AF539275

No. 1842
$21.95

PRINCIPLES & PRACTICE OF ELECTRICAL & ELECTRONICS TROUBLESHOOTING

DANIEL R. TOMAL & DAVID V. GEDEON

TAB BOOKS Inc.
BLUE RIDGE SUMMIT, PA. 17214

"Megger" is a trademark of the James G. Biddle Co., Plymouth Meeting, PA 19462

FIRST EDITION

FIRST PRINTING

Library of Congress Cataloging in Publication Data

Tomal, Daniel R.
Principles and practice of electrical and
electronics troubleshooting.

Includes index.
1. Electric apparatus and appliances—Maintenance and
repair. 2. Electronic apparatus and appliances—
Maintenance and repair. I. Gedeon, David V. II. Title.
TK452.T586 1985 621.3'028'8 84-23980
ISBN 0-8306-0842-7
ISBN 0-8306-1842-2 (pbk.)

Contents

Acknowledgments

We wish to thank the many people and companies who contributed to the development of this book. We express our appreciation to the following people who provided help and assistance: our wives, Sheryl and Annette, for their typing and grammatical review; Ray Tomal and Jack Ward for their suggestions and photos; and Jim George, Pam McRee, and Steven Peltz for their fine illustrations.

We also thank the many companies who provided information, especially Allen-Bradley Company; Apple Computer, Inc.; B & K-Precision; Creswell, Munsell, Fultz, and Zirbel, Inc.; Franklin Electric Company; Global Specialties, Inc.; Hewlett-Packard Company; Ideal Industries, Inc.; McGraw-Edison Company; RCA Corporation; Square D. Company; Tektronics; Texas Instruments, Inc.; and Westinghouse Electric Corporation.

Introduction

New advances in the electrical/electronic field mandate the need to have qualified technicians to service and repair countless numbers of products and devices. This book has been written based on the need for a simple and basic approach to understanding electrical/electronic troubleshooting.

This book explains basic troubleshooting and repair of electric motors, industrial controls, residential/industrial wiring, appliances, radio and stereo, television, digital equipment, and other electrical/electronic products and devices. It is designed for handymen, trade and apprenticeship technicians, and beginning students. The book's unique strength is that it consolidates fundamentals and repair of a wide range of electrical/electronic devices, thereby eliminating the need for people to purchase several books. At the beginning of each chapter, basic theory is reviewed, troubleshooting and repair procedures are then presented, and a series of multiple-choice and essay questions conclude the chapter. This book is not too technical for beginners and consists of a series of logical, short chapters with many illustrations to help ensure the fullest comprehension of the material. Several references and handy troubleshooting guides are presented in the appendices. This book provides basic prerequisite material for the prospective technician that is absolutely essential for more advanced and specialized study in the fascinating and rewarding field of electricity and electronics.

Chapter 1 provides the basic theory and testing of electrical/electronic components such as capacitors, transistors, diodes, silicon-controlled rectifiers, electron tubes, resistors, inductors, and integrated circuits.

Chapter 2 presents the explanation and use of the most commonly used test equipment such as the vom, vtvm, oscilloscope, tube tester, signal generator, transistor checker, CRT restorer, megohmmeter, sweep/

marker generator, noise generator, growler, and signal level meter.

Chapter 3 explains the basic theory and repair of electric motors. Some of the types of motors covered include capacitor, shaded-pole, three-phase, repulsion, universal, synchronous, and stepping.

Chapter 4 explains the theory, basic functions, and servicing of industrial controls such as the bimetallic thermal overload relay, manual starter, magnetic overload relay, limit switch, push-button control station, drum switch, and pneumatic timing relay.

Chapter 5 covers troubleshooting of residential and industrial wiring. The basic theory and fundamentals of wiring are explained, followed by troubleshooting of such circuits as distribution panel, three-way and four-way switch control, wye and delta wiring, polyphase-industrial control wiring, and television and antenna distribution systems.

Chapter 6 explains the troubleshooting and repair of appliances. They include a coffee maker, iron, slicing knife, toaster, can opener, hair blower, and hot water heater.

Chapter 7 presents the fundamentals, repair, and testing procedures for radio and stereo equipment. This equipment includes tape players, AM/FM receivers, multiple receivers, and turntables.

Chapter 8 presents the fundamentals, testing procedures, and repair of black-and-white and color television receivers. This chapter describes a conceptual overview of television troubleshooting rather than a specified make or model.

Chapter 9 presents the fundamentals of digital logic and basic gates; digital integrated circuits; packaging and family types; reading logic and connection diagrams; physical handling (insertion, removal and precautions) of integrated circuits; and troubleshooting associated with small scale integration.

Chapter 10 discusses the fundamentals of sequential and combinational digit circuits such as counters, registers, encoders, multiplexers and demultiplexers, flip-flops, clocks, binary numbers, and troubleshooting techniques associated with medium to large scale integration.

Chapter 11 presents the fundamentals, testing procedures, and repair of microcomputer systems and peripheral equipment. It includes personal microcomputer systems and industrial programmable controllers plus other digital equipment.

Chapter 1

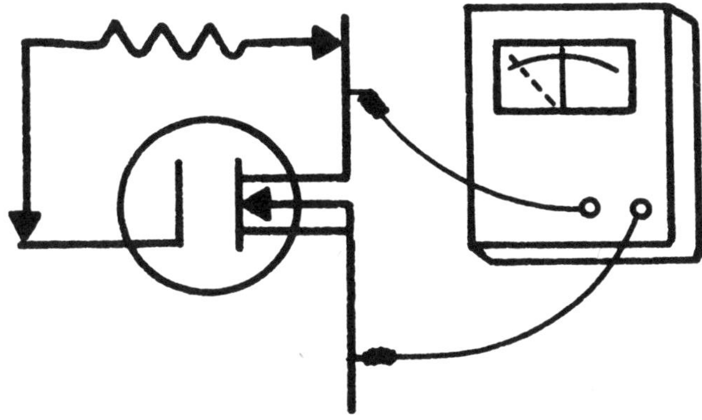

Troubleshooting Electrical/Electronic Components

The need to understand basic testing equipment techniques and to develop a fundamental approach to troubleshooting is essential in servicing electrical/electronic products and devices. Most electrical/electronic products and devices are similar/common types of components such as wire, resistors, capacitors, diodes, transistors, switches, and contacts. The need to understand the basic theory and testing of these components is a prerequisite for technicians. In this chapter, you will learn a basic approach to troubleshooting and fundamental theory and testing procedures for many of the most common components used in electrical/electronic apparatus.

REASONS FOR BREAKDOWNS

Ideally, most people would like electrical/electronic products and devices to be "breakdown proof," but unfortunately this is not the case. Most breakdowns are probably, directly or indirectly, a result of abuse or lack of maintenance.

More specifically, electrical/electronic breakdowns can be categorized by some very basic causes as follows:

1. Heat.
2. Moisture.
3. Dirt and contaminants.
4. Abnormal or excessive movement.
5. Poor installation.
6. Animals and rodents.

Whenever too much heat is applied to electrial or electronic devices, problems usually occur. Heat increases the resistance of circuits, which in turn increases the current. Heat will cause the materials to expand, dry out, crack, blister, and wear down much quicker—sooner or later the device will break down.

Moisture will also cause circuits to draw more current, and eventually break down. Moisture (water and other liquids) causes expansion, warping, quicker wear, and abnormal current flow (short circuits).

Dirt and other contaminants, such as fumes, vapors, abrasives, soot, grease, and oils, are materials that cause electrical/electronic devices to "clog" or "gum" up and operate abnormally until they finally break down.

Abnormal or excessive movement can lead to breakdowns. Vibration and physical abuse are the leading causes of these types of breakdowns.

Poor installation is often the work of an unqualified technician or one who is careless or in a hurry. Failure to tighten a bolt or properly solder a connection results in an electrical/electronic device breaking down too soon.

Many technicians and electricians find a problem that was caused by an animal or rodent. It is not uncommon to find that a rat or other small rodent has chewed on an electrical wire or that an animal of some type found its way into a motor. The effects of breakdown causes are few and are listed as follows:

1. Short.
2. Open.
3. Ground.
4. Mechanical.

Basically a short circuit results when the current takes a direct path across its source. Figure 1-1 illustrates a short circuit in an electric motor caused by a defect in the motor where two wires of the circuit touch and cause a bypassing of the normal current flow.

Short circuits draw more current because the resistance in the circuit decreases, and as a result, the voltage decreases. Typical signs of short circuits are as follows:

1. Blown fuses.
2. Increased heat.
3. Low voltage.
4. High amperage.
5. Smoke.

An open circuit results from an incomplete circuit. Figure 1-2 illustrates an electric motor with an open circuit caused by a break in the motor circuit which prevents the current from flowing in a complete path.

An open circuit will have infinite (unlimited) resistance and zero current since its path has been broken. Typical signs of an open circuit are as follows:

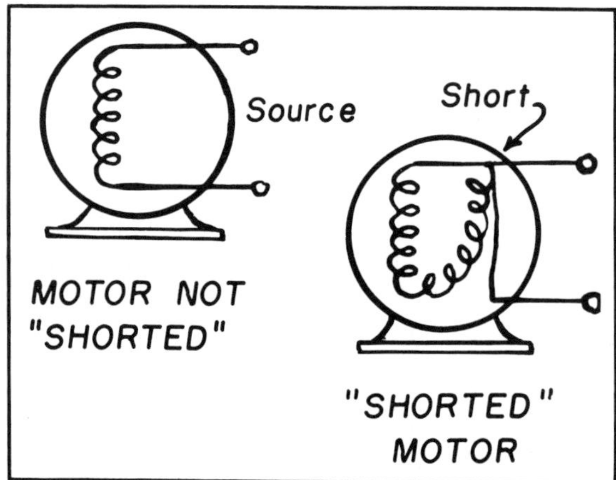

Fig. 1-1. Example of a good circuit and a shorted circuit.

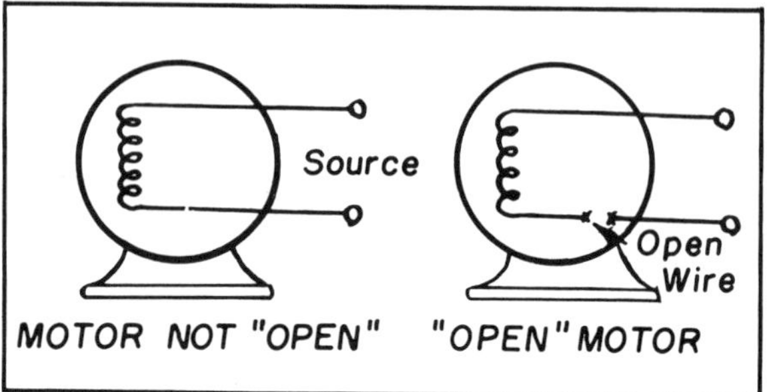

Fig. 1-2. Example of a good circuit and an open circuit.

1. Infinite resistance.
2. Zero amperage.
3. Device completely dead (inoperable).

A ground results when a defect in the insulation or placement of a wire or component causes the current to take an incorrect (abnormal) route in the circuit. Figure 1-3 illustrates an electric motor with a grounded circuit. The ground results when part of the windings make electrical contact with the iron "frame" of the motor.

Common grounds result from wires with poor insulation or misplaced components. Shocks can be obtained from a grounded motor since the frame of the motor has become part of the electrical circuit. Typical signs of a ground are as follows:

1. Abnormal amperage reading.
2. Abnormal voltage reading.
3. Abnormal resistance reading.
4. Shocks.
5. Abnormal circuit performance.

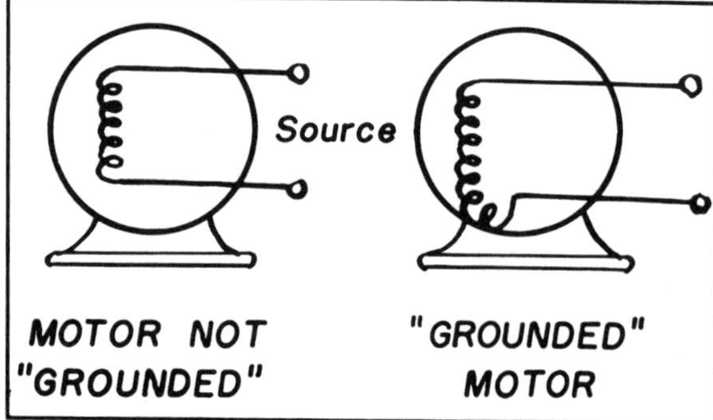

Fig. 1-3. Example of a good circuit and a grounded circuit.

Mechanical problems are a result of too much friction, wear, abuse, or vibration, where the physical part of an electrical/electronic device causes the breakdown. Broken belts, worn bearings, loose bolts, worn contacts, damaged chassis, and broken controls are common examples of mechanical problems. Typical signs of mechanical problems are as follows:

1. Noisy operation.
2. Abnormal operation.
3. Visual inspection.
4. Circuit failure.

The most important tool or instrument a service technician can have is his/her senses. Most troubleshooting problems can be found by the use of one or all of the following senses:

1. Sight.
2. Smell.
3. Touch.
4. Hearing.

Before any sophisticated attempt is used to analyze a problem, first look for an obvious cause. A cracked circuit board, broken wire, burnt or charred component, or any type of damaged item can quickly lead the technician to the problem.

There is not a more common smell than that of a burned transformer. A good technician should easily be able to identify this smell. Also, burned cables, insulation, wires, and components may give direction to the servicer in the troubleshooting operation.

Many technicians rely on their sense of touch to locate component faults. Integrated circuits should never be *hot* when touched by the servicer. A hot integrated circuit would indicate a short in the integrated circuit. Likewise, a hot, smoky motor is a common sign of a short circuit. But, a 10-watt line resistor should not be cold when touched—it should feel warm or hot. This cold resistor would indicate an "open" component.

BASIC METHODS IN TROUBLESHOOTING

There are basic techniques common to all technicians when troubleshooting electrical/electronic devices. Which techniques the servicer uses depends upon what type of defect or symptom exists.

The troubleshooting techniques that will be introduced, and later explained in detail throughout this book, are as follows:

1. Voltage measurements.
2. Amperage measurements.
3. Resistance measurements.
4. Substitution.
5. Bridging.
6. Heat.
7. Freeze.
8. Signal tracing/injection.
9. Component testers, test lamps.
10. Resoldering, adjusting, etc.
11. Bypassing.

The voltage measurement of a circuit is usually taken by using a voltmeter or an oscilloscope. A zero voltage reading may identify an open circuit, while a low voltage reading may indicate a shorted component. Remember, always connect a voltmeter in parallel with the circuit when measuring voltage (Fig. 1-4).

The amperage measurement of a circuit is usually taken by using an ammeter or a "clamp on" ammeter. The ammeter indicates and locates common circuit faults, such as shorts, opens, and grounds. Remember, always connect the ammeter in series with the circuit when measuring current (Fig. 1-5).

An ohmmeter is used to measure the "continuity," resistance of a circuit, or resistance of a component. This technique is very valuable in locating shorts, grounds, and open circuits. Remember: always shut off the power before measuring resistance (Fig. 1-6).

The substitution technique simply means replacing a suspected faulty component with a known good component. This method can save valuable time and frustration for the servicer.

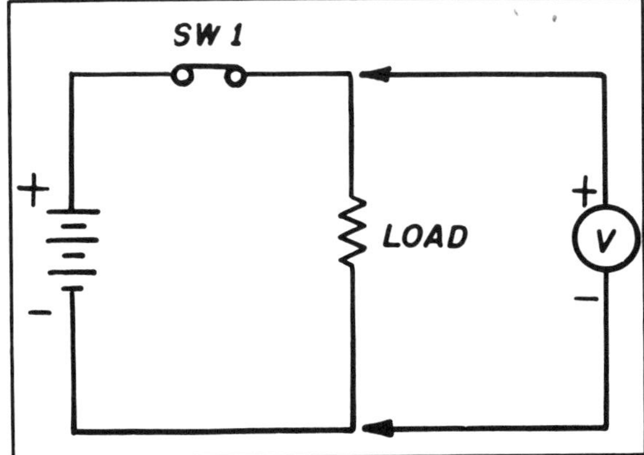

Fig. 1-4. Always connect a voltmeter in parallel with the circuit.

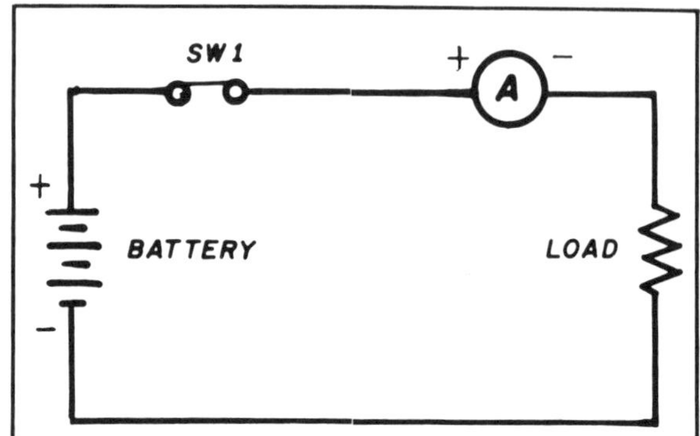

Fig. 1-5. Always connect the ammeter in series with the circuit.

When a technician suspects a component (usually a capacitor) to be faulty, he/she "jumps" or places a known good component across the suspected faulty component from the circuit. This is called "bridging." The servicer can save valuable time and frustration by bridging (Fig. 1-7).

Application of heat is a technique whereby the technician applies heat to a suspected "thermal intermittent" component. This thermal intermittent component breaks down under heat. By applying heat to this expected intermittent component—usually done using a hot blower—the servicer can determine the quality of the component. Do not use enough heat to damage nearby components, particulary plastic components.

The freezing technique is used by the technician to temporarily restore a component to normal operation. The freezing technique gets its name from the use of cold air from a fan or a chemical coolant. The freezing technique cools the suspected thermal intermittent component, thus temporarily restoring the component to normal operation.

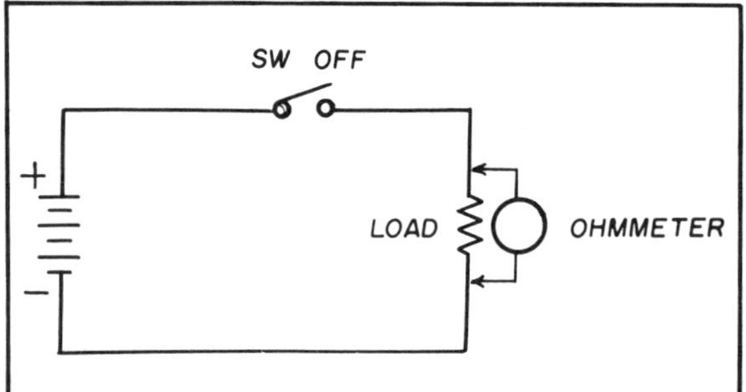

Fig. 1-6. Always turn off the power in the circuit before measuring resistance.

Fig. 1-7. Bridging a component with another.

Signal tracing/injection is most often used in servicing radios. The technician injects a signal into the malfunctioning receiver in order to locate the specific inoperable (dead) stage (Fig. 1-8). A signal is injected into the various points preceding each stage. A tone will be heard at the speaker if the stage is operating. The defective stage will not allow the signal to pass through, and the signal will not be heard at the speaker.

Component testers are instruments used to test the quality of the component. Component testers include megohmmeter, capacitor checker, test lamp, transistor/diode tester, CRT tester, tube tester, and others.

Resoldering, adjusting, and aligning are all techniques used by a technician on suspected problems. Many times, a technician will use these techniques because he/she has a feeling that a problem exists. These techniques may also be used if they have been successful in the past. A poor electrical solder connection is called a cold solder joint.

Bypassing is a technique that a technician may use to

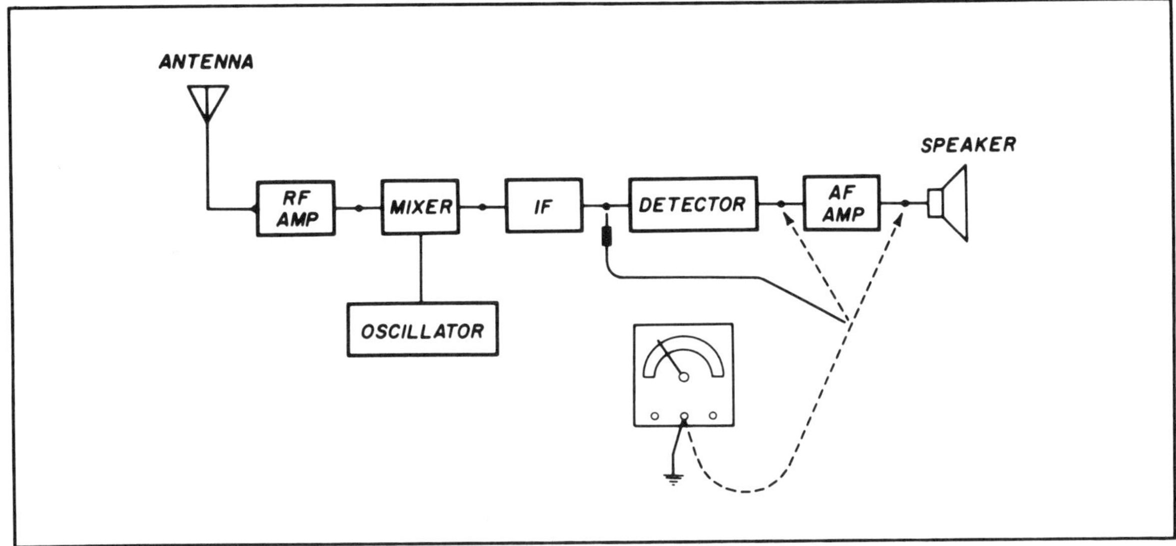

Fig. 1-8. Signal injection process in a radio.

locate a suspected problem. This bypassing technique requires unplugging one of several circuits. By "shutting off' a transistor, its effect on the total performance of the circuit can be observed. Suspected problems can be identified in this manner.

In diagnosing electrical/electronic troubles, it is important that a technician follow a logical systematic procedure to eliminate unnecessary time, tests, and replacement of parts. "Time is money," and a good technician needs a good "cookbook" approach to troubleshooting.

Every technician should know the difference between troubleshooting and repair. These two terms are often confused. Troubleshooting is the process of "problem solving." A problem can be described as a deviation from a standard, such as a malfunctioning or inoperable product. When a technician has identified the cause of a problem, he/she then decides how to repair it. A common mistake is to attempt to repair a product before troubleshooting it. For example, a technician might replace a defective component, only to turn on the product and the component shorts out. "Over-fixing" is not only time-consuming but expensive.

One approach to troubleshooting is: define the problem, investigate the problem (e.g., voltage, amperage, resistance readings), analyze the information, and determine the cause of the problem. Here is a list of questions that should be asked of a customer or operator in determining the cause of a problem:

1. When did the defect occur?
2. What symptoms, noises, smells were noticed when the defect occurred?
3. Under what conditions did the defect occur?
4. Which parts of the product are okay?
5. Where did the defect occur?
6. To what degree is the product defective?

A step-by-step procedure is important when approaching a problem. Remember, do not overlook the obvious! Use common sense when approaching a problem. Figure 1-9 points out the steps one should use in analyzing a problem.

Most troubleshooting procedures can be greatly aided by use of diagrams, schematics, and blueprints. Basically, service diagrams consist of the following:

1. Schematic diagrams.
2. Line drawings and blueprints.
3. Pictorial diagrams.

Most schematic diagrams consist of an electrical or electronic layout of the circuit design. These diagrams present specific component values and data. Schematic diagrams often specify normal operating voltages and current, cold resistance values, signal waveforms, and other information. Figure 1-10 shows an example of a schematic diagram.

Basic line drawings and blueprints show layout

placement of wire/cable and controls. Line drawings are usually used in residential/industrial wiring and controls to aid when installing, locating, and tracing circuits. A blueprint of a home is shown in Fig. 1-11.

A pictorial diagram can serve to be most useful in providing a layout of the location and placement of specific component parts. Many times a schematic diagram will be followed by a pictorial diagram. A pictorial diagram usually shows a "picture" of a circuit. Figure 1-12 shows an example of a pictorial diagram.

Often the success of troubleshooting a device depends upon the availability of servicing diagrams. Many foreign products and equipment are difficult to service when service literature is not available. Often, an electronics technician will refuse to work on a product knowing that a service diagram is not available.

Regardless of the trouble or situation, a good technician will always record a mental or written note of a circuit problem he/she has repaired and use this information in the future.

BASIC COMPONENTS

The ohmmeter is probably one of the most important meters in servicing. This meter is used to measure continuity or resistance in a component or circuit. A component having continuity has resistance near zero. On the other hand, a component having no continuity would have infinite resistance.

When testing basic components, the technician is mostly concerned with the resistance of the component or continuity measurement. For example, when checking a fuse, a good fuse will read zero ohms, but an open (blown) fuse will read infinite resistance (Fig. 1-13).

Like the fuse, when testing an appliance cord, or an extension cord, a good cord will have continuity, and a broken wire (open) will have no continuity (Fig. 1-14).

When testing a switch, the same procedure is used. A single-pole, single-throw switch should have continuity one way and not the other (Fig. 1-15).

Resistors are manufactured in various shapes, sizes, and values. The main purpose of the resistor is to limit current flow and/or reduce voltage. Most resistors are made of carbon or wire and are manufactured in prescribed ohmic values. For example, a 1000 ohm resistor at 10% tolerance would be color coded brown, black, red, and silver. Therefore, using an ohmmeter, the resistor

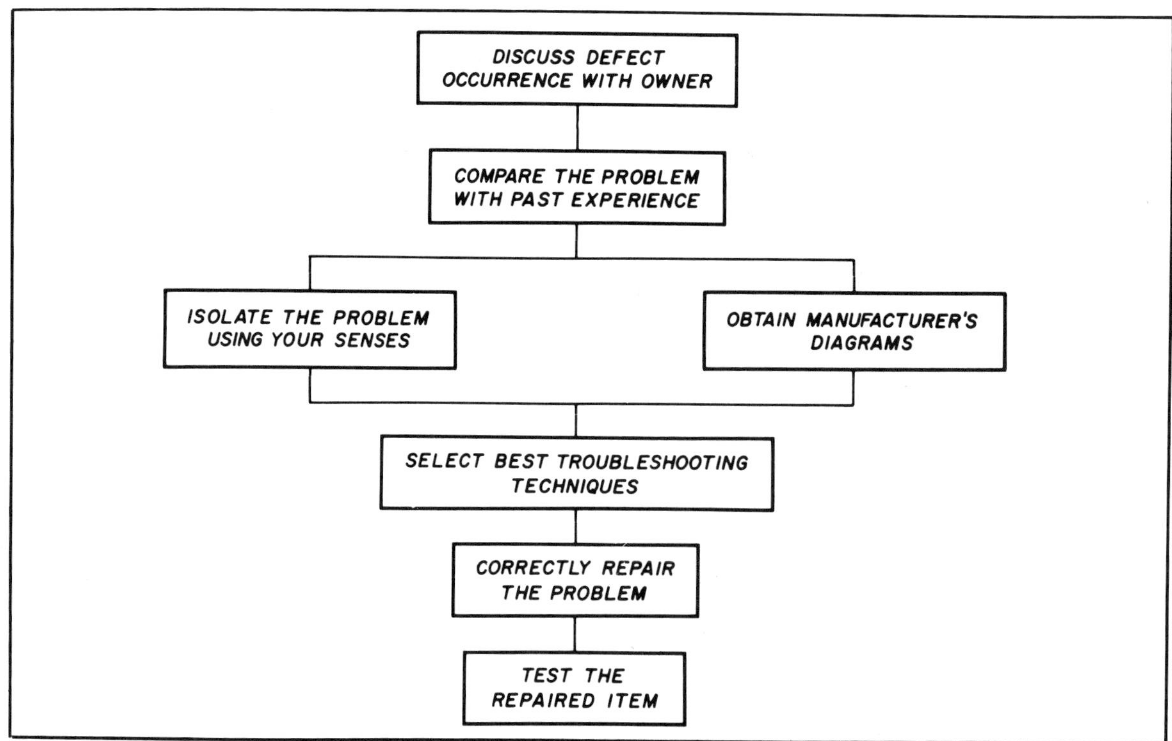

Fig. 1-9. Steps one should use to analyze a problem.

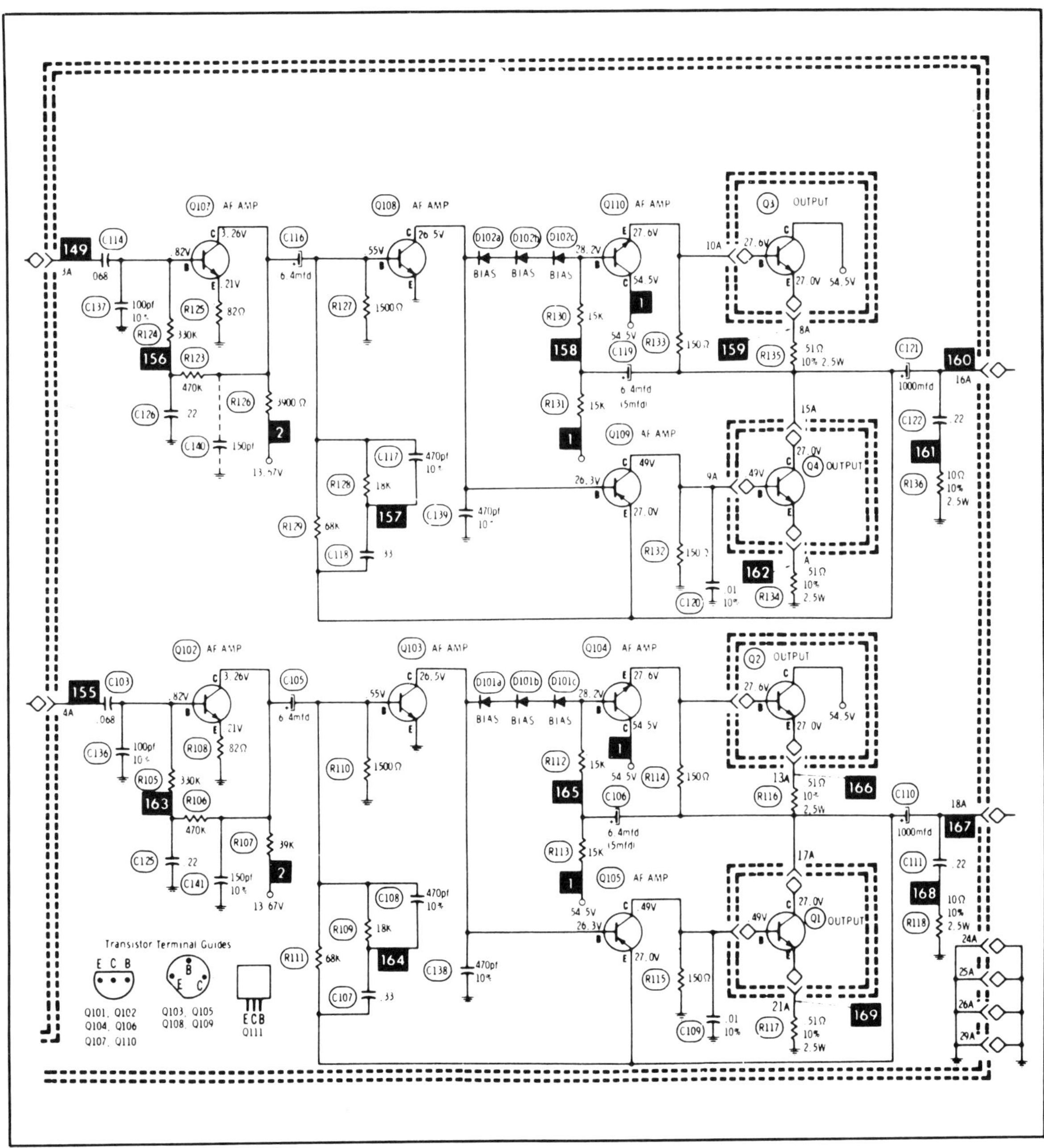

Fig. 1-10. A typical schematic diagram (courtesy Howard W. Sams & Co., Inc.).

should measure between 900 and 1100 ohms (Fig. 1-16). An open resistor would have infinite resistance, and a defective resistor could have any value below 900 ohms or above 1100 ohms.

Resistors are rated in watts that determine the ability of the resistor to absorb the heat that is produced within the resistor. The actual physical size of the resistor determines the watt rating of the resistor. A larger resistor simply can absorb more heat than one of a smaller size.

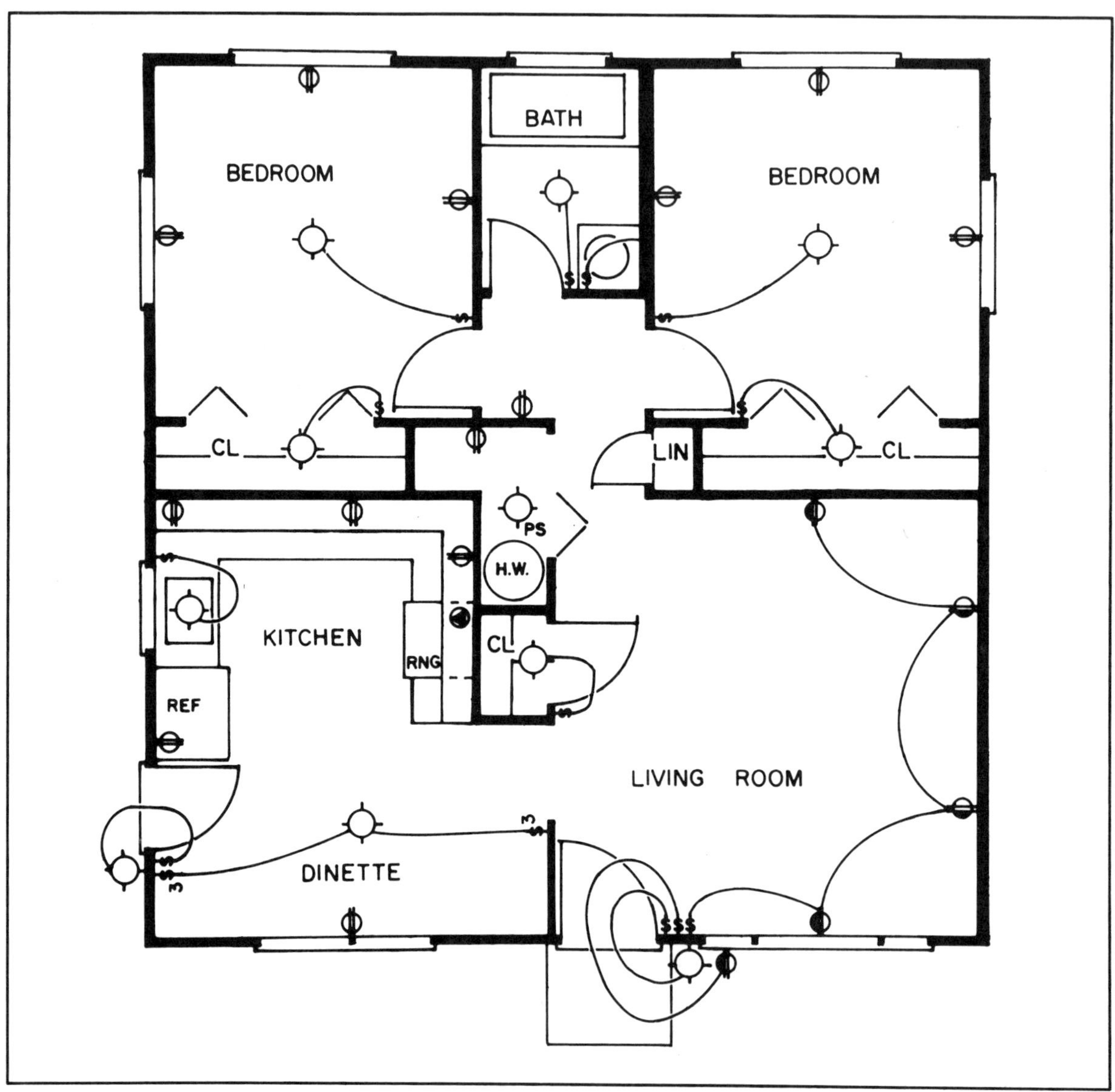

Fig. 1-11. A typical blueprint of a home.

The most common defects of resistors are physical—*cracking* and *charring*. When excessive current and heat tend to increase the resistance, the resistor *opens*. A charred or discolored resistor should be replaced, since the resistor will often check good with an ohmmeter, but break down (open) under voltage in the circuit.

Variable resistors are called potentiometers and can be measured and tested in two simple ways: (1) Use an ohmmeter and measure the value of the potentiometer across the two end terminals. The value should equal the value printed on the potentiometer. Place one lead on the center terminal and the other lead on one of the end terminals. The potentiometer, when turned, should vary the resistance accordingly (Fig. 1-17). (2) The second way to test a potentiometer is to turn the potentiometer while it is in the circuit. If a *scratchy*, rough sound is heard in the speaker the potentiometer needs cleaning or replacing. To clean the potentiometer, turn off the power

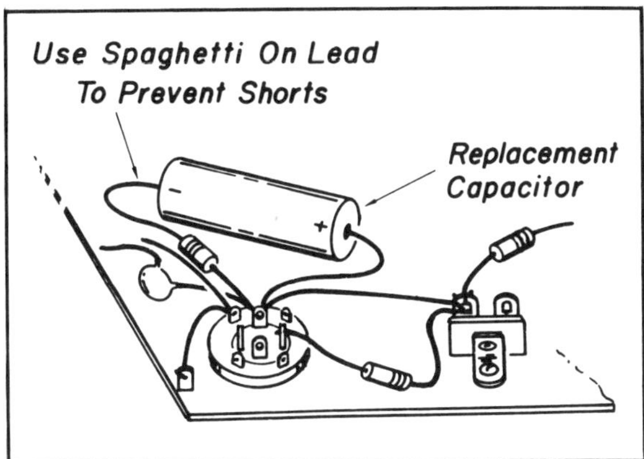

Fig. 1-12. A typical pictorial diagram.

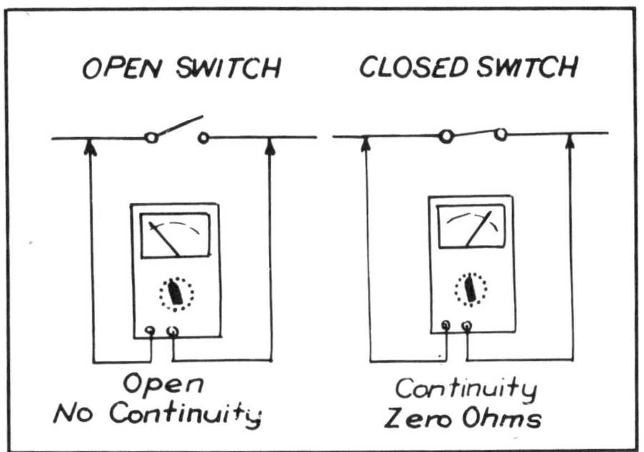

Fig. 1-15. Checking a switch for continuity using an ohmmeter.

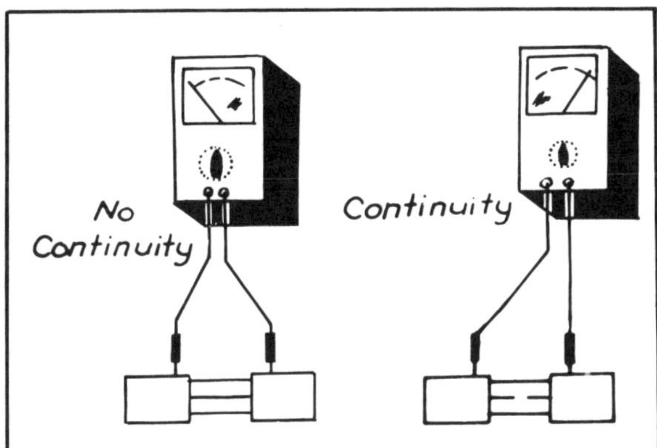

Fig. 1-13. Checking a fuse for continuity using an ohmmeter.

and spray an electronic component cleaner in the sliding contact area while turning the control back and forth.

The correct voltage output of a battery is very important. An excellent battery should exceed its rated value. For example, a new dry cell rated at 1.5 volts dc should measure 1.5 to 1.6 volts. On the other hand, an old, weak battery will read less than 1.5 volts. An automobile battery (lead acid cells) rated at 12 volts dc will often exceed 13 volts when fully charged.

The quality of a receiver is largely determined by the speaker. When checking speakers, first make an inspection. *Rattles* and heavy *vibration* are often signs of a defective speaker. Visually inspect the speaker for cracks, dirt, and other faults. If in doubt about the actual

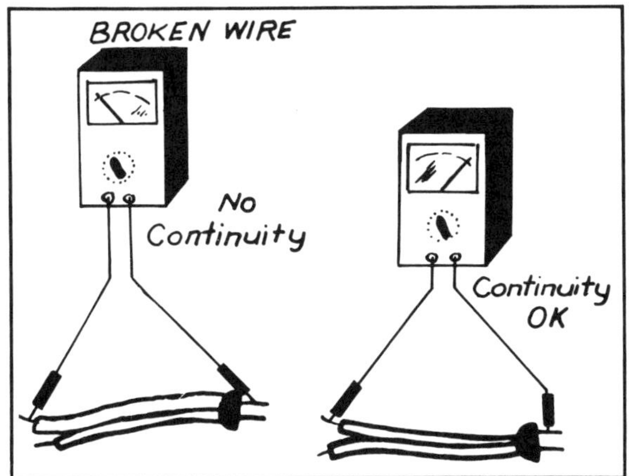

Fig. 1-14. Checking a cord for continuity using an ohmmeter.

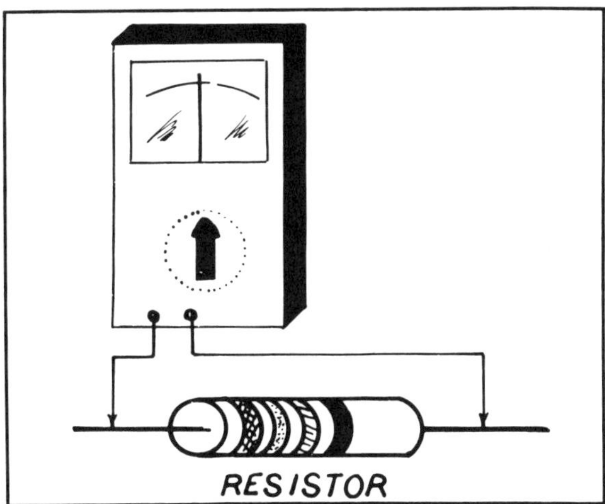

Fig. 1-16. Checking the value of a resistor using an ohmmeter.

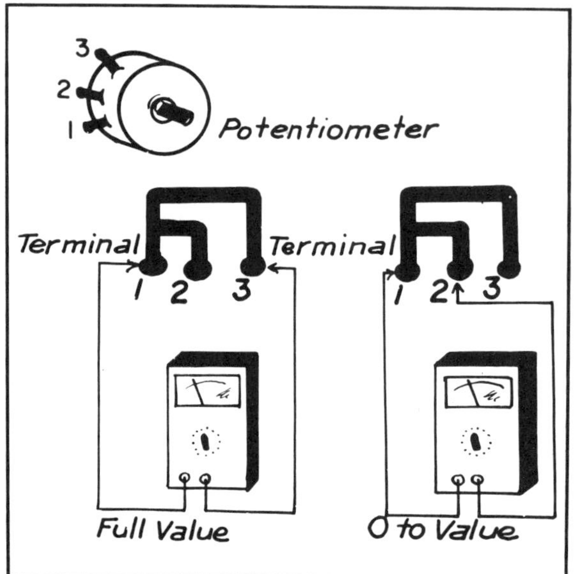

Fig. 1-17. Checking the potentiometer using an ohmmeter.

quality of a speaker, the best testing technique to use is that of substitution. Simply substitute a known good similar value speaker in its place.

When replacing a speaker, it is important to replace the speaker with one having the same impedance and power ratings. Primarily, the voice coil of the speaker determines the impedance and power rating of the speaker. The power rating, measured in watts, determines the maximum power at which the speaker should be operated. The impedance (in ohms) of the speaker is used to *electrically* match the speaker input to the output of the receiver. The impedance of a speaker can be roughly approximated by measuring the resistance of the voice coil with an ohmmeter, then multiplying this value by 1.25

Some common speaker values are 3.2, 4, 8, 10, 16, and 20 ohms. The internal construction of a speaker is shown in Fig. 1-18.

Another method used to check a speaker is to place an ohmmeter across the voice coil terminals. Listen and watch for a small "pop" and movement of the speaker cone. A defective speaker will show no movement or popping sound. Also, this method can be helpful in *phasing* two or more speakers together. Place the ohmmeter on the voice coil terminals and note whether the cone moves *in* or *out*. Reversing the polarity of the ohmmeter will reverse the movement of the cone. Next mark the positive side of each cone when it is in the *out* position (Fig. 1-19). Then, connect each speaker to the audio amplifier while observing the correct polarity. Sound reproduction should be improved since the speaker cones will move in and out together. The in and out movements of the speaker cones keep the speakers *in phase*, or *in step* with each other. If they are out of step, or opposing each other at certain frequencies, the sound waves will be cancelled.

Capacitors are used for hundreds of different purposes. They come in various sizes, shapes, types, and values. Basically, a capacitor is a device that has the ability to store an electrical charge. Capacitors consist of two conducting plates, separated by an insulating *dielectric* material. The value of a capacitor is expressed in

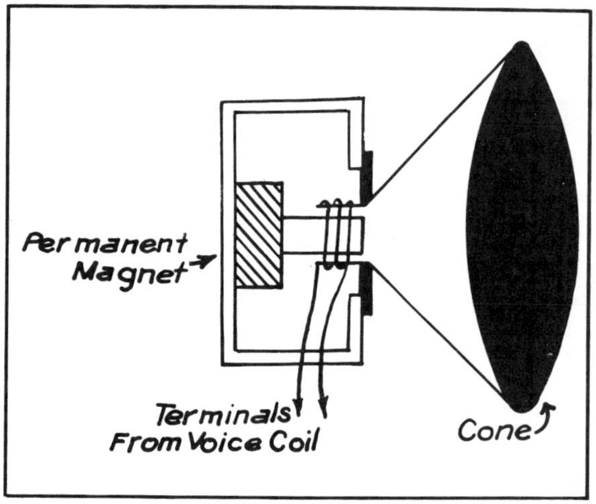

Fig. 1-18. Internal construction of a speaker.

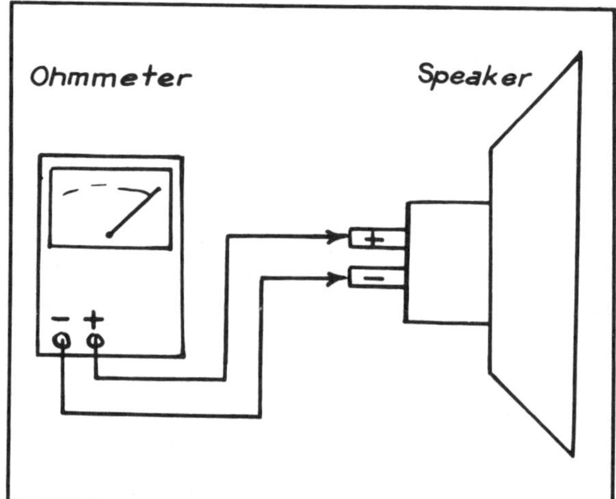

Fig. 1-19. Phasing speakers with an ohmmeter.

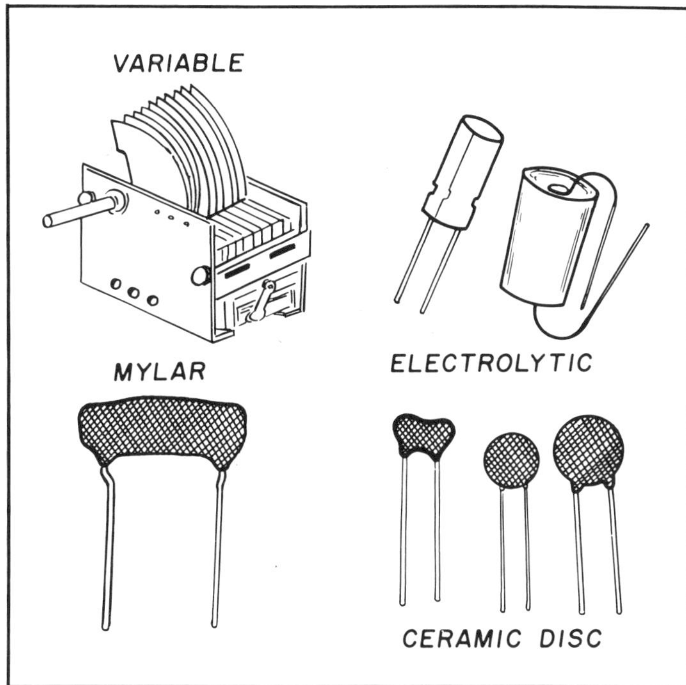

Fig. 1-20. Various types of capacitors.

microfarads (μF), which represent the amount of electrical charge. Figure 1-20 shows some different types of capacitors.

There are few techniques used in troubleshooting capacitors. They are as follows:

1. Resistance measurement (ohmmeter).
2. Capacitance measurement (capacitor checker).
3. Spark test.
4. Bridging.
5. Substitution.

A capacitor below .25 μF should not show a reading on the ohmmeter since it is too small to advance the meter movement. A near zero reading would indicate a shorted capacitor. All capacitors above .25 μF should register on the ohmmeter.

When checking a capacitor, place the ohmmeter on a high scale such as 10,000 ohms and place the leads of the ohmmeter across the leads of the capacitor. First, make sure you have discharged the capacitor by shorting the leads with a piece of wire or a screwdriver. When the leads of the meter have been placed across the capacitor, you should see the needle deflect upward and then slowly drop back down to near zero. Failure to deflect the needle would indicate an *open* capacitor, and failure of the needle to drop down would indicate a *shorted* capacitor (Fig. 1-21).

Another method used to check a capacitor is the *spark test*. Connect the capacitor across a voltage source for just one second. This charges the capacitor. Make sure the voltage you have applied does not exceed the voltage rating specified on the capacitor (Fig. 1-22). After the capacitor has been charged, short its terminals together using a screwdriver or a similar device. A good capacitor will show a spark. Absence of a spark means it is defective.

Fig. 1-21. Checking a capacitor with an ohmmeter.

Fig. 1-22. Checking a capacitor by the spark test.

A *capacitor checker* is a useful device for testing the performance of a capacitor. Besides checking its specific capacitance rating, it can also test for other characteristics of a capacitor, such as leakage and opens.

The *bridging* method is also a good way to check the performance of a capacitor. Usually, filter capacitors that are suspected of being open will first be bridged before being replaced. This is simply a method where the suspected defective capacitor is bridged (or jumped) with a known good capacitor within ±10 percent of the rated value. A noticeable difference in the quality of performance of the product/device (such as radio or TV) under test should result upon bridging the capacitor. In Fig. 1-23 the filter capacitor in this power supply is bridged with a known good filter capacitor.

The *substitution* technique, like the bridging

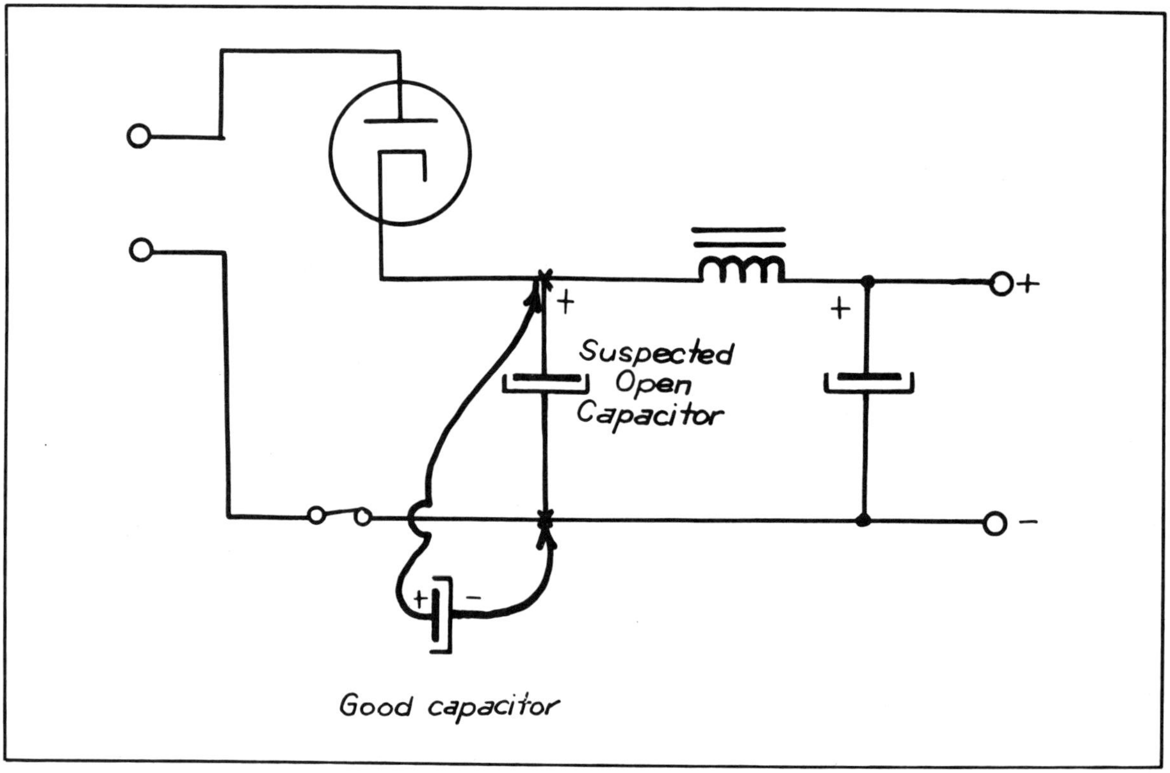

Fig. 1-23. Bridging a known good capacitor across a suspected open capacitor.

method, determines the quality of the capacitor by means of another capacitor. When using the substitution method, simply replace the suspected defective capacitor with a known good capacitor with similar value and rating. The performance of the product/device will indicate the effect of the new capacitor. Remember, never exceed the voltage rating of a capacitor. A capacitor rated at 100 volts should only be replaced with one at 100 volts or more; otherwise, it may be destroyed.

SEMICONDUCTORS

The understanding of basic theory of semiconductors can be a real asset for the technician in troubleshooting these components. One of the first known semiconductors was the crystal detector. This consisted of a lump of crystal galena with a thin wire catwhisker combination under pressure by a spring. This combination rectified current allowing it to flow in only one direction.

Although the galena crystal was unreliable, it was the first step in semiconductor application. The development of modern diodes and transistors started with the basic theory and development of p- and n-type materials.

To develop a p- or n-type material, a crystal material of germanium or silicon is used. Silicon has an atomic number of 14 with 4 (valence) electrons in its outer shell, Germanium has an atomic number of 32 with also 4 (valence) electrons (Fig. 1-24).

In order to form a p-type material, an impurity, galium or indium is added. This impurity is called a trivalent because it has a valence of three. This means that this trivalent has three electrons in its outer shell. When either galium or indium is added to silicon or germanium (both have a valence of four), one valence electron is left unfilled.

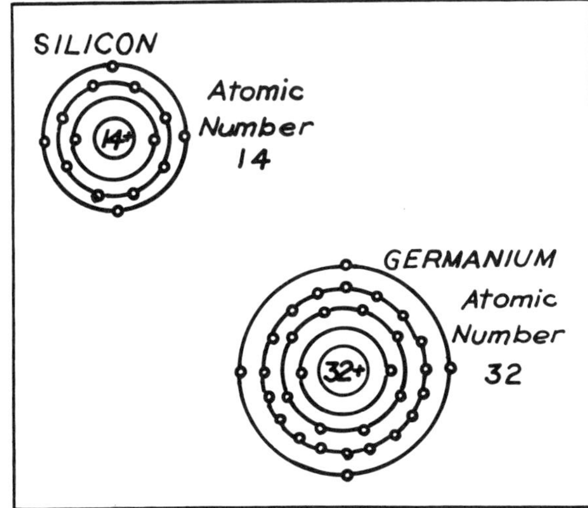

Fig. 1-24. The atomic structure of silicon and germanium.

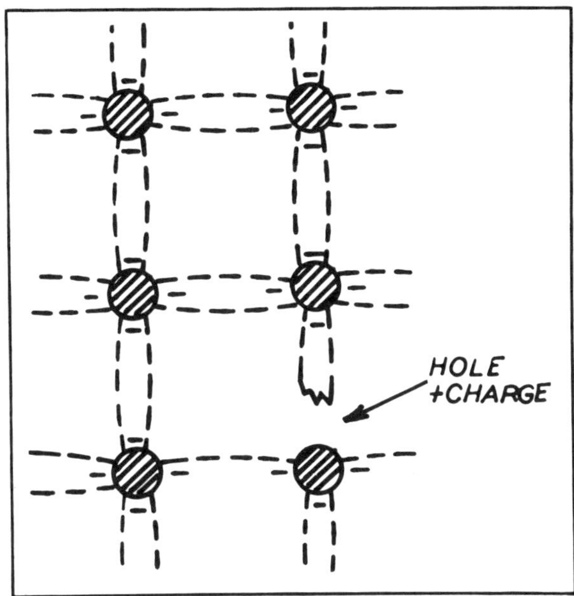

Fig. 1-25. Adding an acceptor impurity to a crystal leaves a hole forming a p-type material.

The unfilled gap called a *hole* takes on a positive charge and forms a p-type material. The impurity which created a hole is referred to as *acceptor* impurity (Fig. 1-25).

To form an n-type material an impurity of arsenic or antimony is added. This impurity, called a pentavalent because it has five valence electrons in the outer shell, when added to germanium or silicon joins with the four valence electrons adding one free electron. This free electron constitutes a *negative* charge in the atom and, therefore, this impurity is given the name of a *donor* (Fig. 1-26).

When a p- or n-type material is pieced together, a p-n junction is formed. This p-n junction is called a *diode* because it allows current to pass in only one direction. When a battery is hooked up to a diode negative to negative and positive to positive, current flows through the diode. This is called *forward bias* and is shown in Fig. 1-27.

When a diode is forward biased the negative charge of the battery *repels* the negative charges in the diode, forcing them into the positive side of the diode. The

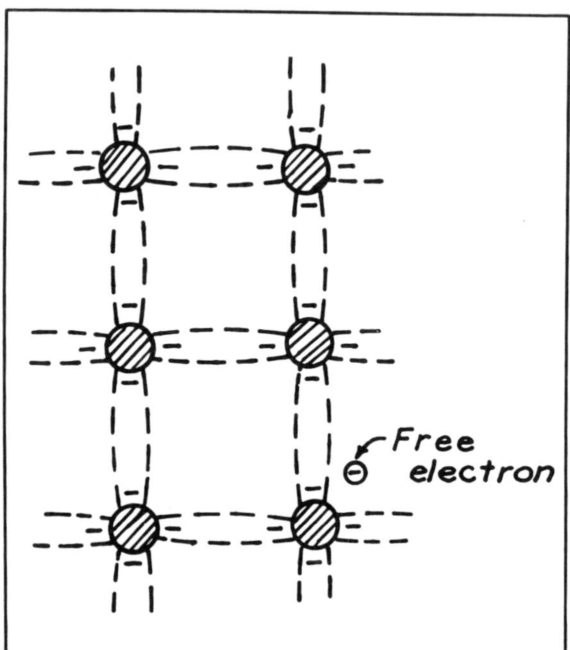

Fig. 1-26. Adding a donor impurity to a crystal adds an extra electron forming an n-type material.

positive force of the battery pulls these electrons on through the diode. This completes the circuit.

When the battery is reversed, the diode is reverse biased, because the positive force of the battery and the positive holes of the diode attract each other, allowing no current flow to take place. A reversed biased diode is shown in Fig. 1-28.

The p (positive) side of a diode is called the anode, and the n (negative) side is called the cathode. It is important for the technician to know which side is which. The arrow indicates the p side. The line indicates the n side. A line or dot placed on the diode by the manufacturer indicates the cathode side. Notice that the manufacturer marks the cathode side (negative) of the diode with a positive mark or band. When the diode is connected to the positive side of the power source, it will be reverse biased. Figure 1-29 shows various diode configuration.

To test a diode, the technician can use either an ohmmeter or a diode/transistor checker. When checking a diode with an ohmmeter, place the selector switch on R × 100 and place the leads across the diode. In the *forward bias* direction, the ohmmeter should read less than 100 ohms. In the *reverse bias* direction, the ohmmeter should read about 5000 ohms. Figure 1-30 shows the correct testing of a diode with an ohmmeter.

Most types of diodes can be checked using an ohmmeter. Keep in mind that the actual resistance value of the diode is not very important as long as you get a *low/high* reading when the ohmmeter polarity is switched. If any doubt exists after the ohmmeter check, it is recommended that a new diode be substituted. Also, keep in mind that a low resistance reading both ways is common when the diode is tested in the circuit. To be sure that the diode is good, unsolder one lead and check

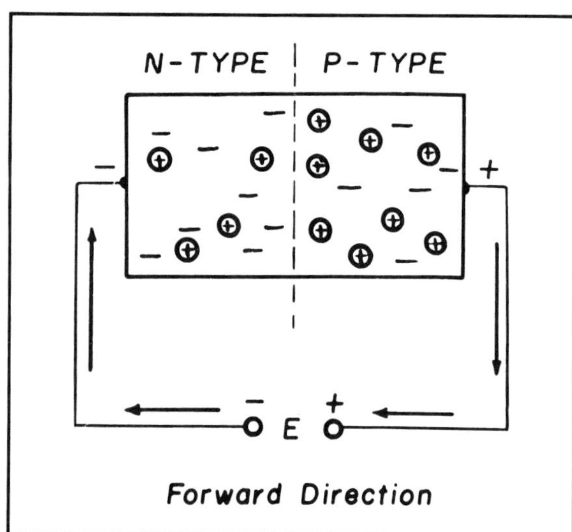

Fig. 1-27. Forward biased diode.

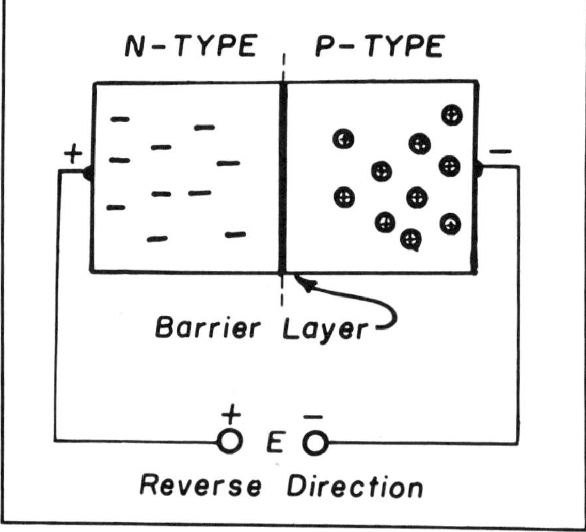

Fig. 1-28. Reverse biased diode.

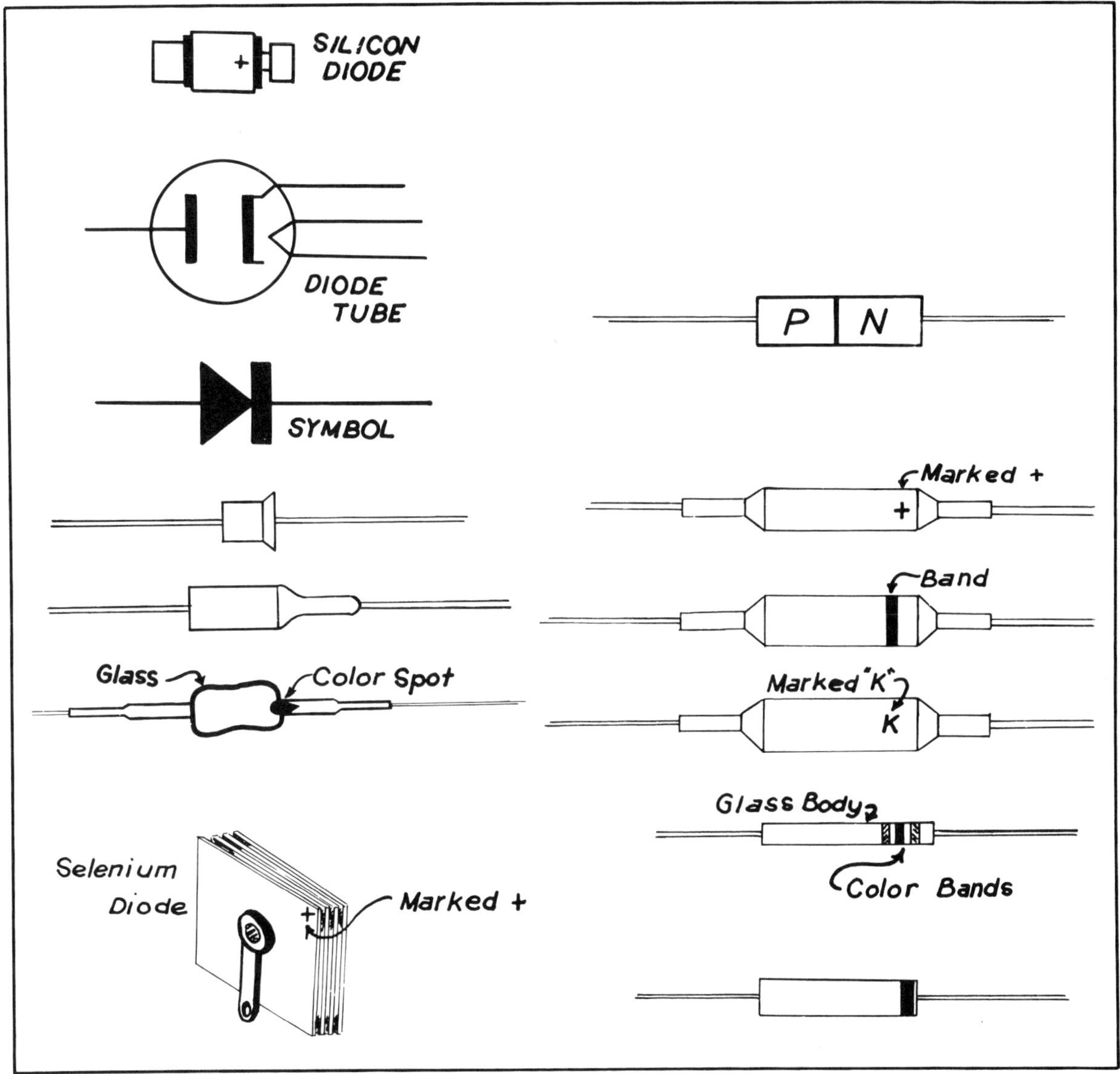

Fig. 1-29. Various diode configurations.

the diode again with the ohmmeter. Keep in mind when replacing a diode that a diode will only block a certain amount of voltage in the reverse direction. This is called the prv rating [peak inverse (reverse) voltage]. Never exceed this voltage rating, or the diode will be destroyed.

Although there are numerous types of diodes—zener, light-emitting, photoconductive, varactor, tunnel—each has its own unique features. When in doubt of the diode's quality, the best testing method to use is substitution.

The transistor is actually made up of two diodes back to back. A transistor is either an npn or a pnp combination. The first section is called the emitter. The middle section is called the base. The last section is called the collector. Figure 1-31 shows examples of both types of transistors.

A technician should understand the reason why a transistor amplifies. Figure 1-32 shows an npn transistor where the emitter/base is forward biased, so it has low resistance to current flow. Also, the collector/base has

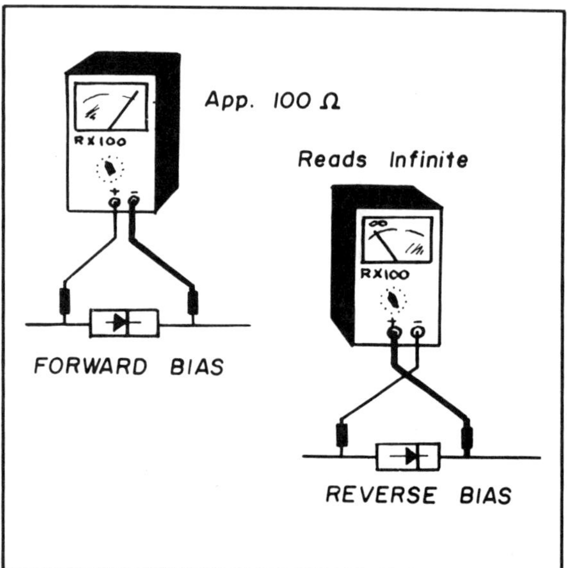

Fig. 1-30. Checking a diode with an ohmmeter.

high resistance to current flow or is reverse biased. The negative potential of the battery forces the electrons in the emitter to flow into the base region. Very few of these electrons bond with positive holes since most of them continue on through to the collector region. This is due to the strong positive attraction of the battery. The electrons complete the circuit by returning to the battery supply. Keep in mind that new positive holes are being pulled into the base region from the battery when electrons fill the holes.

Since the collector region has a higher value of resistance than the emitter region any current change in the emitter will cause a greater proportional change in the collector. A signal passing through this transistor will be amplified.

The amount of signal amplification can be controlled by regulating the amount of electron flow into the base region. The amount of electrons supplied to the base region determines the amount of electrons available to the collector region. The regulation of electrons into the base region is called *biasing*. In a transistor, the forward bias (or emitter-to-base bias) determines the amplification of the transistor. The forward biasing of a transistor can be controlled by increasing or decreasing the voltage or resistance at the emitter/base region (Fig. 1-32).

The basic operation of current flow in the pnp transistor is very similar to that of the npn except that instead of current flow by electrons, the current flow in the pnp is completed by holes. The positive force from the battery forces the positive holes from the emitter through the base/collector region and back to the negative side of the battery. Here again as with electrons in the npn transistor, a small number of holes bond with electrons in the base region, but the majority of these holes continue on into the collector region. Conduction takes place by *hole* current from emitter to collector. Electric flow is opposite to hole flow (conventional theory). Therefore, electron flow in this circuit is considered to travel in the opposite direction, or from collector to emitter. Don't let this explanation of *hole* current confuse you; basically, the main function of both transistors in the circuit is the same. Both transistors will amplify (see Fig. 1-33).

The three basic circuit configurations that transistors can be connected to are the common base, common

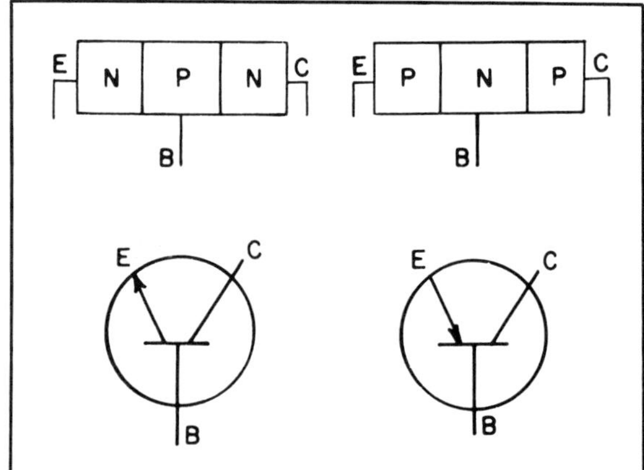

Fig. 1-31. The three sections of a transistor.

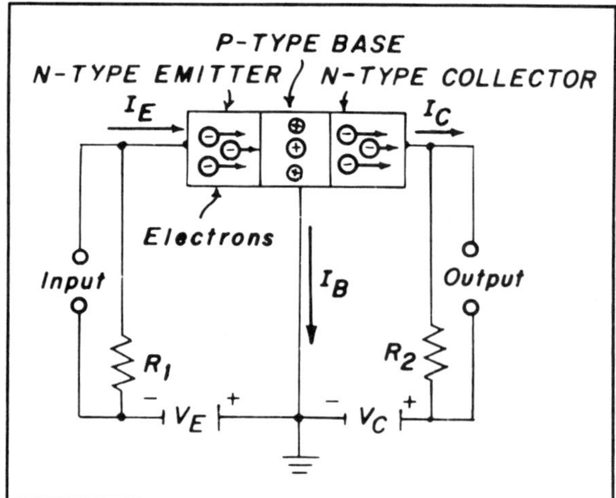

Fig. 1-32. Electron flow in an npn transistor.

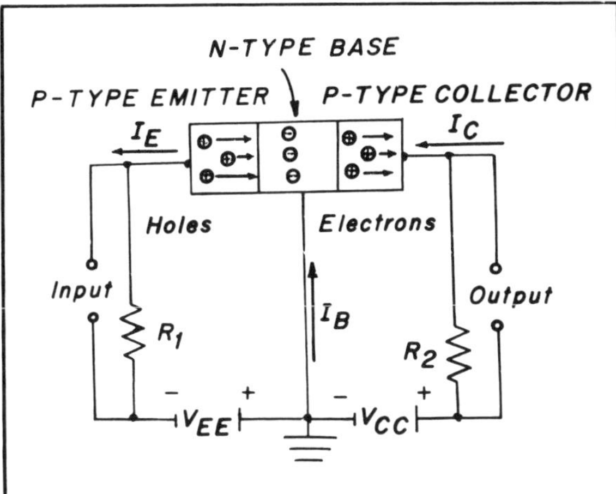

Fig. 1-33. Electron flow in a pnp transistor.

emitter, and common collector. Each circuit configuration has its own unique characteristics. Figure 1-34 and Table 1-1 illustrate these basic differences.

The circuit operation and troubleshooting tips and pointers will be discussed in more detail in a later chapter.

Transistors are usually checked by either a transistor checker or by an ohmmeter. Figure 1-35 illustrates how to check a transistor with an ohmmeter. Keep in mind that a transistor is actually two diodes back to back and, therefore, can be checked accordingly. To test a transistor for a short or open, simply connect the positive lead of the ohmmeter (R × 100) to the base and the negative lead to the emitter of an npn transistor. Now the transistor is forward biased, and a low resistance reading should be read. Reversing the leads will reverse bias the emitter/base regions, and the ohmmeter will show a high

Table 1-1. Three Basic Transistor Circuit Configuration Characteristics

Transistor Function	Common Base	Common Emitter	Common Collector
E gain	High	High	Low
I gain	Low	High	High
Input Z	Low	Moderate	High
Output Z	High	High	Low
Power gain	Moderate	High	Moderate

resistance reading. The base/collector regions are checked in the same way. Remember, a low/high resistance reading should always take place. Two highs indicate an *open* transistor; two lows indicate a *shorted* transistor (out of circuit test).

Transistors can be checked *in* or *out* of the circuit in this manner. It is recommended that a transistor that has been checked defective in the circuit be taken out of the circuit and checked again before being replaced.

Using the ohmmeter is another way to help determine the emitter and collector leads and/or determine the quality of the transistor. First locate the emitter/collector leads by either the manufacturer's pictorial reference or through use of the high/low readings on the ohmmeter. Place one lead of the ohmmeter on the emitter and one on the collector. Some kind of reading should result on the ohmmeter scale. Now short the base lead to the emitter; the resistance on the meter scale should increase. When the base lead is shorted to the collector, the resistance on the meter scale should decrease (Fig. 1-36).

The FET (field-effect transistor) is a special transistor used frequently in electronic circuitry. Although

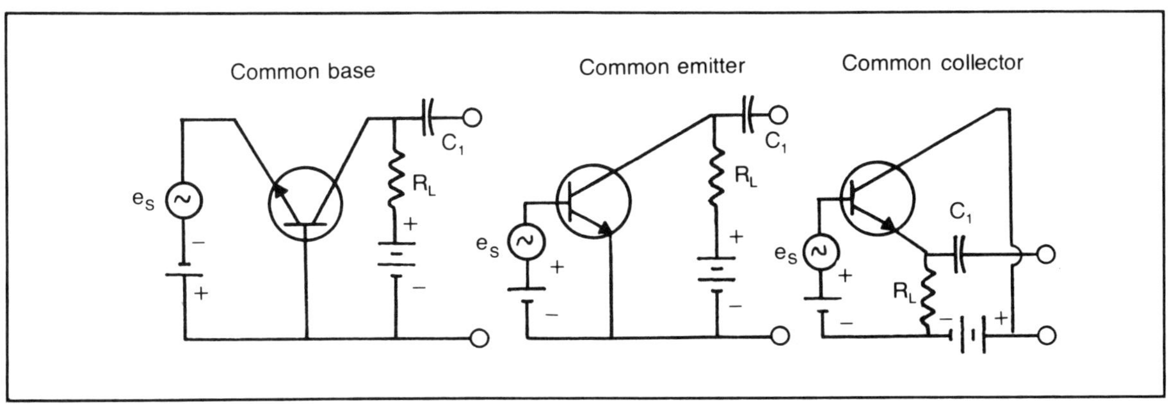

Fig. 1-34. Three basic transistor circuit configurations.

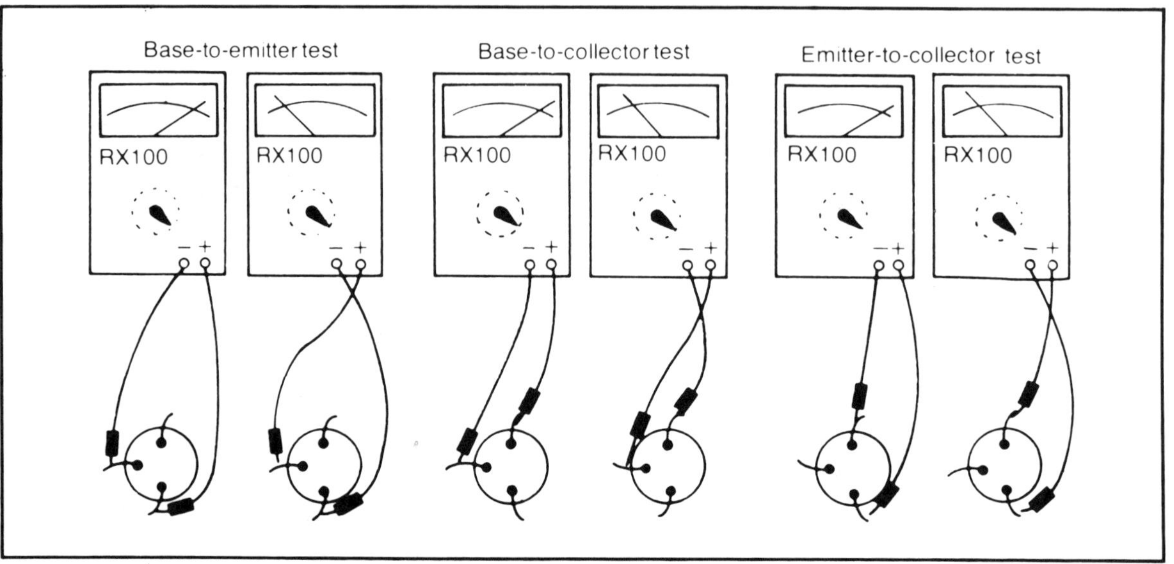

Fig. 1-35. Checking a transistor for shorts or opens using an ohmmeter.

its outward appearance is similar to the bipolar transistor (npn's and pnp's), construction is different. The FET consists of three terminals—source, gate, and drain—which correspond to the emitter, base, and collector of the bipolar transistor (Fig. 1-37).

Current flows between the source and drain, the "resistive" (i.e., semiconductor-substrate channel) parts of the FET. The gate is a diode-junction that is reverse biased, rather than forward biased as in the bipolar transistor. Therefore, the gate has very high resistance allowing for a very high input impedance desired in many circuits.

The field-effect transistor that is a junction-gate type is called a JFET (junction field-effect transistor). The JFET can be checked with an ohmmeter like the bipolar transistor. The ohmmeter (R×100) shows diode action (high/low resistance) between the source and gate and likewise between the drain and gate. Two high readings indicate an open transistor, and two low readings indicate a short. The ohmmeter reading between the source and

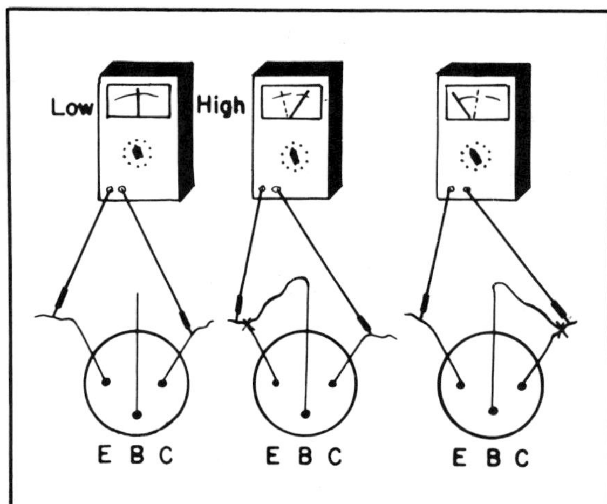

Fig. 1-36. Checking the quality of a transistor with an ohmmeter.

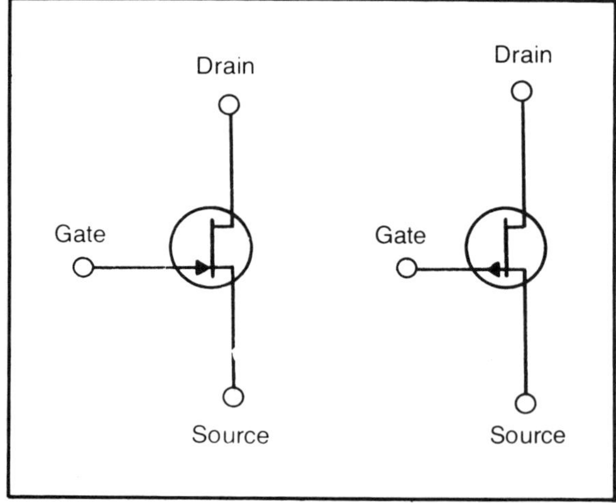

Fig. 1-37. Schematic symbols for N and P channels field effect transistors.

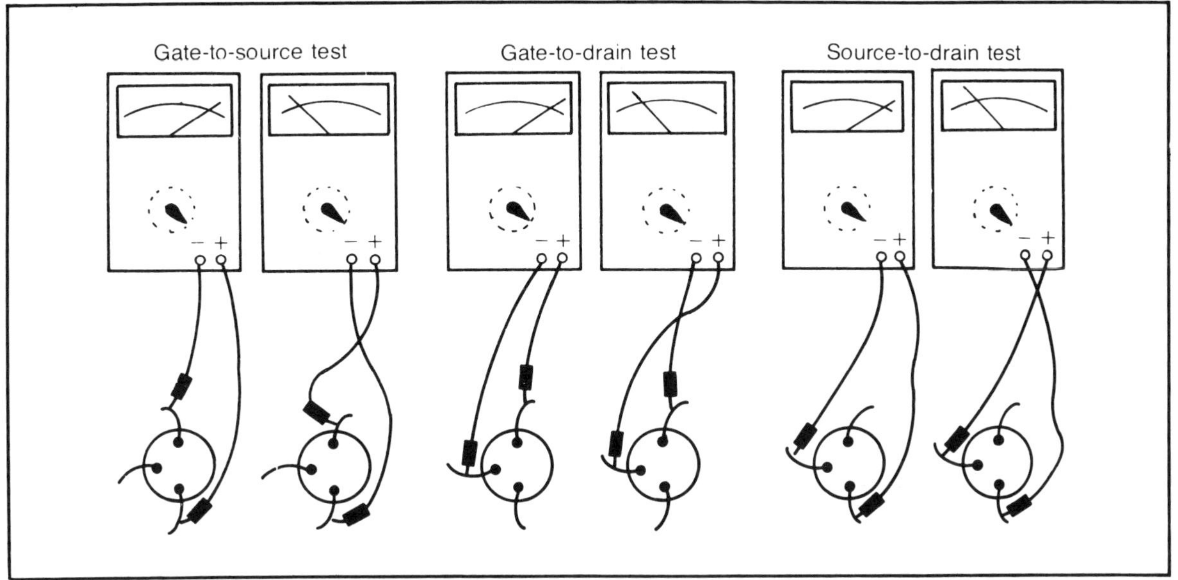

Fig. 1-38. Checking a JFET for opens and shorts using an ohmmeter.

drain shows a high resistance in either polarity in a good transistor. Two lows indicate a short (Fig. 1-38).

A MOSFET is another transistor called a metal-oxide-semiconductor field-effect transistor. This transistor is referred to as an IGFET type (insulated-gate field-effect transistor) because the gate is electrically insulated from the source-drain channel (i.e., semiconductor substrate) by a thin layer of silicon dioxide (Fig. 1-39). The MOSFET can be a P-channel or N-channel type. The current flow in a P-channel is reduced by applying a positive voltage and increased by applying a negative voltage. Moreover, three basic types of MOSFETs include the "enhancement," "depletion," and "depletion enhancement." The enhancement type conducts from source to drain when forward-biased and remains "cut off" (i.e., no current flow) at zero bias. The depletion type conducts at zero bias, is reversed with forward bias, and is reduced with reverse bias. With enough reverse

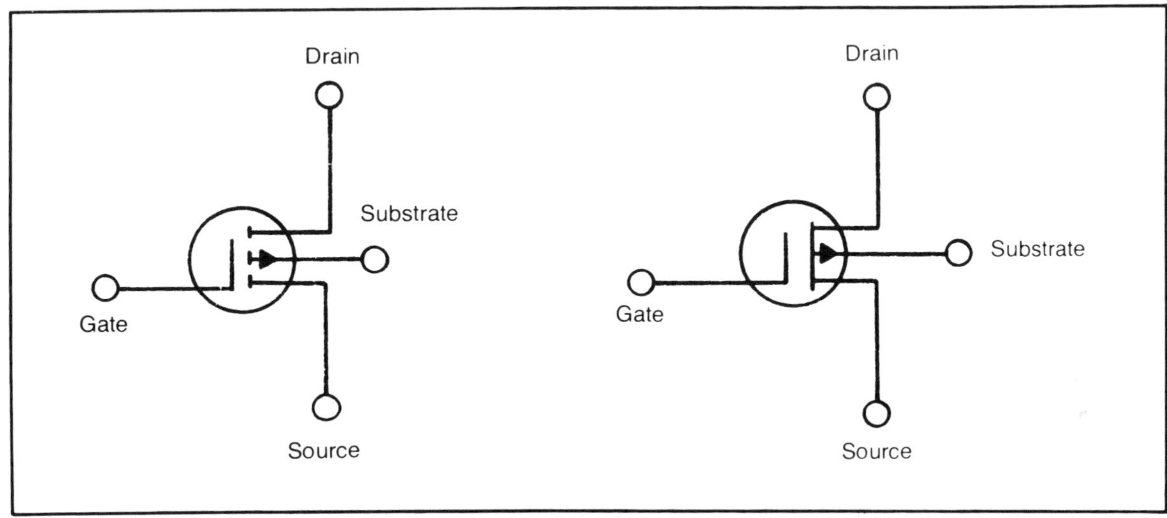

Fig. 1-39. Schematic symbol for enhancement-type MOSFET. Schematic symbol for depletion-type MOSFET.

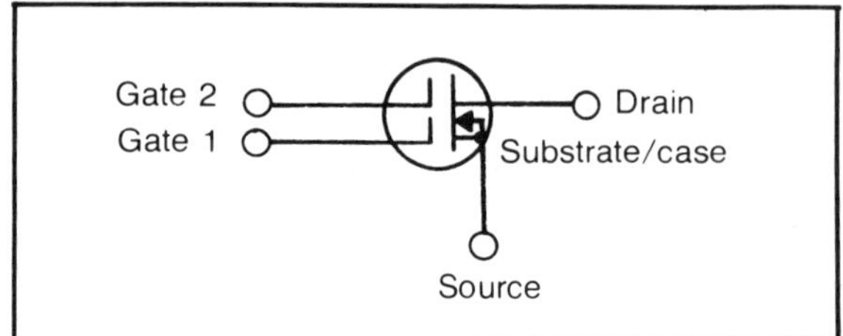

Fig. 1-40. Schematic symbol for dual-gate N-channel depletion MOSFET.

bias, it can be cut off. The "depletion-enhancement" type has some conduction with zero bias. Current is reduced with negative bias and increased with positive bias. MOSFETs have high input impedance. They are also sensitive to static electricity and must be carefully handled.

For this reason, the gate is kept shorted to the source by twisting their leads together during shipping and handling or by a spiral shorting-spring. Dual-gate protected MOSFETs help overcome this problem by reducing the internal load. Figure 1-40 shows the schematic symbol for this type MOSFET. The dual-gate MOSFET acts as a single-gate MOSFET by connecting the gate leads together. The MOSFET can be checked by using an ohmmeter. There should be zero resistance between the gate and source or drain. A reading on the ohmmeter indicates a short. To check the drain-source condition, place a 15 K resistor from the gate to drain. With the ohmmeter leads on the source and drain, a change in resistance suggests the MOSFET is okay. Note, however, the best test is by substitution or by use of test instruments (Fig. 1-41).

There are several different troubleshooting techniques. Many of these can be used to *directly* or *indirectly* test the performance of a transistor. Other techniques besides resistance and component checkers are as follows:

1. Voltage measurements.
2. Heat/freeze.
3. Signal tracing.
4. Substitution.
5. Transistor *cutoff*.

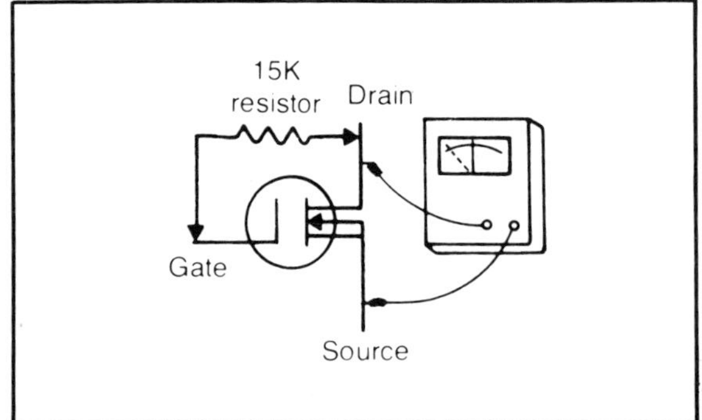

Fig. 1-41. Checking a MOSFET for opens and shorts using an ohmmeter.

Voltage readings can be very useful in determining transistor circuit action. For example, the transistor in Fig. 1-42 is given the manufacturer's specified normal operating voltages. If the transistor were *open* or not conducting, the transistor would not draw current and the voltage at the collector would not be 6 volts but the full source voltage of 10 volts. If the transistor were shorted, the transistor would draw excessive current. This would *load down* the circuit. Therefore, if the voltage at the collector were low, this could very well indicate a shorted transistor or faulty bias resistor.

Often transistors can be checked by simply applying heat from a hot blower, such as a hair dryer. If the heat causes the transistor to break down, and applying a chemical "freeze" mist coolant or *cool air* from a fan will restore the transistor to normal operation, the transistor can be diagnosed as defective. This technique is commonly used in locating *thermal intermittent* transistors. Thermal intermittent transistors are defective transistors that usually break down after lengthy operation. Temperature rise increases the current in the transistor; this increased current conduction in turn produces more

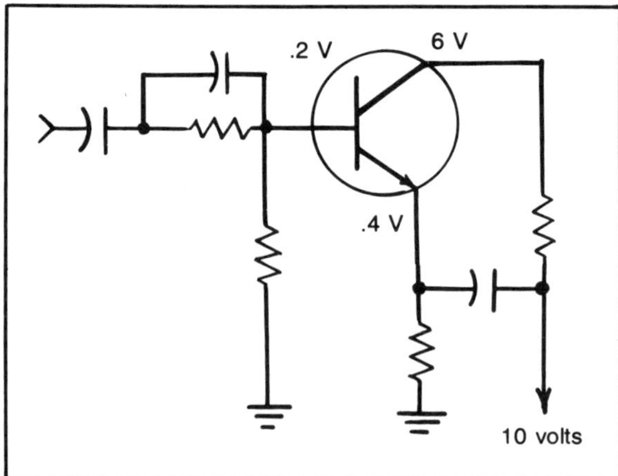

Fig. 1-42. Typical operating voltages of a transistor.

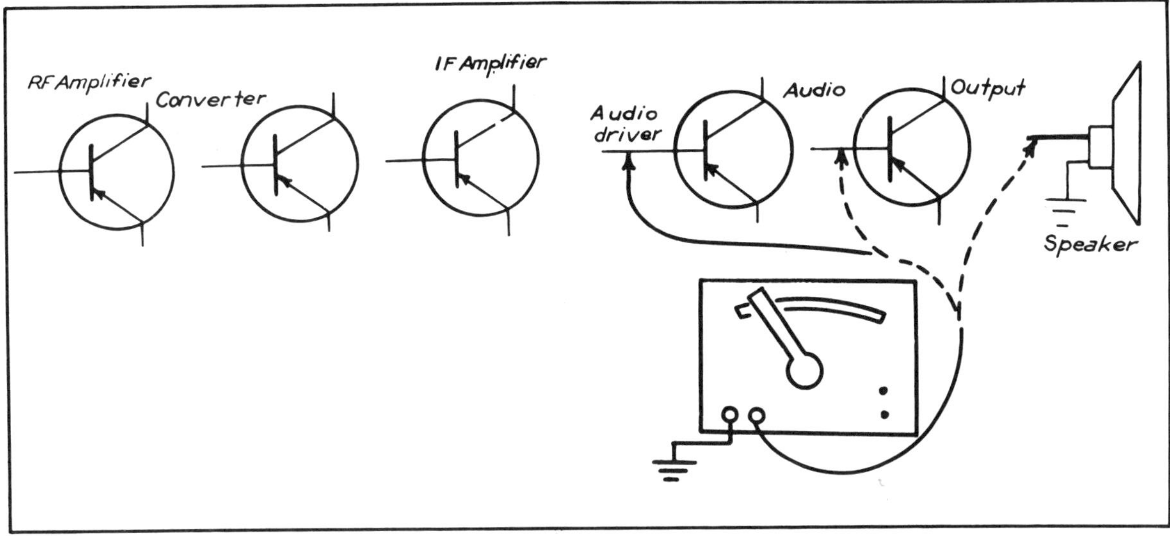

Fig. 1-43. Signal tracing of a radio receiver.

heat that causes the transistor to draw even more current. Eventually the transistor destroys itself. This principle is called *thermal runaway*. Remember, do not apply too much heat to the transistor. The transistor may be good and could be damaged by the excessive heat.

Signal tracing can also be used to isolate a particular defective transistor. For example, if a signal is injected into each stage of a malfunctioning transistor receiver starting at the speaker and working backward, the defective (open) transistor would prevent the signal from reaching the speaker. The signal injection points are shown in Fig. 1-43.

The transistor substitution technique can be effective in determining a defective transistor. Keep in mind that when substituting a transistor be sure a similar transistor is used. Many technicians prefer to "tack on" the substitution transistor to the copper clad side of the circuit board first to see if the suspected transistor really is bad. This can save the technician valuable time.

Another technique a technician often uses to determine if a transistor is operating is to short the base to emitter which *cuts off* the transistor (Fig. 1-44). A noticeable difference in the overall operation of the equipment should result if the transistor is working. If no noticeable difference is indicated, the transistor is most likely defective. Use caution when you do this test and make sure you *don't* short the base to the collector, since this will cause the transistor to draw excessive current and destroy itself.

The transistor cutoff technique can be compared to locating a bad sparkplug wire on your car. While the automobile is idling, each plug wire is lifted off for a second and the idling operation of the engine is noticed. If the removed sparkplug wire causes the engine to idle *rougher*, the plug is good, but if the performance of the engine is unchanged, the plug wire is probably defective.

It should be mentioned at this point that when transistors are replaced, certain precautions should be noted. These precautions are as follows:

1. Never overheat transistors.
2. Use a heatsink.

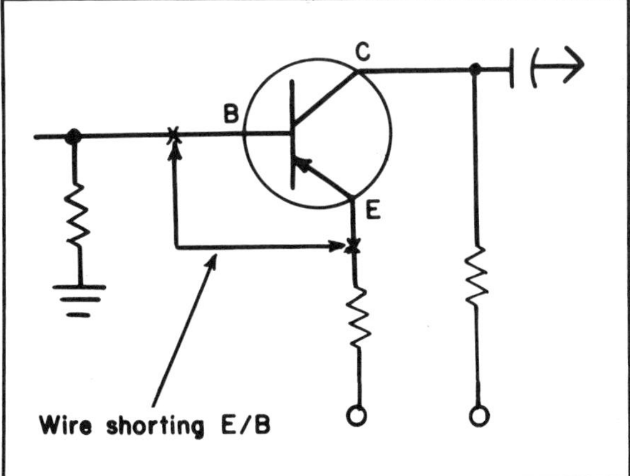

Fig. 1-44. Cutting off a transistor by shorting the emitter to base.

3. Use a 35-watt solder iron.
4. Use an exact or recommended replacement type.
5. Identify E,B,C positions.

The SCR (silicon-controlled rectifier) is formed when three diodes are arranged back to back (Fig. 1-45). The SCR acts like an electronic switch and consists of an anode, gate, and cathode. The SCR acts like a rectifier, except that conduction in the forward direction can only take place when sufficient voltage *triggers* the gate, at which time the SCR conducts as long as sufficient holding voltage is maintained.

The SCR is a popular device used in burglar alarms and automatic control circuits. It can best be checked by substitution or an ohmmeter.

When checking an SCR with an ohmmeter, place the selector knob on the R × 10,000 scale. With the negative lead connected to the cathode and the positive lead to the anode a good SCR should read over 1 megohm. Zero ohms would indicate a shorted SCR. To see if the gate junction is operating, short the gate lead to the anode—the meter should read zero ohms.

INTEGRATED CIRCUITS

The actual theory of IC (integrated circuits) is not the intention of the author, but rather to present basic troubleshooting techniques.

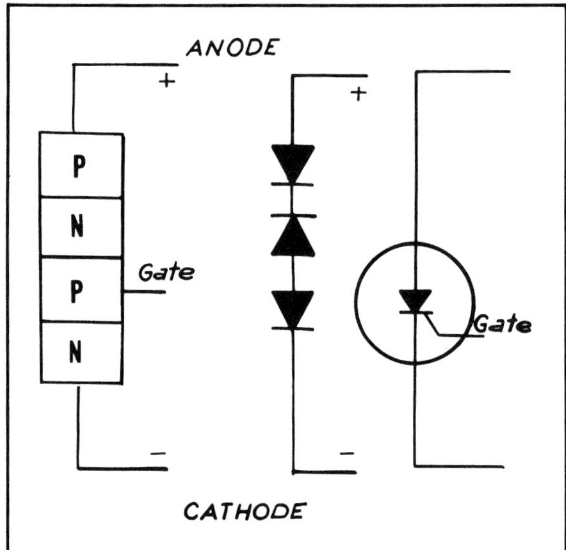

Fig. 1-45. The construction of an SCR (silicon controlled rectifier).

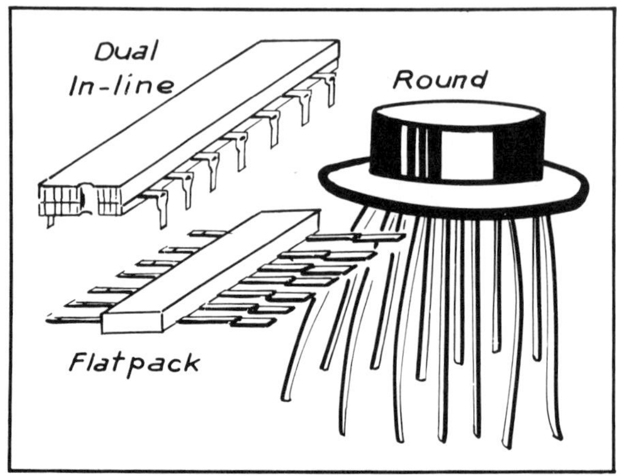

Fig. 1-46. The three basic configurations of an IC integrated circuit.

The basic three types of integrated circuits are as follows:

1. Dual-in line type.
2. Round type.
3. Flat type

Figure 1-46 illustrates these three basic types. The IC basically consists of many micro-size components. One small IC may consist of several resistors, capacitors, diodes, and transistors, all connected into a micro circuit. They are hermetically sealed in a ceramic or plastic package. Figure 1-47 shows an example of a typical IC diagram.

Although there are various shapes, types, and sizes of integrated circuits, the troubleshooting techniques employed are common to all of them.

1. Use of senses.
2. Heat/freeze.
3. Voltage check.
4. Capacitor bypass.
5. Substitution.
6. Logic pulse probe.

The first step in troubleshooting an IC is to use your *senses*. Look for obvious problems such as corroded, defective, or damaged pins, sockets, or solder connection. Make sure the IC is completely inserted in its socket. Check the component identification number of the IC with those of the manufacturer to make sure the correct IC is in the circuit and is correctly positioned.

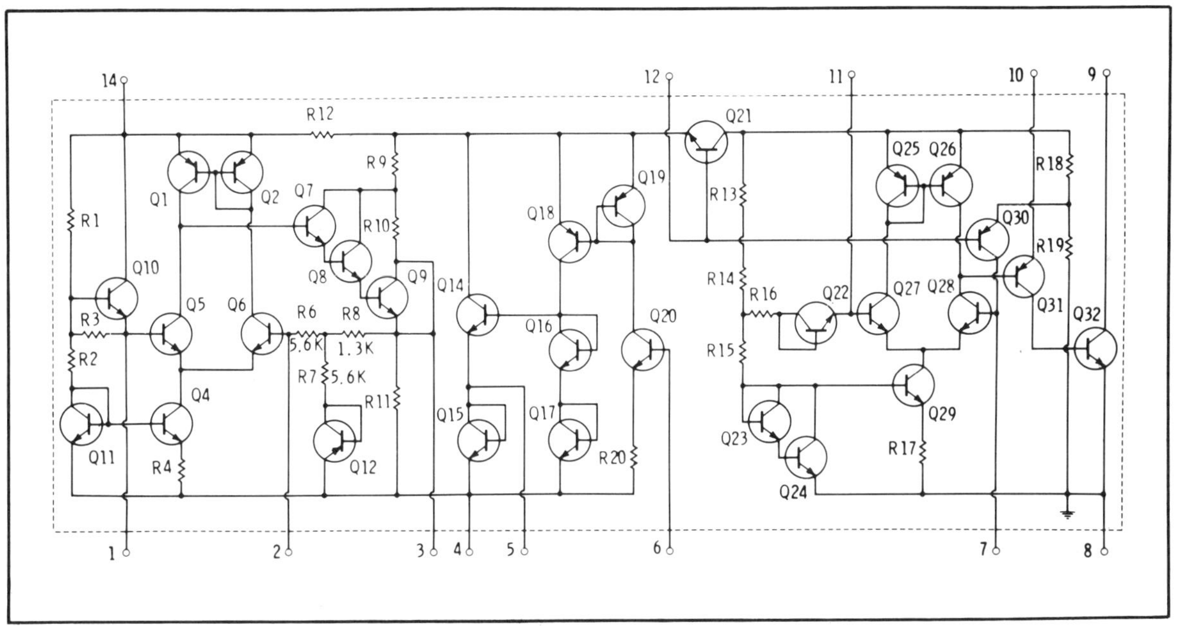

Fig. 1-47. Typical internal diagram of an IC audio preamp and a 1 watt audio output using ECG 1043 (courtesy Sylvania, Inc.).

Touch is one technique that many service technicians use. Using your finger while the circuit is in operation, touch the top insulated case of the IC and note its temperature. A hot IC is a good indication of a defective or *shorted* component. Most IC components should feel cool to warm when touched.

Heat and freeze is another technique often used to check for a defective IC. As stated before, a suspected thermal intermittent component can be checked by first heating the component with a hot blower—noting the performance of the circuit—and then cooling or freezing the component. The defective thermal intermittent IC should break down when heated but operate again when cooled off.

Voltage checks can be performed easily with a voltmeter. Simply measure the voltage of each pin of the IC and compare it with that of the manufacturer's operating voltages. An incorrect voltage reading probably indicates a faulty IC (Fig. 1-48).

Sometimes a suspected faulty IC can be "jumped" by using a capacitor. The *capacitor bypasses* the signal around the IC (Fig. 1-49). If the signal increases when the IC is jumped with the capacitor, the IC is probably defective.

Any suspected IC should be replaced by a similar known good IC. This technique of *substitution* saves valuable time for the service technician. Realistically speak-

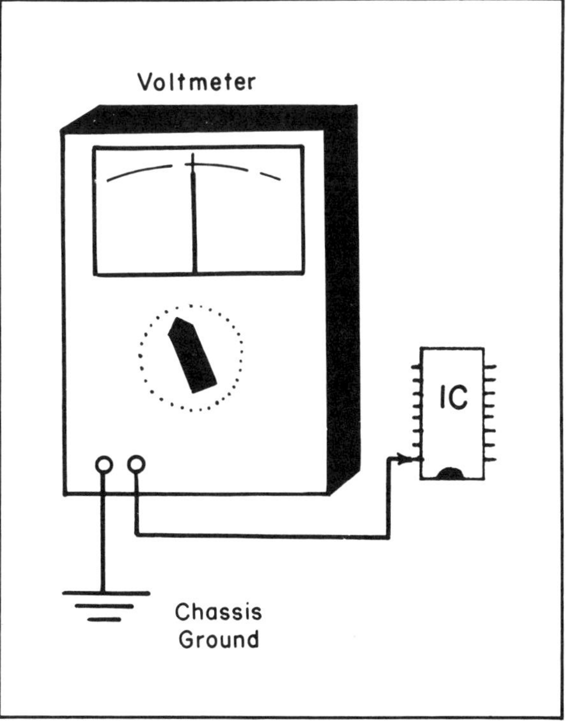

Fig. 1-48. Checking voltages on each pin of an IC.

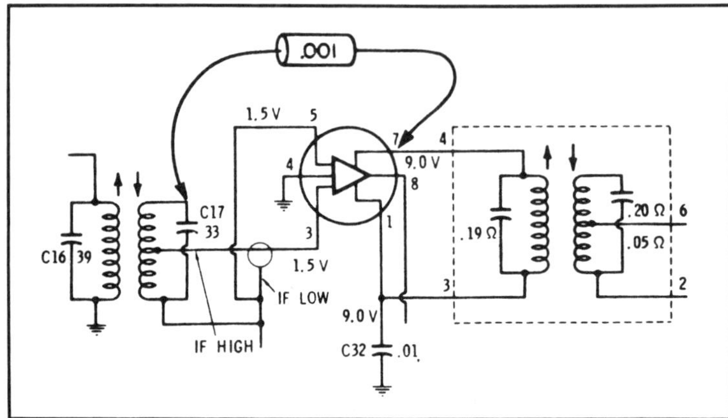

Fig. 1-49. Bypassing an IC with a capacitor (courtesy Motorola, Inc.).

ing, the reason technicians don't solely use this technique is because it would require having a large inventory of ICs on hand which would be costly. Also keep in mind that if the root of the problem is not a bad IC, replacing a bad IC with a good one could destroy the good one.

The *logic digital IC probe* is probably one of the most important test instruments used by service technicians. This small, hand-held probe is generally used to test logic pulses and levels. The probe contains a complex circuit which identifies, through use of LEDs (high, low, medium, open), operating logic level responses. Like the voltmeter, the logic probe is applied to each IC pin or test point. The response is compared to the manufacturer's data.

When a defective IC has been found, replace it by using the following few tips.

1. Order the exact replacement.
2. Insert or position the IC exactly as the original IC. It is extremely easy to put an IC in backwards! Always identify the number 1 pin of the IC; this is often indicated by the manufacturer placing a small dot next to this pin.
3. When inserting a dual in-line 16 pin IC into the socket, it is easy to miss and smash at least one of the 16 pins. Make sure all 16 pins are aligned properly before pressing the IC fully down into the socket.
4. Never overheat an IC. If the IC has to be soldered into the circuit, use a small 35-watt soldering iron.

ELECTRON TUBES

Since electron tubes are rarely used today, little mention of their theory will be discussed. Basically, besides the CRT (cathode ray tube) the most common tubes are the diode, triode, tetrode, pentode, gas filled, and multi-element. Figure 1-50 gives the symbol for each of these tubes.

The diode tube consists of a negative cathode and a positive plate. In Figure 1-51, the negative cathode when heated gives off electrons and conduction takes place. The process of giving off electrons from the cathode is called *thermionic emission*. When the polarity is re-

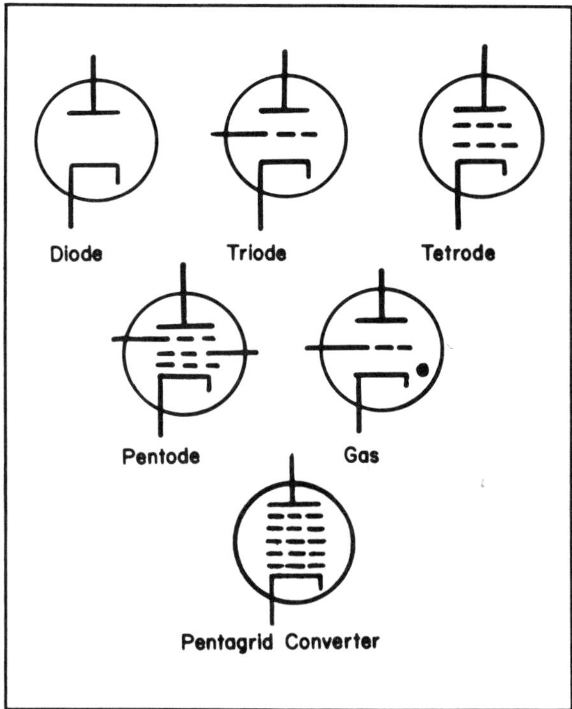

Fig. 1-50. Common types of electron tubes.

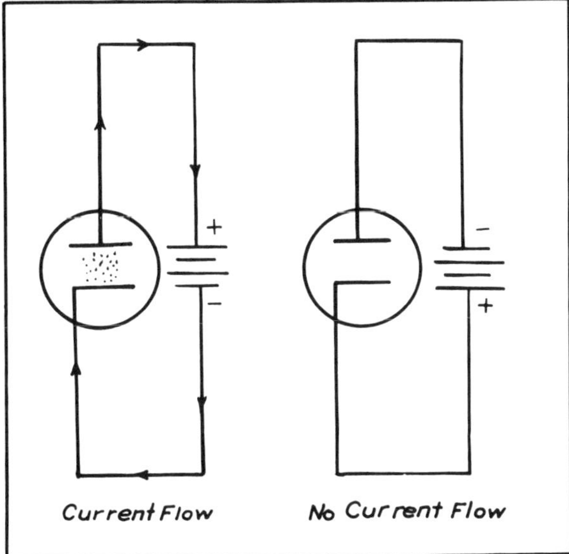

Fig. 1-51. Conduction in a diode tube.

versed, no thermionic emission takes place, and no current flows. This valve-like action serves as a one-way gate and is, therefore, used as a rectifier.

The amount of electrons that reach the plate from the cathode in the triode tube is controlled by placing a fine-mesh wire called a grid. This control grid is made negative in relation to the cathode. The more negative the grid, the less the current flows; the less negative the grid, the more the current flows. *Cutoff* is the point where the grid is made too negative and current flow stops. *Saturation* is the point where the grid is at the least negative point and current flow is at a maximum. Figure 1-52 shows a triode tube with a bias battery connected

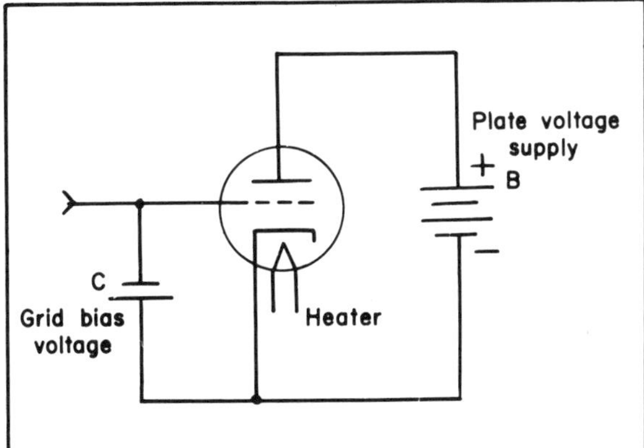

Fig. 1-52. Typical circuit of a triode tube.

between grid and cathode (note negative side of battery connected to grid).

In order to prevent *interelectrode capacitance,* an undesired effect in the triode, a second grid, called a *screen grid* is added in the tetrode tube (Fig. 1-50).

When increased performance under certain applications is desired, a third grid is added. It is called a *suppressor grid* (Fig. 1-50). The pentode which contains this grid eliminates secondary emission (uncontrolled acceleration electrons near the plate) by controlling these accelerated electrons.

Power tubes are largely to provide high power amplification. Gas tubes are often filled with nitrogen or mercury vapor and are used in high current applications. The *thyratron* tube is a common example of a gas tube. Multielement tubes are tubes consisting of two or more *tubes* which are enclosed in the same glass envelope. The *pentagrid* converter is a common example of a multi-element tube. It consists of both the local oscillator and mixer stages in a receiver (Fig. 1-50).

The few techniques used to test electron tubes are as follows:

1. Tapping.
2. Visual.
3. Tube checker.
4. Substitution.

Fig. 1-53. Determining the quality of a tube by tapping on it.

To examine the quality of a tube, while the circuit is operating, use the plastic end of a screwdriver. Gently tap the top of each tube while listening or watching the performance of the circuit (Fig. 1-53). If any type of noise is heard or seen while tapping on the tube, or if there is any change in the picture, the tube is probably defective. Keep in mind that a loose connection or poor soldering connection in the same area could cause the same problem.

Another way to quickly examine the quality of some tubes is to visually see if the heater is lit. If the heater is *open*, no glow will be produced, and the tube will not operate. An ohmmeter can also be used to check the heater. A good heater will read near zero ohms, but an open heater will read infinity (Fig. 1-54).

Although the tube tester can be a helpful instrument for the service technician, it can also be a handicap. For example, there is nothing more comical than to see a "do it yourselfer" trying to play technician by pulling out every tube from a TV set and taking them all to the local drugstore to be tested on a tube tester. Undoubtedly, the tube tester will indicate that many of the tubes need to be replaced but will often fail to pinpoint the one tube that is causing the problem. Valuable time and money can be saved by understanding the function of each tube in a circuit and how to properly use a tube tester. For example, tube testers cannot match the operating conditions of the circuit. They cannot adequately measure interelectrode capacitance. Also, oscillators, limiters, and high voltage tubes (where characteristic curves are critical) are difficult to test on the tube tester. The best word of advice to follow is "when in doubt, replace the tube."

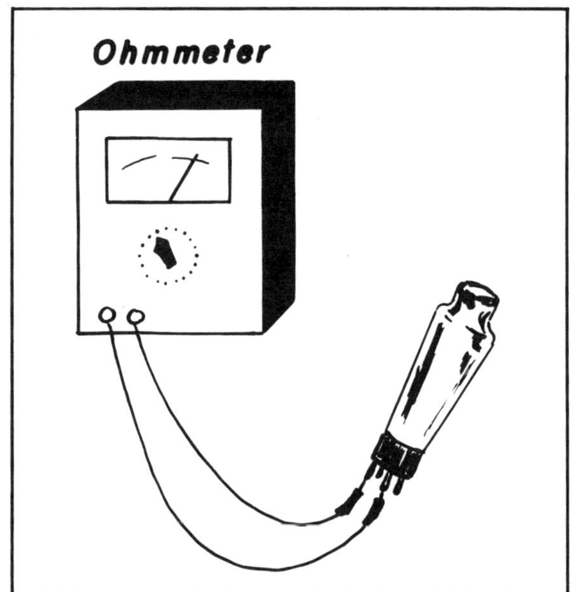

Fig. 1-54. Checking the heater of a tube with an ohmmeter.

Substitution of a known tube for a suspected defective tube can be a time-saver for the service technician. Remember, if the tube failed because of a circuit problem, the substitution of a new tube will only ruin the new tube. For example, when servicing a superheterodyne receiver, if a 35W4 rectifier tube is found shorted, also look for a shorted filter capacitor. Most likely the filter capacitor caused the rectifier tube to short. Also, before replacing any tube, it is a good idea to visually inspect the area for charred resistors or any other problem that could have caused the tube to fail.

Unlike the transistor, which could theoretically last forever, the life of an electron tube is limited because the cathode simply wears down and, with time, emits less and less electrons. Also, mechanical vibration, excessive heat, and current all contribute to tube breakdowns.

When replacing a tube, be sure to use the exact replacement or recommended substitute and make sure the tube socket is clean and free of corrosion. Also, be careful not to bend the tube pins.

SELF-EXAMINATION

Select the best answer:

1. Which of the following is not an example of a source of breakdown?
   - A. Heat.
   - B. Moisture.
   - C. Poor installation.
   - D. Animals and rodents.
   - E. None of the above.
2. Which of the following is not one of the senses commonly used by service technicians?
   - A. Sight.
   - B. Hearing.
   - C. Touch.
   - D. Taste.
   - E. Smell.
3. A hot smoky product or device is often a sign of:
   - A. A short.
   - B. A ground.
   - C. An open.
   - D. All of the above.
   - E. None of the above.

4. A circuit that has infinite resistance is called __________ circuit.
   A. A short.
   B. A ground.
   C. An open.
   D. All of the above.
   E. A or C.
5. Voltage measurements are often taken using a voltmeter or:
   A. An ammeter.
   B. An oscilloscope.
   C. An ohmmeter.
   D. A watt meter.
   E. None of the above.
6. Signal injection/tracing is a method commonly used in troubleshooting:
   A. Electric motors.
   B. Residential wiring.
   C. Industrial wiring.
   D. Radio.
   E. Any of the above.
7. A technique whereby a suspected defective component is replaced by a "good" component is called:
   A. Bypassing.
   B. Substitution.
   C. Bridging.
   D. Both B and C.
8. A cold solder connection can best be repaired by:
   A. Substitution.
   B. Bridging.
   C. Resoldering.
   D. Cooling.
   E. Freezing.
9. When using a "step-by-step" analysis approach to troubleshooting, the first step should be:
   A. Discussion of defect with customer.
   B. Acquisition of service information.
   C. Selection of troubleshooting technique.
   D. Repairing of the problem.
   E. All of the above.
10. The type of diagram that illustrates the component parts of a product or device is called a:
   A. Line drawing.
   B. Schematic diagram.
   C. Blueprint.
   D. Pictorial diagram.
   E. Schematic print.
11. A component having no continuity would have __________ resistance.
   A. Zero.
   B. Infinite.
   C. Both A and B.
   D. None of the above.
   E. Small.
12. A good fuse will have __________ resistance.
   A. Zero.
   B. Infinite.
   C. Small.
   D. Both A and B.
   E. None of the above.
13. The physical size of a resistor that determines the ability of the resistor to absorb heat is rated in:
   A. Ohms.
   B. Volts.
   C. Watts.
   D. Farads.
   E. None of the above.
14. The new fully charged lead acid storage battery should measure:
   A. Over 12 volts.
   B. 2 volts.
   C. 11 volts.
   D. 12 volts.
   E. None of the above.
15. Capacitors can be tested by:
   A. An ohmmeter.
   B. A spark test.
   C. Bridging.
   D. Only B and C.
   E. All A, B, and C.
16. In order to make a p-type crystal:
   A. A pentavalent of galium is added.
   B. A trivalent of indium is added.
   C. A pentavalent of antimony is added.
   D. A trivalent of arsenic is added.
   E. None of the above.
17. The term acceptor is referred to in the:
   A. Addition of pentavalent to the crystal.
   B. Addition of trivalent to the crystal.
   C. Both A and B.
   D. None of the above.
18. Actually a transistor is:
   A. 1 diode back to back.

B. 2 diodes back to back.
C. 3 diodes back to back.
D. 4 diodes back to back.
E. None of the above.

19. High voltage gain and low current gain are characteristics of the:

A. Common base circuit.
B. Common emitter circuit.
C. Common collector circuit.
D. Both A and C.
E. None of the above.

20. If the operating voltage at the collector of the transistor was much lower than normal, one could suspect:

A. A defective filter.
B. An open resistor.
C. An open transistor.
D. A shorted transistor.
E. None of the above.

21. To cut off a transistor for troubleshooting purposes:

A. Short E and B.
B. Short B and C.
C. Short gate to anode.
D. Either A or B.
E. None of the above.

22. The SCR is considered to be ________ diodes back to back consisting of an anode, cathode, and ________.

A. 2, plate.
B. 3, gate.
C. 4, base.
D. 2, emitter.
E. 3, plate.

23. When checking thermal intermittent integrated circuits the suggested best troubleshooting technique is:

A. Voltage.
B. Resistance.
C. Heat/freeze.
D. Current.
E. Bridging.

24. The tube that contains 3 grids is the:

A. Triode.
B. Tetrode.
C. Pentode.
D. Multielement.
E. Power.

25. Which of the following techniques is not common in troubleshooting electron tubes?

A. Tapping.
B. Tube checker.
C. Bridging.
D. Substitution.
E. Both A and C.

26. The gate in the field-effect transistor is generally:

A. Reverse biased
B. Forward biased
C. Zero biased
D. None of the above

27. The MOSFET is often referred to as a:

A. FET.
B. Bipolar transistor.
C. IGFET.
D. SCR.

28. The "depletion" MOSFET conducts at:

A. Forward bias.
B. Reverse bias.
C. Zero bias.
D. None of the above.

29. Current is reduced in the "depletion-enhancement" MOSFET with:

A. Positive bias.
B. Negative bias.
C. Zero bias.
D. None of the above.

30. The dual-gate MOSFET acts as a single-gate MOSFET by connecting the ________ leads together.

A. Gate.
B. Drain.
C. Source.
D. Emitter.

QUESTIONS AND PROBLEMS

1. List and explain six reasons for breakdowns.
2. List and explain four senses commonly used in troubleshooting.
3. What are the effects of breakdown causes? Also list how each one is different.
4. What are the characteristics of a short circuit?
5. What are the characteristics of an open circuit?
6. What are the characteristics of a grounded circuit?
7. What are the characteristics of a mechanical problem in a circuit?

8. Explain the difference between the terms bridging and substitution.
9. Explain the technique of signal tracing.
10. List and explain different types of service diagrams.
11. Explain the techniques in troubleshooting capacitors.
12. Name several different types of capacitors.
13. Explain the structure of a diode.
14. Explain how to test a diode.
15. What is a diode crystal detector?
16. Explain the structure of a transistor.
17. Explain how to test a transistor.
18. Explain how to test an SCR.
19. Explain the various techniques used to troubleshoot transistors.
20. Explain the various techniques used to troubleshoot in integrated circuits.
21. Explain the difference between several types of electron tubes.
22. Explain the few basic ways to check an electron tube.
23. Why should tube testers be used with caution in testing tubes?
24. What is a CRT?
25. What is a thyratron tube?
26. Explain how to check the field-effect transistor.
27. What is a MOSFET?
28. Explain the different types of MOSFETs.
29. Explain how to check a MOSFET.
30. What special consideration must be taken when shipping or handling the MOSFET?

Chapter 2

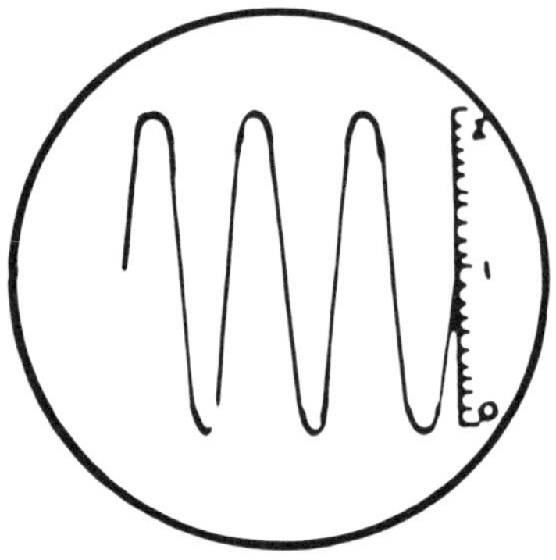

Test Equipment

Most electrical/electronic products and devices could not be effectively serviced without the aid of test equipment. There are hundreds of different types of test equipment used today. The use of this test equipment helps the service technician's speed and accuracy in troubleshooting and isolating the problem. In this chapter, some of the most popular types of test equipment used by technicians and servicemen are presented.

VOM, VTVM, DIGITAL MULTIMETER

The most commonly used voms (volt ohm milliammeter) measure ac/dc voltages, current and resistance values. Most of these meters are portable and require no external power source for operation (Fig. 2-1). They often provide many other features, such as measurement of microvolts and decibels. Overload protection is usually provided by an internal fuse or circuit breaker. Since this meter is often portable, it has become a very popular meter for electricians and electronic technicians, especially those in industry.

One disadvantage of the vom is that the impedance of the meter itself can, under certain conditions, "load down" the circuit and affect the measurement of voltage. Therefore, it is not uncommon for the voltage measurement to sometimes be inaccurate. Small, inaccurate measurements of voltage do not usually present a problem for industrial electricians. It does, however, present a problem for the electronic technician. Small inaccuracies in measuring voltage may greatly affect the electronic diagnosis.

The solid-state FET vom overcomes this problem of "loading down" a circuit by its high input impedance and precision-regulated internal power supply. This vom is a versatile, portable meter for field and factory servicing and design testing.

A dual-triode is commonly used in the vtvm (vacuum tube voltmeter) circuit. The meter needs an external source of power, and so is less portable than the vom. The vtvm has high input impedance and is fairly accurate. The vtvm (Fig. 2-2) tends to be a very popular meter for "bench servicing." Some vtvm's are designed with solid-state devices which have increased their flexibility.

The digital multimeter (Fig. 2-3) tends to be the most popular meter for electronic technicians who need extreme accuracy as in laboratory work and digital equipment testing and servicing. This meter uses circuits that produce numerical readouts using LEDs, eliminating the need to interpret the needle postion used on analog-type meters. Nickel-cadmium batteries are

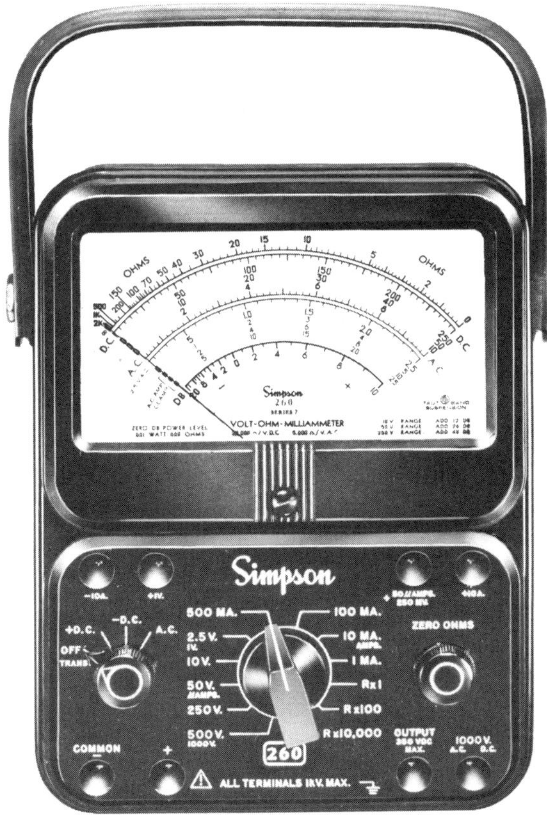

Fig. 2-1. A typical vom (courtesy Simpson Electric Co.).

generally used in this meter and need to be regularly recharged.

OSCILLOSCOPE

The basic advantage of the oscilloscope is that it provides a visible display of the waveform being measured. Most oscilloscopes use electrostatic deflection. The beam sent from the electron gun is deflected vertically or horizontally by pairs of vertical and horizontal plates. Figure 2-4 illustrates the position of these plates.

Although the oscilloscope is largely used to measure peak to peak voltage, other measurements that can be taken are frequency, time periods, waveslopes, phase angles, and frequency response. Figure 2-5 illustrates a typical oscilloscope.

The basic operating controls and their functions in an oscilloscope are as follows:

1. *Intensity*—controls the brightness of the electron beam.
2. *Focus*—adjusts the sharpness of the beam.
3. *Vertical Center Control*—controls the vertical positioning of the electron beam.
4. *Horizontal Center Control*—controls the horizontal positioning of the electron beam.
5. *Vertical Gain*—adjusts the height of the waveform.
6. *Horizontal Gain*—adjusts the width of the waveform.
7. *Sweep Control*—adjusts the frequency of the horizontal sweep oscillator.
8. *Sync Selector*—permits use of an internal or external synchronization.
9. *Z-Axis*—varies the intensity of modulation of the trace.
10. *Calibration Scale*—permits a scale for measurement of voltage waveforms.

The basic setup for an oscilloscope consists of the following:

1. Turn intensity, focus, gain, sync amplitude, to minimum.

Fig. 2-2. A typical vtvm (courtesy B & K-Precision/Dynascan Corp.).

Fig. 2-3. A typical solid-state digital multimeter (courtesy B & K-Precision/Dynascan Corp.).

2. Turn vertical/horizontal controls to mid-range.
3. Turn on scope and adjust intensity control to minimum brightness.
4. Allow one to two minutes for the scope to warm up and then adjust the focus control for a sharp trace.
5. Center the trace signal by adjusting the vertical/horizontal controls.
6. Connect a 6.3 volt ac source to the vertical input for calibration.
7. 6.3 volts rms equals 9 volts peak voltage or 18 volts peak to peak and, therefore, should be adjusted to display 1.8 divisions on the CRT screen (Fig. 2-6).
8. Adjust the sync control until a stationary pattern displaying three sine waves appears.
9. Now the oscilloscope is "set-up" and calibrated.

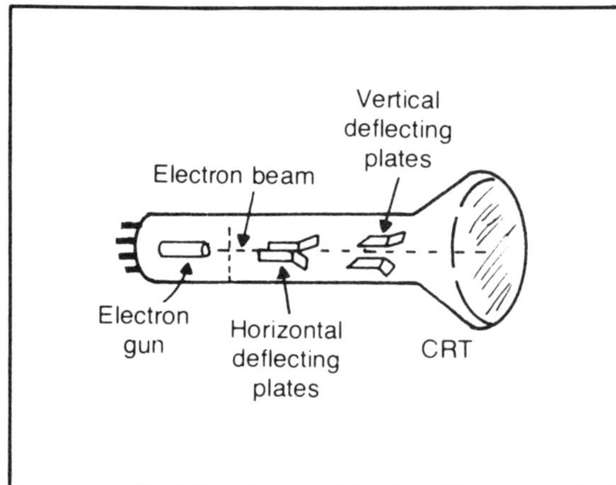

Fig. 2-4. The vertical and horizontal deflection plates of an oscilloscope.

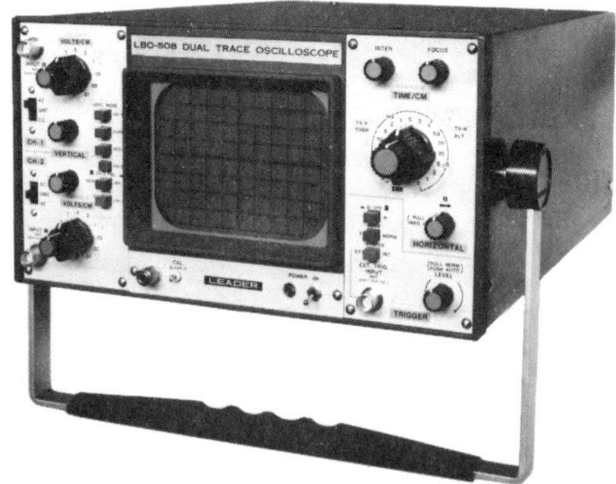

Fig. 2-5. A typical oscilloscope (courtesy Leader Instruments Corp.).

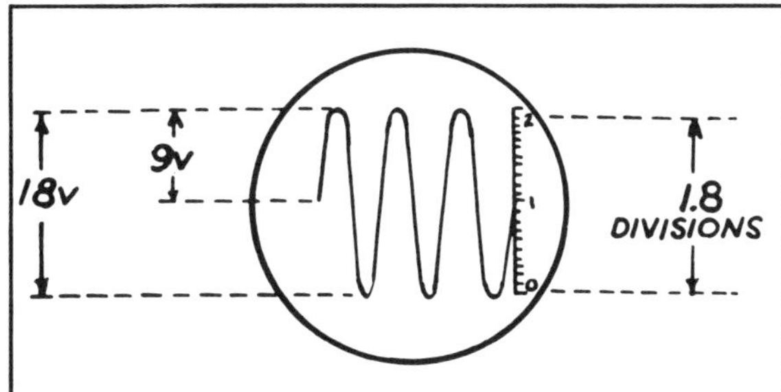

Fig. 2-6. A calibrated display of an oscilloscope.

Each division will now equal 10 volts peak. The vertical attenuator can be used to multiply the divisions by 0.1, 1, 10, etc., as desired for proper measurements.

When calibrating an oscilloscope with an internal calibrator, the scope can be calibrated by adjusting a fixed pattern of 1 volt peak to peak.

When troubleshooting with an oscilloscope, three basic accessory probes are commonly used.

1. Low-capacitance probe.
2. Demodulation probe/rf probe.
3. Voltage divider probe.

The low-capacitance probe is generally used when measuring high frequency or high impedance circuits. The "loading effect" is reduced by using this probe, which increases the accuracy of measurement.

The demodulation probe/rf probe is often used when measuring rf signals where the signal must be detected before being displayed on the scope.

The voltage divider probe is used when the voltage being measured is desired to be "stepped down" (reduced). The usual voltage division ratio is 10:1 or 100:1.

COMPONENT/EQUIPMENT TESTERS

The basic component/equipment testers commonly used are as follows:

1. Tube tester.
2. CRT tester.
3. Transistor tester.
4. Capacitor tester.
5. Frequency counter.
6. Color generator.
7. Megohmmeter.
8. Voltage tester.
9. Growler.
10. Test lamp.
11. Neon voltage tester.
12. Clamp-on ammeter.
13. Signal level meter.

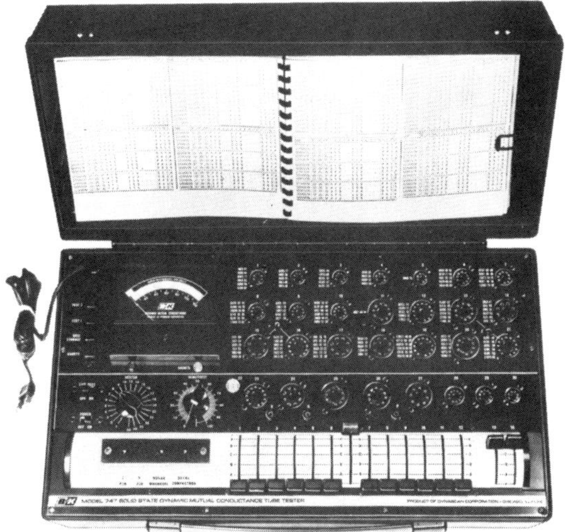
Fig. 2-7. A typical tube tester (courtesy B & K-Precision/ Dynascan Corp.).

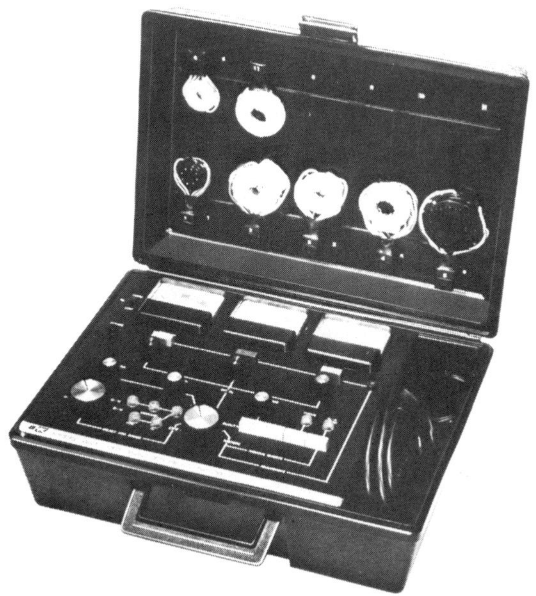
Fig. 2-8. A typical CRT restorer (courtesy B & K-Precision/ Dynascan Corp.).

The tube tester is a fairly accurate way of testing electron tubes. However, it should not be depended upon exclusively for complete accuracy. Tube testers cannot match the operating circuit conditions where the tube is used, such as in a TV or radio (Fig. 2-7).

The CRT tester checks the performance of the CRT. Also, it usually contains a "restorer" circuit. The restorer circuit prolongs the life of the CRT by its ability to rejuvenate the CRT (Fig. 2-8).

Transistor testers are fairly accurate checkers of diodes and transistors. They also have the capability of checking the performance of these components "in" or "out" of circuit. They also measure transistor leakage and Beta and automatically identify emitter, base, and collector leads (Fig. 2-9).

Capacitor testers not only check the quality of the capacitor but also determine the value of unknown capacitors. Many capacitor testers have the ability to check most capacitors "in" or "out" of circuit, which can speed up the servicing time of the product (Fig. 2-10). Also, the capacitor tester can identify power factor values, leakage, opens, problems that are often common to capacitors. Keep in mind that the actual capacitance value of the capacitor can only be accurately measured when the capacitor is out of the circuit. Remember, never touch the terminals of the capacitor tester when the voltage is turned up! Severe shocks can result.

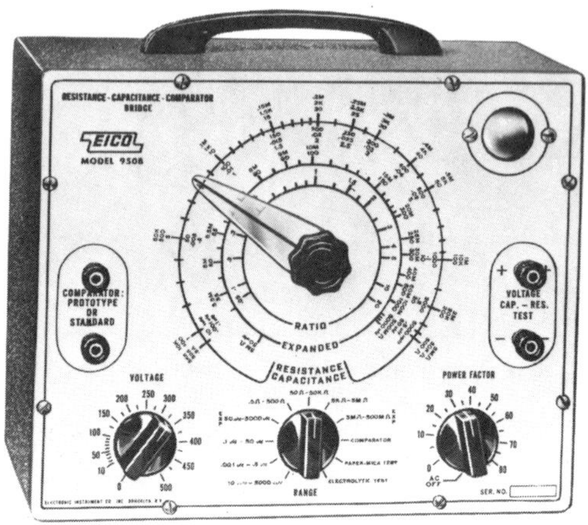

Fig. 2-10. A resistance-capacitance-comparator bridge (courtesy Eico Electronic Instrument Co., Inc.).

Fig. 2-9. A typical transistor checker (courtesy Sencore, Inc.).

Frequency counters are used to measure the frequency in hertz of an electronic product. Frequency counters are often used to adjust the frequency of radio receivers and transmitters. It also is a very valuable instrument used in research and experimentation (Fig. 2-11).

The color generator produces various test patterns which can be used in analyzing and converging color television (Fig. 2-12). It provides such patterns as purity,

Fig. 2-11. A typical frequency counter (courtesy B & K-Precision/Dynascan Corp.).

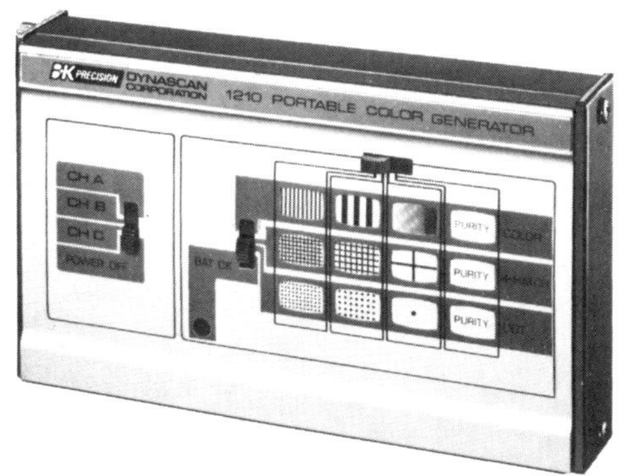

Fig. 2-12. A digital IC color generator (courtesy B & K-Precision/Dynascan Corp.).

dots, crosshatch, gated rainbow, horizontal/vertical lines and other test patterns. These meters are small and lightweight and can easily be carried for in-home servicing.

A megohmmeter is an insulation-resistance meter. It is used to check the electrical resistance of an insulator by indicating the resistance on a scale as it supplies a voltage. The megohmmeter provides the voltage source by either a self contained hand operated generator, by battery or by power supply. The quality, measured in resistance, of an insulator is determined by its ability to withstand the voltage without leakage. Figure 2-13 illustrates a "Megger"® which is a popular hand-operated generator megohmmeter.

Voltage testers are commonly used by electricians in measuring ac voltages (Fig. 2-14). These testers are portable, easy to use, and can be easily carried in the electrician's back pocket.

Growlers consist of two kinds—internal and external—and are used to test armatures and stators of electric motors, generators, and other equipment. Some growlers have a built-in meter that tests for defective coils, reverse coils, or other problems. The growler is highly sensitive, very durable, and can be easily transported. A typical internal growler is shown in Fig. 2-15.

A test lamp is a simple test device used to check continuity. This device is sometimes preferred to the ohmmeter because the illuminance of the bulb allows the technician to watch the test lamp leads and device under service rather than on the scale of the ohmmeter. Figure 2-16 shows an example of the construction of a typical test lamp.

Fig. 2-13. A hand-operated megohmmeter (courtesy Biddle Instruments).

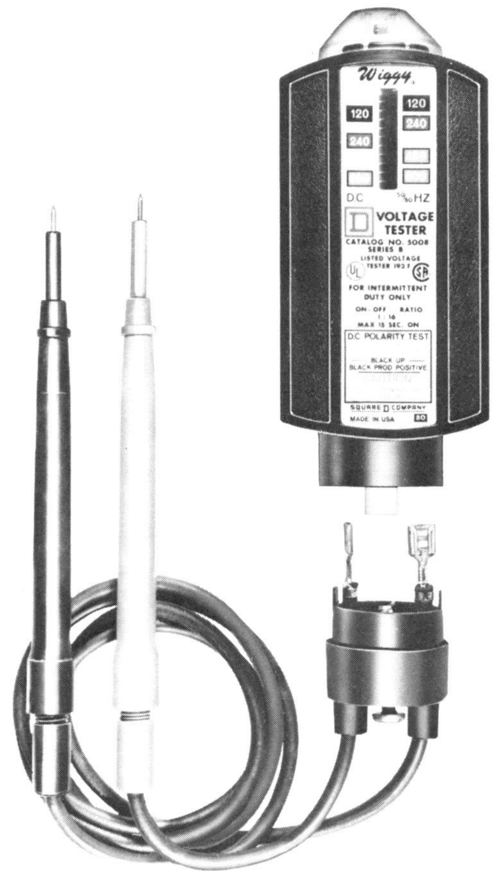

Fig. 2-14. A voltage tester for an electrician (courtesy Square D. Co.).

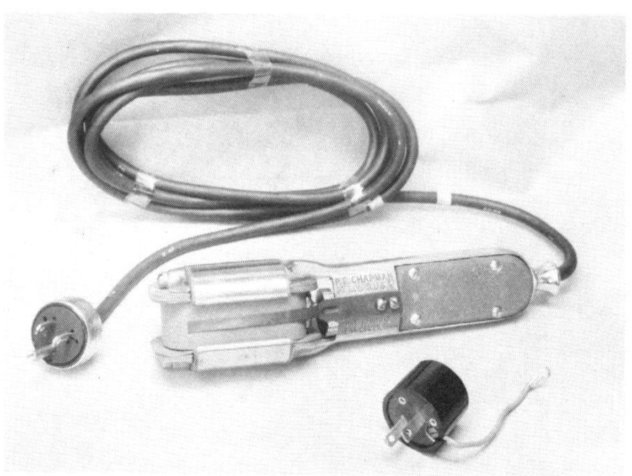

Fig. 2-15. A typical internal growler (courtesy P. E. Chapman Electrical Works, Inc.).

The neon voltage tester is used to check the presence of voltage in a circuit. This instrument is most often used in troubleshooting housewiring. The tester is very simple to operate, inexpensive, easy to carry and uses a neon bulb to indicate the presence of voltage rather than with a meter. Figure 2-17 shows a typical example of a neon voltage tester.

The most common ammeter used by electricians in industry is the clamp-on ammeter. This meter contains transformer jaws that clamp around a conductor without interrupting the circuit. These meters are portable and rugged. Figure 2-18 illustrates a typical clamp-on ammeter.

An instrument that is used in troubleshooting/installing TV distribution systems is the signal level meter (i.e., field strength meter); see Fig. 2-19. This

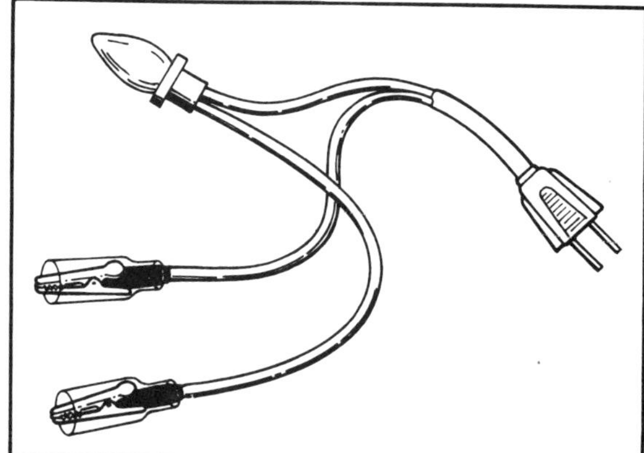

Fig. 2-16. A test lamp.

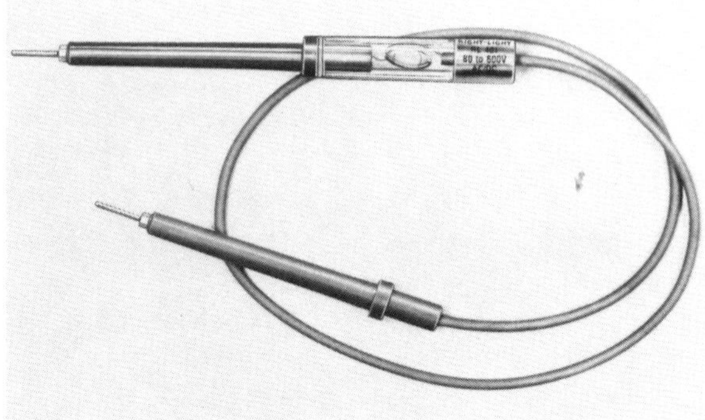

Fig. 2-17. A neon voltage tester (courtesy ETCON Corp.).

meter provides accurate VHF-UHF measurements and features a built-in loudspeaker. This meter is a tuned-rf voltmeter with a microvolt or decibel scale. This meter has many uses, its most common being to position antennas and measure signal levels. Other uses include

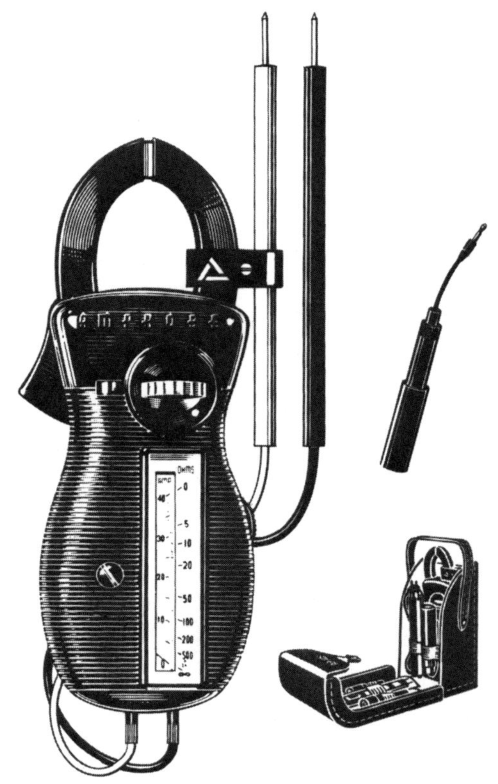

Fig. 2-18. A clamp-on type ammeter (courtesy Amprobe Instrument, Division of SOS Consolidated, Inc.).

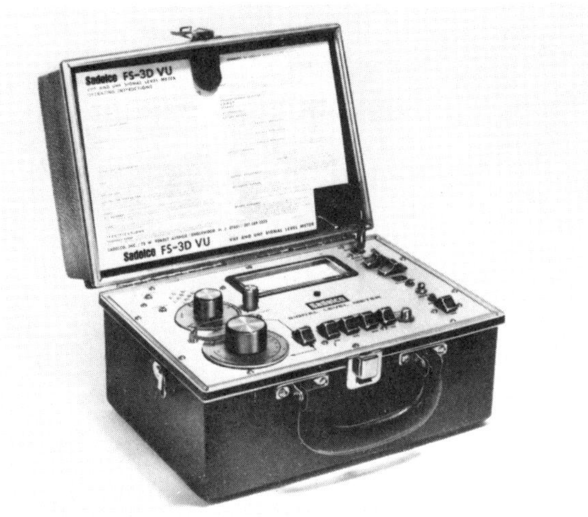

Fig. 2-19. A typical signal level meter used to troubleshoot/install TV distribution systems (courtesy Winegard).

measuring insertion loss, amplifier gain, automatic gain control settings, subchannel signals, return loss (i.e., standing wave ratio) and noise levels.

SIGNAL GENERATORS

The basic kinds of generators used in troubleshooting are as follows:

1. Af/rf generators.
2. Marker generator.
3. Sweep generator.
4. Noise generator.
5. Logic pulsers and logic probes.

The af/rf generators are commonly used in aligning

Fig. 2-20. A typical rf signal generator (courtesy B & K-Precision/Dynascan Corp.).

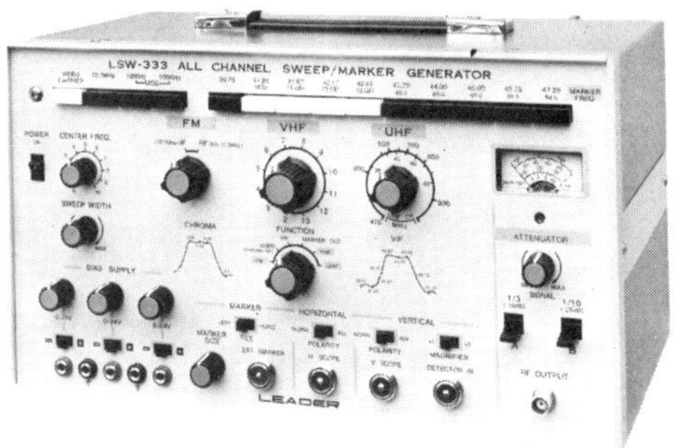

Fig. 2-21. A sweep/marker generator (courtesy Leader Instrument).

and troubleshooting radio receivers. The af generator provides signals in the audio range while the rf generator provides signals in the radio frequency range. Both meters provide sine wave or square wave outputs, contain built-in attenuators, and provide low distortion outputs. Figure 2-20 shows an example of a typical rf signal generator.

The marker generator provides an unmodulated frequency, which is used in identifying frequencies on a response curve used in television alignment.

The sweep generator is also used in television alignment. It provides an fm signal at a desired range of frequencies. The sweep generator and marker generator are usually built-in together. Figure 2-21 shows a typical example of a sweep/marker generator.

A noise generator is a small hand-held signal probe that is handy in signal tracing radio receivers (Fig. 2-22). This small generator sends out a broad-band signal from af to rf. This typical frequency range is from 1 kHz through 30 MHz.

Logic pulsers and logic probes are hand-held, portable instruments used to diagnose logic states in digital circuitry. The logic pulser operates like a signal injector, and the logic probe acts like a signal tracer in audio and radio frequency circuits.

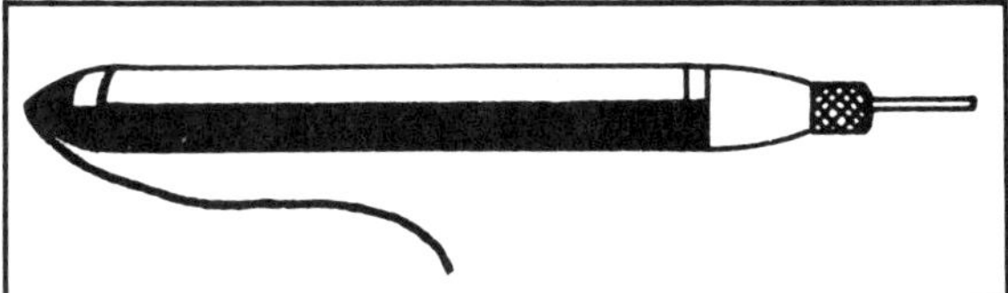

Fig. 2-22. A typical noise generator.

The logic pulser (Fig. 2-23) provides a one-shot positive and negative pulse by depressing the pulse button. A continuous pulse train can be injected by keeping the pulse button depressed. Other types of logic pulsers have additional features such as automatic polarity selection and pulse variations.

The logic probe (Fig. 2-24) is used to diagnose the logic states (e.g., Hi or Lo) by comparing the voltage to reference thresholds for the desired logic family. Logic probes offer speedy testing and eliminate the need for expensive/bulky oscilloscopes or voltmeters. There are many different types of logic probes offering such applications as memory, overload protection, and very high speed testing. Logic monitors are similar to logic probes but offer the ability to clip on to the IC pins, test points, or bus line when multi-point testing is desired. Two common types of monitors are the 16-channel and 40-channel. These instruments have a wide variety of applications in troubleshooting microprocessors and process control systems.

Fig. 2-23. A hand-held digital logic pulser (courtesy Global Specialties Corp.).

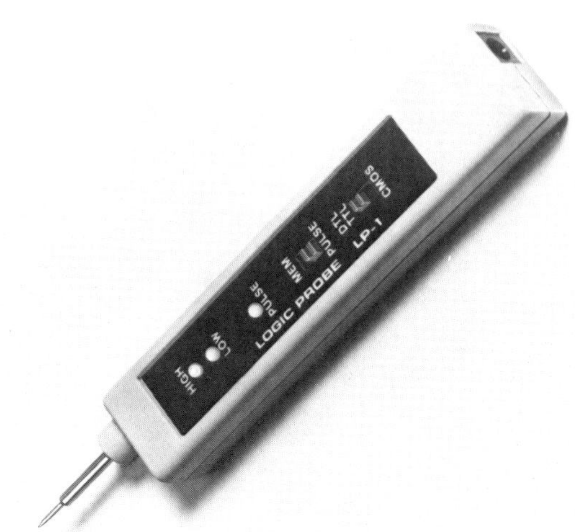

Fig. 2-24. A hand-held digital logic probe (courtesy Global Specialties Corp.).

SELF-EXAMINATION

Select the best answer:

1. Which meter uses a dual triode tube and requires external power for operation?
   A. Tvm.
   B. Vom.
   C. Vtvm.
   D. FETVM.
   E. None of the above.
2. The oscilloscope is largely used to measure:
   A. Rms voltage.
   B. Average voltage.
   C. Effective voltage.
   D. Peak to peak voltage.
   E. None of the above.
3. The control that adjusts the frequency of the horizontal sweep oscillator is the:
   A. Horizontal gain.
   B. Z-Axis.
   C. Attenuator.
   D. Sync selector.
   E. Sweep control.
4. The necessary probe commonly used on the oscilloscope when measuring high frequency or high impedance circuits is the:
   A. Demodulator probe.
   B. Low-capacitance probe.

C. Voltage divider.
D. Noise probe.
E. None of the above.

5. The Control on the oscilloscope which adjusts the height of the waveform is the:
   A. Sweep control.
   B. Z-Axis.
   C. Sync selector.
   D. Horizontal gain.
   E. Vertical gain.
6. The meter often used in testing an armature for short circuits:
   A. Vom.
   B. Vtvm.
   C. Megohmmeter.
   D. Growler.
   E. None of the above.
7. A meter device that tests insulation resistance is the:
   A. Transistor checker.
   B. Voltage tester.
   C. Gauss meter.
   D. Megohmmeter.
   E. None of the above.
8. A small hand-held probe generator used in signal tracing receivers is the:
   A. Af generator.
   B. Rf generator.
   C. Sweep generator.
   D. Marker generator.
   E. Noise generator.
9. The accessory probe used on the oscilloscope to detect the signal is the:
   A. Voltage divider.
   B. Low-capacitance probe.
   C. Af probe.
   D. Direct probe.
   E. Demodulator probe.
10. The control on the oscilloscope that permits use of an internal or external sync is the:
    A. Sweep control.
    B. Z-Axis.
    C. Focus.
    D. Sync selector.
    E. Intensity.

QUESTIONS AND PROBLEMS

1. Explain the difference between a vom and a vtvm.
2. What is a tvm and an FETVM?
3. Explain the procedure in "setting up" an oscilloscope for operation.
4. Explain how to calibrate an oscilloscope.
5. What is a low-capacitance probe?
6. What is a voltage divider probe?
7. What is a demodulator/rf probe?
8. Explain what a megohmmeter is used for.
9. Where is the marker generator used?
10. What is a neon voltage tester?

Chapter 3

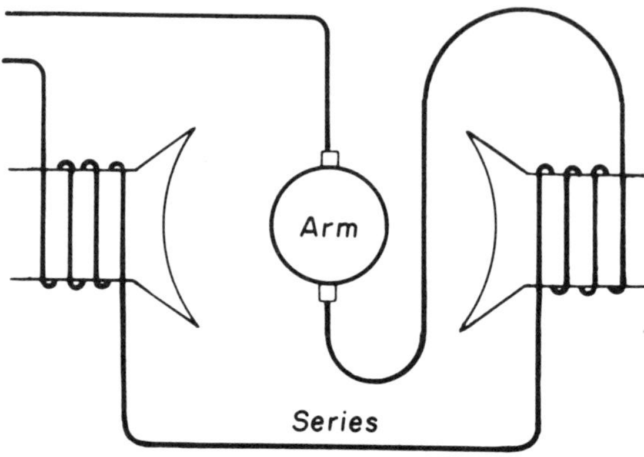

Troubleshooting Electric Motors

Electric motors are among the most widely used electrical devices in domestic, business and industrial applications today. They can be found in such products as refrigerators, computers, and air conditioners, as well as business machines, automobile, and industrial process controls. Electric motors, indeed, play an essential role in turning the "wheels of industry."

With the advent of high technology, the need for electric motors becomes even greater. High technology has demanded new designs and capabilities of the motor. In this chapter, a review of the basic principles of operation, types of motors, and methods in troubleshooting and repairing are presented.

FUNDAMENTALS

The construction and theory of the electric motor is very similar to that of the generator. An electric motor is a device that changes electrical energy into mechanical energy. The generator does the complete opposite. A simple dc generator can be converted into an electric motor by connecting a battery to the terminals of the brushes. Figure 3-1 illustrates a simple electric motor.

The current supplied to the armature from the battery makes it an electromagnet. The armature has a north and a south pole adjacent to its respective field poles as illustrated. Therefore, the armature rotates because like poles repel each other, as illustrated in Fig. 3-2.

The armature continues to rotate since the commutator continuously changes the polarity of the armature poles. This type of motor is called a repulsion-type motor. To increase the efficiency of the electric motor, many coils of wire are used in the armature and field poles. Adding these coils of wire makes the motor more powerful and run much smoother.

The basic parts common to most electric motors include the armature, field windings, end plates, bearings, frame, brushes, starting switch, and base. Figure 3-3 shows some of the common parts of a motor.

Basically, all electric motors operate either on the principle of repulsion or induction. The repulsion-type motor, as you already know, works on the principle of two like magnetic fields opposing each other. The magnetic field of the armature opposes the magnetic field of the stationary field windings and causes the armature to rotate. The commutator continually reverses the polarity of the armature, and the armature continues to rotate. All dc motors and a few ac motors operate on the principle of repulsion and need an armature, commutator, and set of brushes.

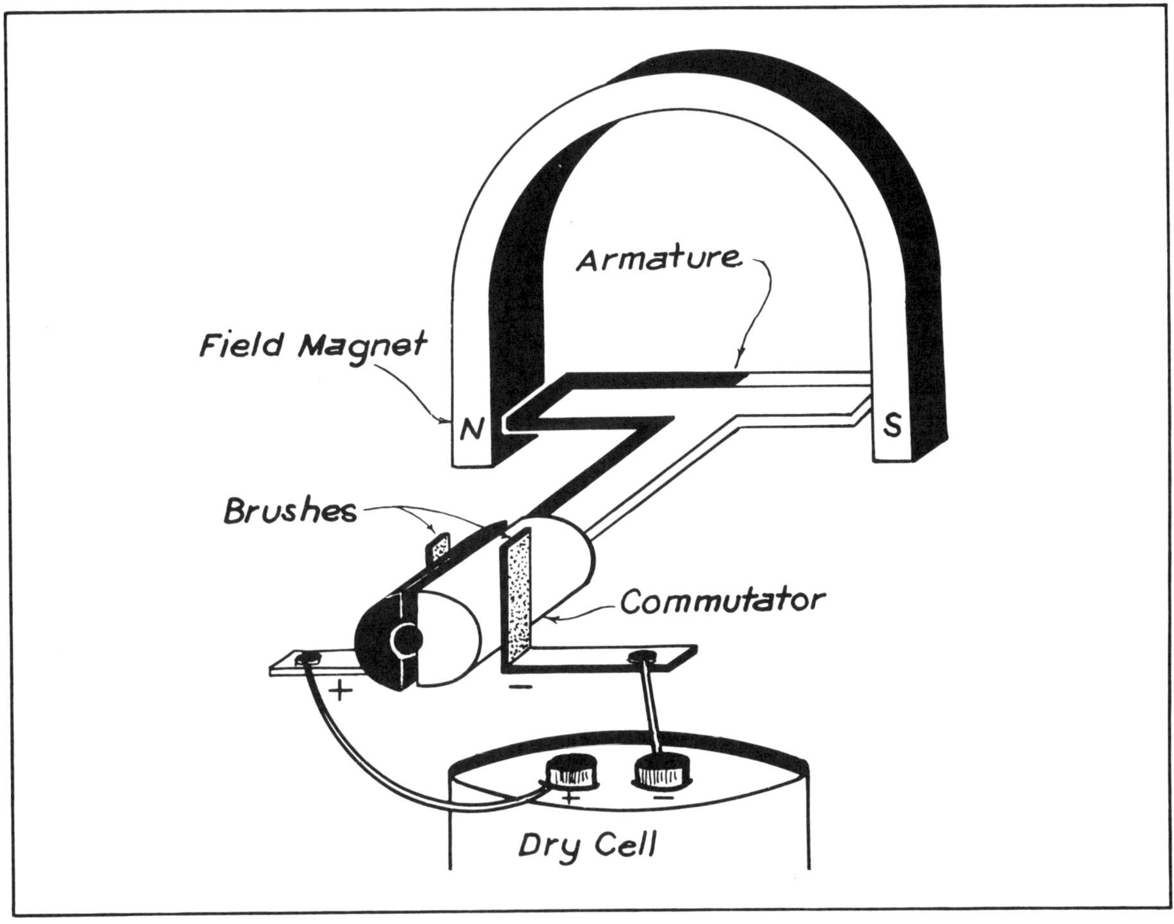

Fig. 3-1. The construction of a simplified electric motor.

The induction motor, as you might guess, works on the principle of induction. Almost all induction motors operate on ac. The induction motor has a rotor that consists of a laminated iron cylinder containing copper bars buried in slots. The appearance of this rotor resembles a squirrel cage and, therefore, is called a squirrel cage rotor. When ac is applied to the stator field windings, a current is induced into the rotor because of induction. This induced current produces a rotor field polarity that is opposite to the stator field windings. The rotor will not begin rotating by itself; therefore, most single-phase motors require a starting winding and switch. Three-phase motors do not require a starting switch since each phase is displaced 120°. Also, the induction motor does not need an armature, commutator, or set of brushes to operate.

There are many different types and classes of electric motors, each having its own particular characteristics and abilities. Recent technological developments have increased the supply of electric motors of varying capabilities. Some of the most common types of electric motors are split phase, capacitor, shaded pole, repulsion, dc, synchronous, universal, polyphase, gear motor, and stepping motor.

Split-Phase. A split-phase motor is an ac, single-phase, induction motor and is generally operated on 115 volts. This motor uses a squirrel cage rotor and operates on the principle of induction. It is used on many devices, such as washing machines, water pumps, refrigerators, and fans. Size generally ranges from 1/20 to 1/2 horsepower. Figure 3-4 illustrates a split-phase motor.

The split-phase motor has two field windings—a running winding and a starting winding. It is called a split-phase motor because the starting winding coils are

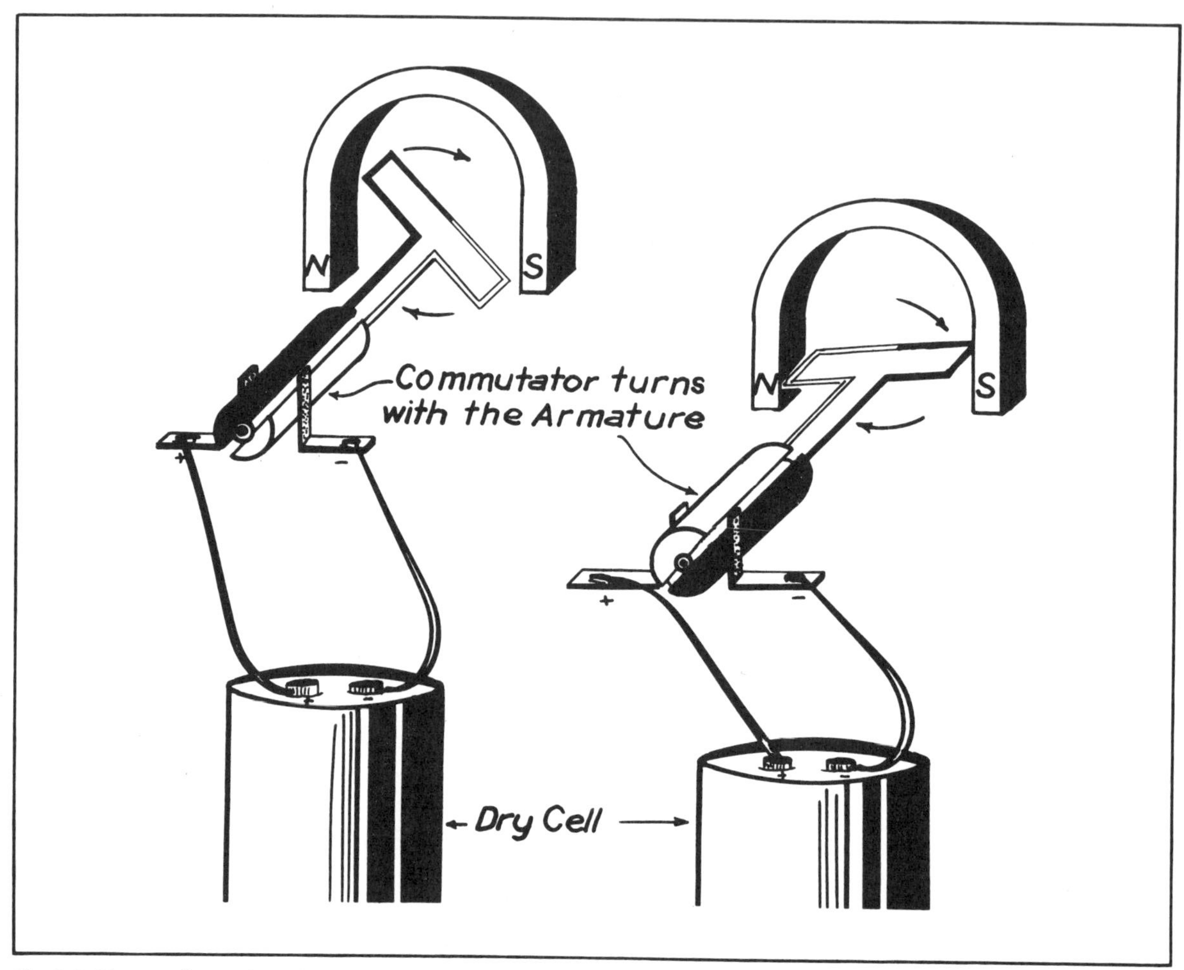

Fig. 3-2. The rotating action of a simplified electric motor.

displaced 90° from the main running windings. Figure 3-5 illustrates the two windings of a split-phase motor.

The starting winding, which is connected parallel to the running winding, is composed of fine, insulated copper wire. The starting windings, or *auxiliary* windings, are responsible for starting the motor and are usually left in the circuit for only a split second. After the motor has reached approximately 75 percent of its speed, a centrifugal switch disconnects the starting windings from the circuit. The main running windings take over. Figure 3-6 shows a picture of the centrifugal mechanism (governor) of a centrifugal switch. Also, this figure shows a rotor with the centrifugal mechanism in place.

The stationary part of the centrifugal switch contains two contacts where the starting windings are connected in the circuit or disconnected from the circuit. Figure 3-7 shows a picture of the stationary part of the centrifugal switch.

Capacitor. The capacitor-type motor is an ac, single-phase, induction motor. It is almost identical in construction to the split-phase motor except that it contains one or more capacitors and generally ranges in size from a fraction of a horsepower to 20 horsepower. Figure 3-8 illustrates a capacitor type motor.

The capacitor is a device that has the ability to store electricity—it is rated in farads, microfarads, and picofarads. The most common types of capacitors used in electric motors are the paper capacitor and the electrolyte capacitor. There are basically three kinds of capacitor motors as follows:

1. Capacitor start.

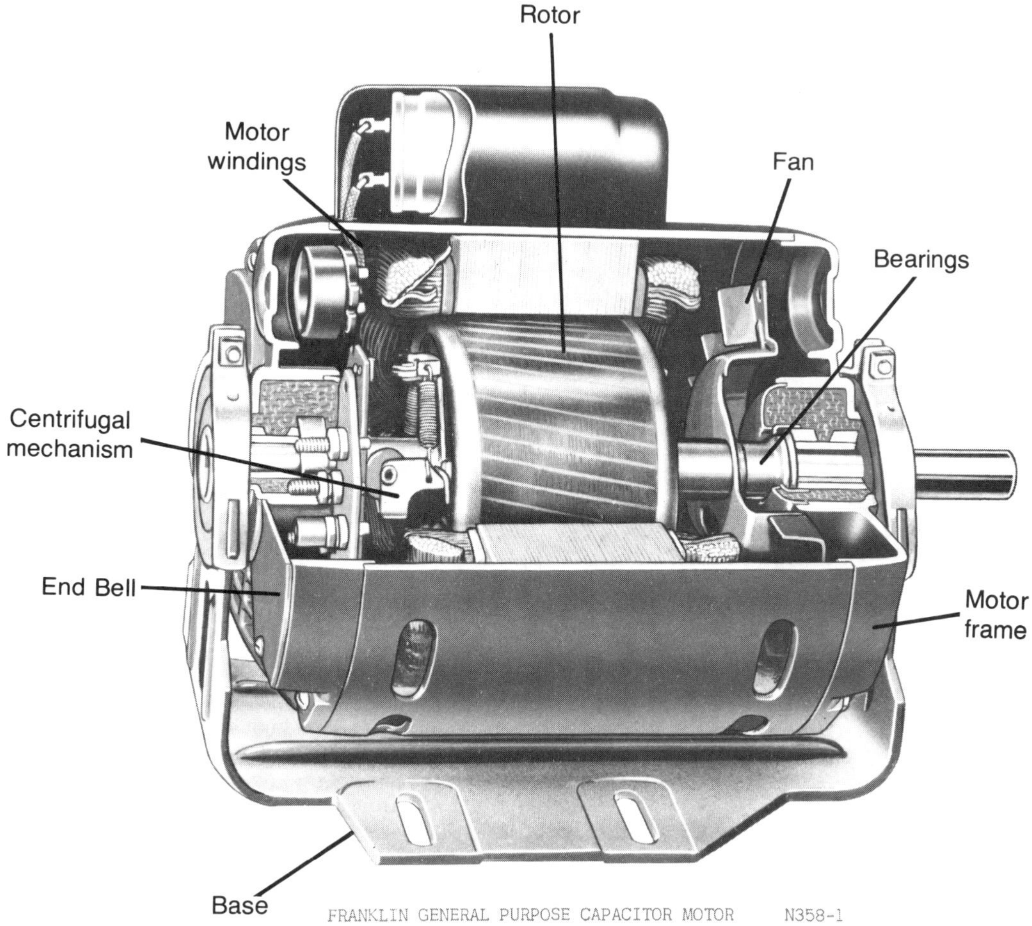

Rotor
Motor windings
Fan
Bearings
Centrifugal mechanism
End Bell
Motor frame
Base
FRANKLIN GENERAL PURPOSE CAPACITOR MOTOR N358-1

Fig. 3-3. The common parts of an electric motor (courtesy Franklin Electric Co.).

2. Capacitor start/run.
3. Two-value capacitor/run motors.

The *capacitor-start motor* uses a capacitor that is placed in series with the starting winding. When the motor is turned on, the capacitor causes the current in the starting winding to lead the current in the running winding. This effect induces a current in the rotor, and it begins to rotate. Figure 3-9 shows a schematic diagram of a capacitor-start motor.

In the *capacitor-start/run motor*, the capacitor and starting winding is left in the circuit at all times. This type of motor is a very quiet and smooth-running motor. It is often used in fans, refrigerators, and air conditioners, where the noise level is to be at a minimum.

The *two-value capacitor-type motor* is also a very quiet motor. It is very similar to the capacitor-start/run

Fig. 3-4. A typical split-phase motor (courtesy Franklin Electric Co.).

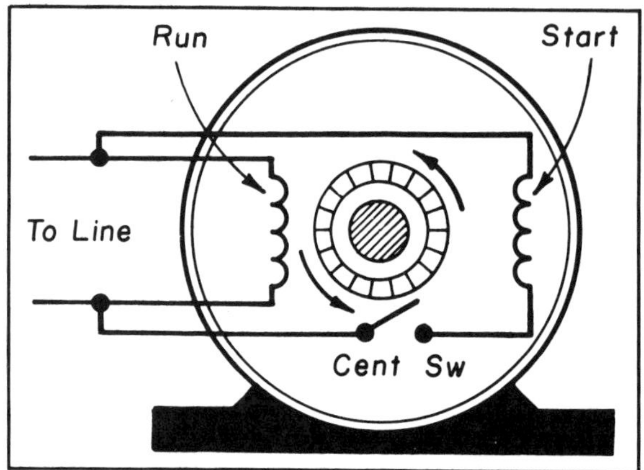

Fig. 3-5. The run and start windings of a split-phase motor.

Fig. 3-6. Typical centrifugal mechanism assembly of a centrifugal switch (courtesy Franklin Electric Co.).

Fig. 3-7. The stationary part of a centrifugal switch (courtesy Franklin Electric Co.).

Fig. 3-8. Typical capacitor-type motor (courtesy Franklin Electric Co.).

motor, except that it usually has two capacitors of different values. The high-value capacitor is used to start the motor, and then the low-value capacitor replaces the high value capacitor once the motor has started running. Also, the two-value capacitor motors is often used on compressors where high starting torque is needed. More than one speed usually can be obtained from this type of motor. Figure 3-10 shows a schematic diagram of a two-value capacitor motor.

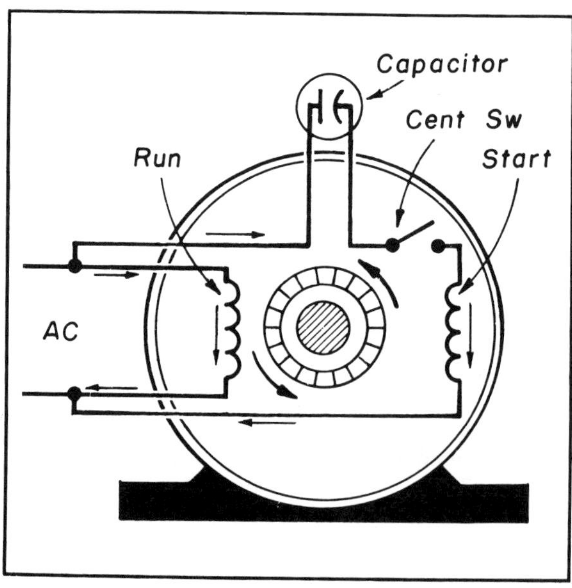

Fig. 3-9. The internal circuit of a capacitor-start motor.

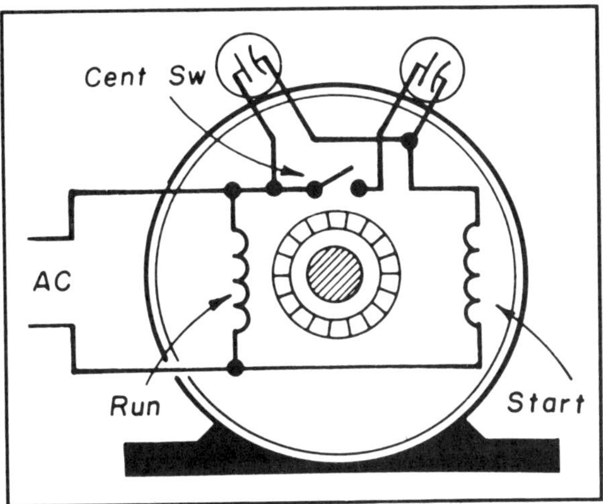

Fig. 3-10. Internal view of a typical two-value capacitor motor.

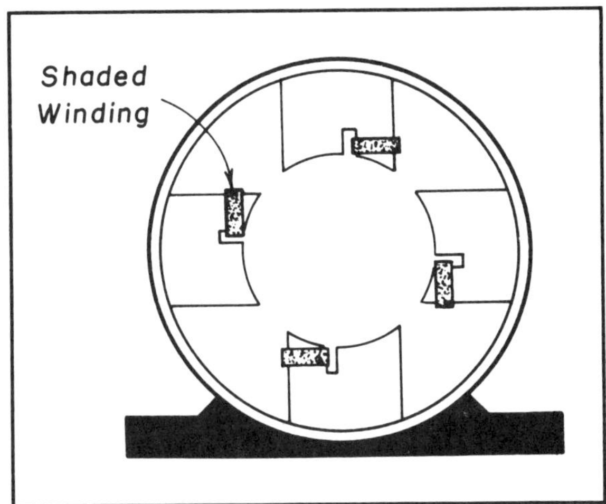

Fig. 3-12. The shading winding of a shaded-pole motor.

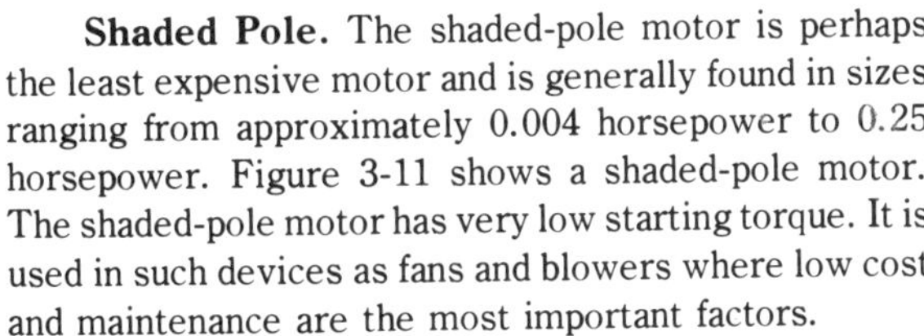

Shaded Pole. The shaded-pole motor is perhaps the least expensive motor and is generally found in sizes ranging from approximately 0.004 horsepower to 0.25 horsepower. Figure 3-11 shows a shaded-pole motor. The shaded-pole motor has very low starting torque. It is used in such devices as fans and blowers where low cost and maintenance are the most important factors.

The shaded-pole motor is a simple ac, single-phase, induction motor and requires very low maintenance. The rotor of the shaded-pole motor is considered a typical squirrel cage type. Its poles tend to project past the laminated iron cylinder and are often referred to as salient poles. The shaded-pole motor does not have a starting winding that is similar to most single-phase induction motors, but rather it has heavy solid copper loops that act as the starting winding (Fig. 3-12).

When current is applied to the shaded-pole motor, the copper loops or *shading coils* produce a magnetic field that is out of phase with the main field windings. This magnetic field induces a current in the rotor, and it begins to rotate. After sufficient speed is reached, the main field windings take over, and the rotor continues to rotate. Figure 3-13 shows the two windings of the shaded-pole motor.

Repulsion. Basically, repulsion motors can be narrowed down to two kinds as follows:

1. Repulsion motor.

Fig. 3-11. Typical shaded-pole motor (courtesy Franklin Electric Co.).

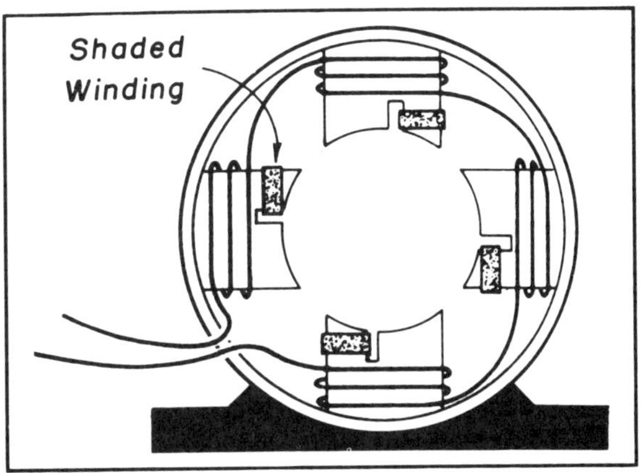

Fig. 3-13. The running and starting coils of a shaded-pole motor.

2. Repulsion start-induction run motor.

As you may recall, the repulsion motor requires an armature, commutator, and a set of brushes and operates on the principle of repulsion of like poles. It is very similar to a dc series motor, and its size varies from 0.5 horsepower to 10 horsepower. The repulsion motor has excellent starting torque and variable speed control characteristics. It is commonly used on such devices as compressors, air conditioners, and pumps where high starting torque is required. The speed of the repulsion motor can often be changed by shifting the brush holder. This will cause the rotor poles to be induced closer or farther away, changing the speed of the motor.

The repulsion start-induction run motor starts on the principle of repulsion; then once it has started rotating, it continues to run by induction. The brushes and commutator are only used during the starting of the motor. Once this type of motor has started running, manually lifting up the brushes would not affect the running performance of the motor. On other types of repulsion start-induction run motors, the entire brush assembly is lifted up by a centrifugal switch assembly. These motors are more complicated in construction, but they eliminate unnecessary brush-wearing.

Direct Current. Direct current motors range in size from motors with a fraction of a horsepower to motors of several thousand horsepower. They are used quite extensively in elevators where excellent starting torque and speed regulation are required.

There are three types of dc motors as follows:

1. Series.
2. Parallel.
3. Compound.

The basic difference in the types of dc motors is the type of circuit connection that is made between the field windings and the armature. In the series motor, the armature and field windings are connected in series. This series motor has the ability to start under heavy loads, and it varies its speed according to its load. Series motors without a load would run wild and literally destroy themselves. Series motors are generally used in automobile starters, cranes, and hoists where high torque is needed at low speeds. Figure 3-14 shows a simple series motor circuit.

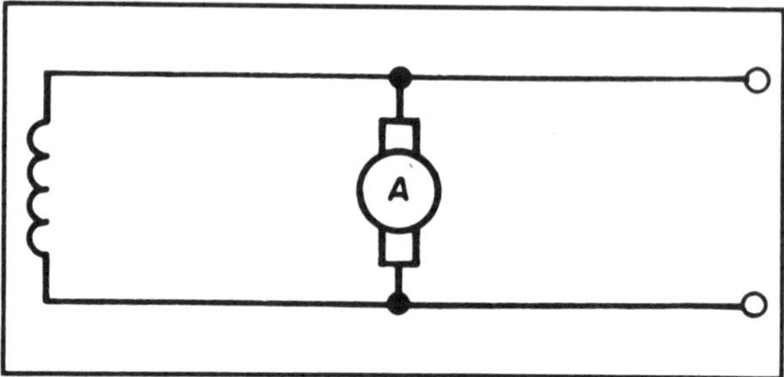

Fig. 3-15. A simple parallel motor circuit.

In a shunt, or parallel, motor, the armature and field windings are connected in parallel. This motor maintains a constant speed under varying loads but does not have as good starting torque as a series motor. Figure 3-15 shows an example of a shunt motor circuit. Shunt motors are generally used in pumps, and elevators, where constant speeds under varying loads are needed.

The compound motor, or *series-parallel motor*, has its armature and field windings connected in a compound circuit. Both the field windings and armature are connected in series and in parallel. Figure 3-16 shows an example of a compound motor circuit.

As you might expect, compound motors combine the features of the series and shunt motors. They offer fairly good starting torque and fairly good speed regulations. Compound motors are generally used in factories to drive large power equipment where good starting and stalling torque are needed.

Universal. Universal motors can operate either on

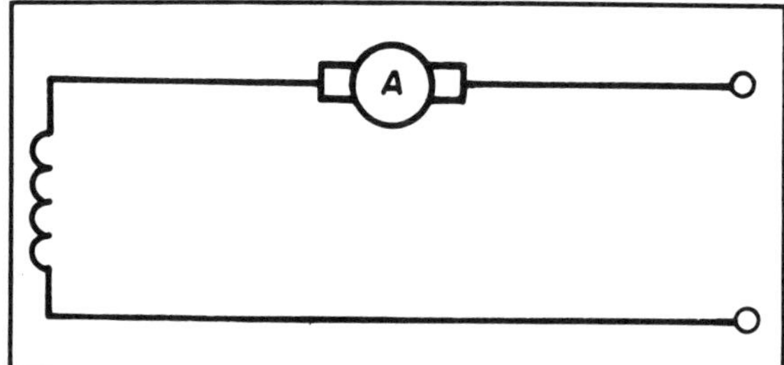

Fig. 3-14. A simple series motor circuit.

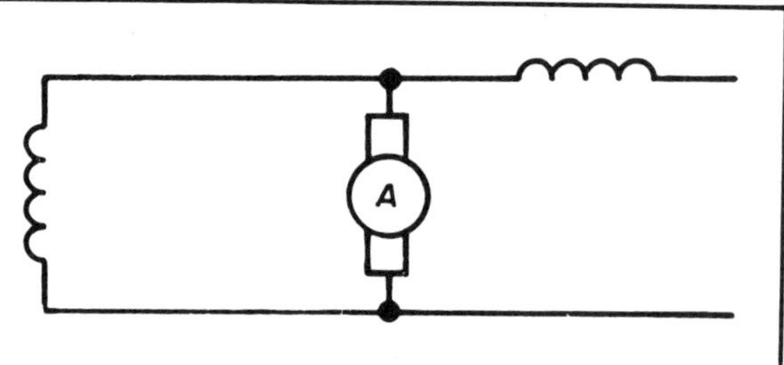

Fig. 3-16. A simple compound motor circuit.

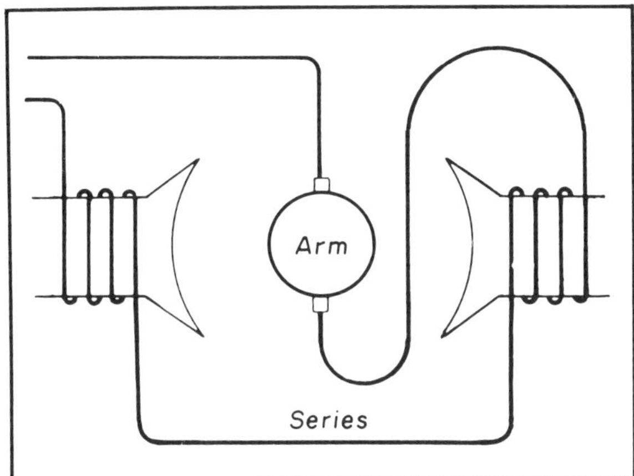

Fig. 3-17. A simple series circuit of a universal motor.

ac or dc. They generally come in a small fraction of a horsepower.

The universal motor is a series-wound motor. It has excellent starting torque and variable speed characteristics. Universal motors are generally used in vacuum cleaners, sewing machines, food mixers, fans, hair dryers, and other appliances in the home. Figure 3-17 illustrates a simple series circuit of the universal motor.

Polyphase. The basic polyphase motor used today is an ac, three-phase induction motor. These motors range from a fraction of a horsepower to several thousand horsepower. Most three-phase motors are found in industry and range from 10 to 100 horsepower in size. Figure 3-18 shows a polyphase motor.

Three-phase motors require little maintenance and repair and are very rugged in construction. The three-phase motor contains a number of coils divided into each separate windings called phases. Each phase has an equal number of coils in its group. The three groups of coils or phases are arranged either in a star or a delta connection. Figure 3-19 displays a diagram of a star and a delta connection.

When a three-phase current is applied to the stator windings, a rotating magnetic field is induced within the metal bars of the squirrel cage rotor. An induced magnetic field will cause the rotor to rotate. The continuous flow of three-phase current, which is 120° apart, maintains the rotor revolving because of induction. Three-phase motors have varying degrees of torque, speed, size, and enclosures—therefore, they are quite ver-

Fig. 3-18. A typical three-phase motor (courtesy Franklin Electric Co.).

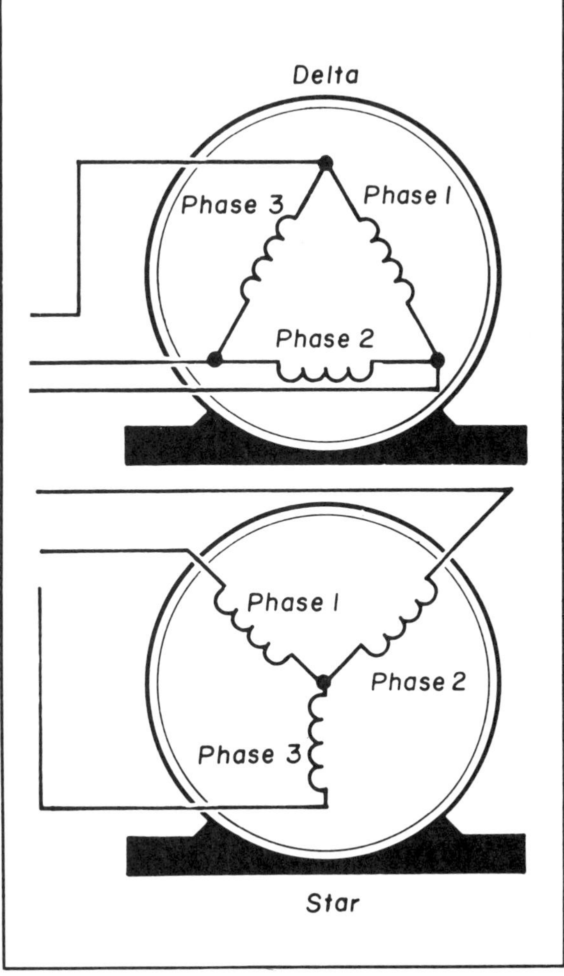

Fig. 3-19. A simplified diagram of a star and delta motor connection.

satile. They are generally used in driving industrial equipment.

Synchronous. Synchronous motors are induction motors that operate at a constant synchronous speed. This synchronous speed is determined by the frequency of the power source and the number of motor poles. Synchronous motors have a wide variety of shapes, sizes, and applications. They are made as small as a fraction of a horsepower for a small electric clock or as large as 3000 horsepower to drive a steel mill.

Synchronous motors are only suitable for operating on alternating current. The speed is constant and will not change with varying loads. The basic principle of the synchronous motor is that the rotor, which contains salient poles, rotates in step with the rotating magnetic field. The rotor tends to "lock in" with the rotating magnetic field and remains in constant rotation, uninterrupted by varying loads. Some rotors of synchronous motors need to be excited by dc; others do not. The excitation of the rotor forms definite poles in the rotor, which lock in step with the rotating magnetic field. Often, this type of motor will have a small dc generator attached to the shaft of the motor where it supplies dc to the rotor.

Gearmotor. The gearmotor is a special motor used for reduced speed and power applications (Fig. 3-20). Gearmotors eliminate the need for external chains and belts and offer precise speed reduction ratios and efficiency. The gearmotor can be either an induction or repulsion type. The selection of a gearmotor is generally determined by the speed and torque output rather than by horsepower rating. Many other factors can determine the selection of a gearmotor, such as the mounting, load, and braking requirements.

Three special kinds of gears used in the gearmotor are the spur, helical, and worm. The spur gear offers high power application but is noiser than the other two types. The helical gear is less noisy than the spur gear and transmits a near-constant motion. The worm gear has minimal noise, and high ratios can be obtained although it is less efficient than the other two types. Gears are manufactured from both metallic and non-metallic material. Nonmetallic gears are more quiet but offer less strength than metallic gears.

Fig. 3-20. Cross section of a gearmotor (courtesy Bodine Electric Co.).

Fig. 3-21. Dc stepping motor (courtesy Superior Electric Co.).

Stepping. The stepping motor is used in digital motion control applications (Fig. 3-21). Some of these applications include computer printers, medical X-ray equipment, photo typesetting, business machines, and industrial process controls. Stepping motors offer fixed and precise movement rather than continuous running motion as do conventional motors. The operation of the stepping motor is based on the theory of induction. Figure 3-22 illustrates the basic operation of a four-step sequence stepping motor. The motor's shaft moves one step (i.e., one-quarter turn) each time a switch is activated. One full revolution is achieved after four steps have been taken.

The drive system used for the stepping motor contains a pulse source and a translator. The pulse source is generally a computer or microprocessor. The translator, energized by a dc power supply, converts the digital pulses into the proper switching sequence for the stepping motor (Fig. 3-23). The stepping motor, in turn, converts the electrical information into mechanical motion to operate the load.

REPAIR AND TESTING PROCEDURES

In diagnosing motor troubles, it is important that the

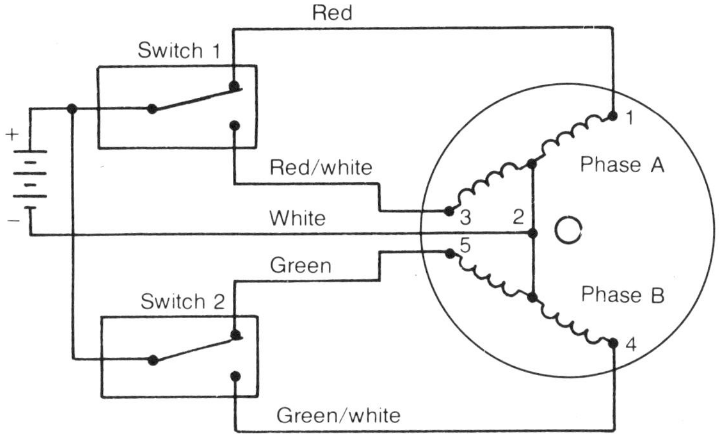

SWITCHING SEQUENCE*

STEP	SWITCH #1	SWITCH #2
1	1	5
2	1	4
3	3	4
4	3	5
1	1	5

*To reverse direction, read chart up from bottom.

Fig. 3-22. Four-step switching sequence (courtesy Superior Electric Co.).

repairman follow a logical, systematic procedure in order to eliminate wasteful time, tests, and replacement of parts. Most of the common motor troubles can be easily checked by using simple test equipment. It is also important that the troubleshooter have a complete understanding of the use of this equipment to analyze and repair the motor.

The ideal procedure in analyzing motor troubles should begin with a hearing/visual inspection. First, inspect the motor for obvious troubles, such as broken end bells, frames, or burned lead wires. Either one of these problems may prevent the motor from running. A noisy motor may be an obvious sign of defective bearings. Check the motor for defective bearings by turning the shaft and trying to move the shaft up and down. A shaft that does not rotate or does move up and down probably indicates bad bearings.

Fig. 3-23. Translator used to convert pulses into switching sequence for the stepping motor (courtesy Superior Electric Co.).

The basic techniques used in troubleshooting electric motors are as follows:

1. Test lamp.
2. Amperage measurements.
3. Growler.
4. Megohmmeter.

Before the repairman tries to run the motor, he should first test the motor for such faulty circuits as grounds, shorts, and opens. As you recall, a ground results from the windings making electrical contact with any iron part of the motor. Common grounds result from wire with poor insulation contacting the stator and end bells. A motor that has a grounded winding may blow fuses, run hot, or lack power. Shocks can be obtained from a grounded motor; therefore, extreme care should always be taken when testing a grounded motor. To test the motor for a ground, connect one lead of the test lamp to one of the motor leads. Connect the other test lamp lead to the stator or frame of the motor. If the lamp lights, this indicates that the motor is grounded. Figure 3-24 illustrates this test procedure.

Here again, as you recall, an open circuit results from a break in the motor circuit which prevents current from flowing in a complete path. A motor will not run with an open circuit. Many times if one of the three phases of a motor is open, the motor has no action; it simply hums. To determine if the motor has an open circuit, connect the test lamp leads to the lead wires of the motor. If the lamp fails to light, the motor has an open circuit. If the lamp lights, the circuit is complete. Figure 3-25 illustrates this test procedure.

A short circuit in a motor is caused by a defect in the motor circuit where two wires of the circuit connect and cause a bypassing of the normal current flow. An ammeter (use *clamp-on* type) can often detect a shorted motor. If the amperage reading exceeds the rated ampere value stated on the name plate of the motor, the motor may be shorted. Keep in mind that other factors, such as low line voltage or an overloaded motor, can cause the motor to draw excessive current. A hot, smoky motor that blows fuses is probably shorted. Also, a motor with a short circuit may heat rapidly, run hot, fail to start, or run slow. A growling noise often accompanies a shorted motor. If power is applied to a single-phase motor and the motor just *hums*, spin the shaft with your hand. If the motor starts running, the problem is in the starting circuit. *However*, if the motor starts, but runs unevenly, slows down, then starts again, the problem is in the running circuit.

Fig. 3-24. Using a test lamp to check a motor for grounds.

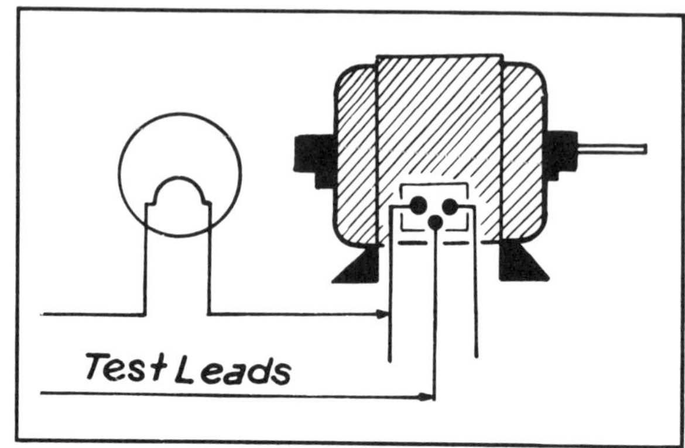

Fig. 3-25. Using a test lamp to check a motor for opens.

Besides using a test lamp, grounds and opens can best be checked by using a megohmmeter (Fig. 3-26). To test a motor for a ground, connect one lead of the megohmmeter to the motor frame and the other to one of the motor terminals. A grounded motor will have a zero or near zero reading. To test for an open circuit, connect the megohmmeter to each pair of phases of the motor. An open motor will show a high reading on the megohmmeter. An ohmmeter can also be used for testing a motor for grounds and opens.

Another way to check field windings for a short is dismantling the motor and applying a small voltage to the stator winding. Each coil now becomes an electromagnet. Place a screwdriver shank by each coil and slowly draw it away noting the magnetic pull. Each coil should have the same amount of pull. The coil with the least magnetic pull probably is shorted. Also, if you touch each coil and find that one is hotter than the rest, the hottest coil is probably shorted.

Before dismantling the motor, mark the two end

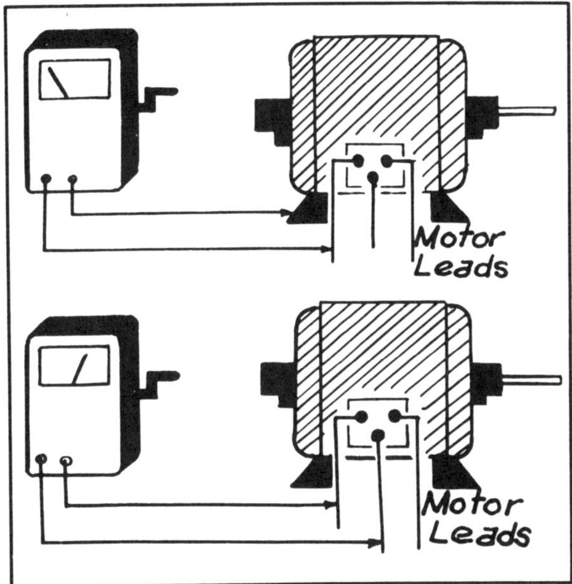

Fig. 3-26. Using a megohmmeter to test a motor for grounds or opens.

bells and frame in reference to each other. Generally, two punch marks indicate the front end of the motor, and one punch mark indicates the rear of the motor. Marking the motor enables the technician to correctly reassemble the motor. The front end shaft should also be marked. This can be done with a punch mark or by scratching an X on the end of the shaft. The base should also be marked in reference to the front end bell of the motor. Many technicians scratch a mark on the rotor shaft by using a knife or small file, indicating the correct position of the rotor. This scratch or mark is usually placed on the front end shaft closest to the front end bell.

To correct a ground in a motor, it is usually necessary to dismantle the motor and trace the windings to locate the part of the circuit that is making contact with the metal of the motor. After locating and correcting the problem, it is generally common practice to clean the windings if they appear dirty or charred. They should be cleaned with a solvent. Reinsulate the windings by spraying or brushing the windings with a coat of epoxy or other air-drying insulating enamel. If it appears that the grounded motor was caused by moisture, dry out the motor in a warm oven or with a fan.

Common causes of open circuits are a defective or poorly aligned centrifugal switch, a defective capacitor, or a broken wire in the motor circuit. In locating the open circuit of a motor that has a capacitor, it is generally a good idea to check the capacitor first. There are several ways to check the condition of a capacitor. One way is to substitute it with a new capacitor having the same rating. If the open circuit no longer exists, the capacitor was at fault. Another method of checking a capacitor is by the spark test. Connect the capacitor across the terminals of a 115 volt line for just a second. After removing the 115 volt line, use a screwdriver blade to short the two terminals of the capacitor (Fig. 3-27). A good capacitor will show a spark. Absence of a spark indicates a defective capacitor.

To test the capacitor for a ground, a simple test lamp can be used. Connect one of the leads of the test lamp to one of the terminals of the capacitor. Connect the other test lamp lead to the metal case of the capacitor. If the lamp lights, the capacitor is grounded and should be discarded (Fig. 3-28). Other methods used to test a capacitor include the use of an ohmmeter, a capacitor tester, or a combination ammeter and voltmeter.

The centrifugal switch often causes a single-phase motor to be open. The switch should be checked to see if the contacts are closing. If the contacts are not closing, washers may need to be added to the rotor shaft to correct the problem. Also, check the condition of the centrifugal switch, since it may be in bad condition and need to be replaced.

The motor windings should also be inspected for possible breaks. One or more broken wires may cause

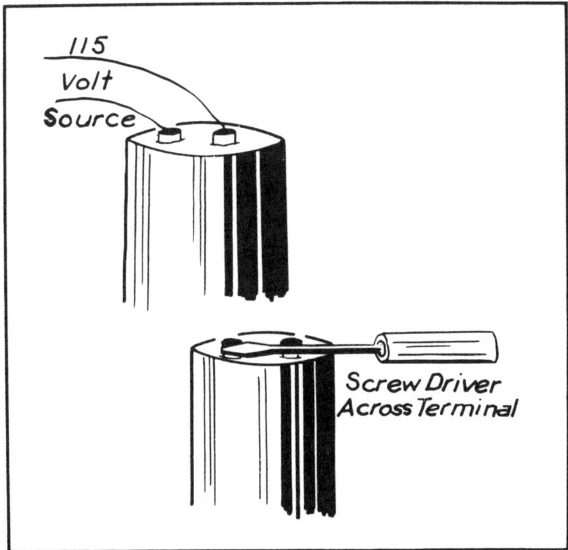

Fig. 3-27. A spark test is performed by using the blade of a screwdriver to short the terminals of a charged capacitor.

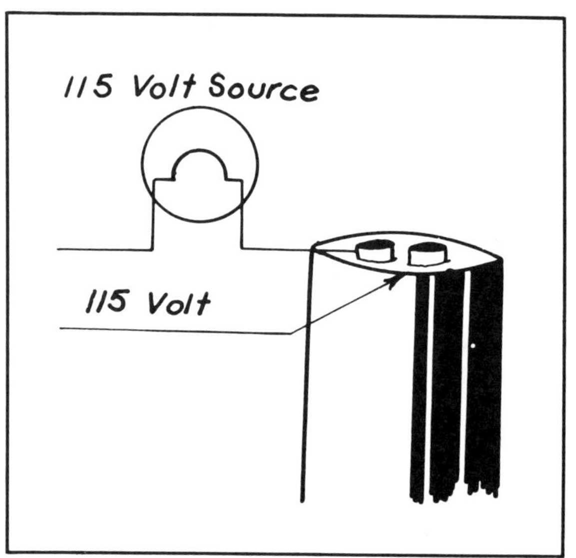

Fig. 3-28. Using a test lamp to check a capacitor for a ground.

the open circuit. If the windings are badly burned or broken and beyond repair, it may be necessary to replace the windings of the motor.

A short in the stator windings can be checked with an internal growler. Place the growler on the laminations of the stator at one end of a coil. Together, the growler and the coils of the stator act as a transformer. The coils of the growler act as the primary circuit, and the coils of the stator act as the secondary circuit. The growler, which has a built-in feeler blade, vibrates excessively when placed on the shorted coil (Fig. 3-29). When a motor has been tested as shorted, it should either be discarded or the windings should be replaced. Rewinding a motor is beyond the scope of this book.

A motor that has a shorted armature may jerk, severely vibrate, hum, growl, fail to run, or blow a fuse. The shorted coil of an armature can often be identified by its discoloration and insulation breakdown.

The armature of a motor can be checked for shorts by using an external growler. Place the armature on the growler with a strip of metal placed on top of the armature. Rotate the armature. If the metal strip rapidly vibrates, the armature is shorted. Figure 3-30 illustrates this principle using an internal growler and a hacksaw blade.

The armature can be tested for a ground by the use of a test lamp. Place one lead of the test lamp on the commutator and the other lead on the armature shaft. If the lamp lights, the armature is grounded (Fig. 3-31).

Bad bearings in a motor can cause the motor to run noisily, hot, or not at all. Generally, bad bearings cannot be saved by cleaning or reconditioning. They are usually replaced. If a ball bearing fails to rotate smoothly, it should be replaced. To remove a ball bearing, a gear puller or a special bearing-removing tool that knocks off the bearing is commonly used. To install a new ball bearing, a drift punch or a piece of round stock is often

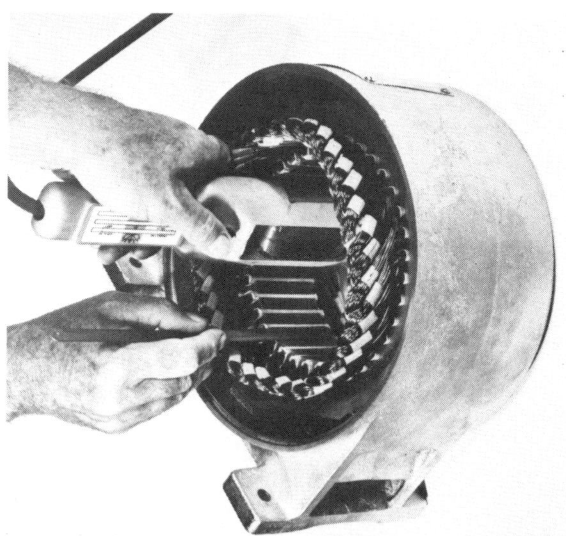
Fig. 3-29. Checking a stator for a shorted coil by using an internal growler (courtesy Crown Industrial Products Co.).

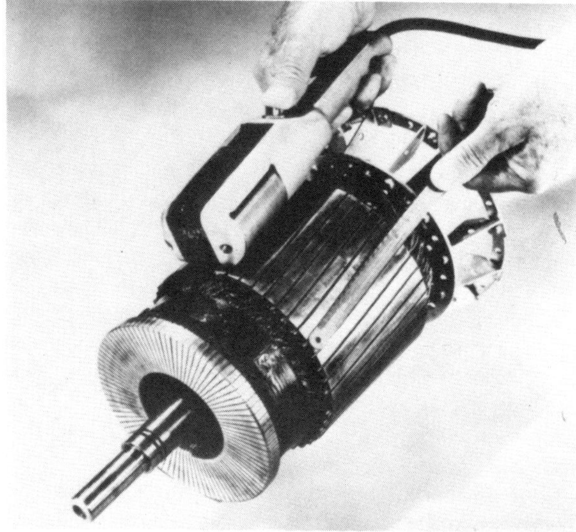
Fig. 3-30. Checking an armature for a short by using an internal growler (courtesy Crown Industrial Products Co.).

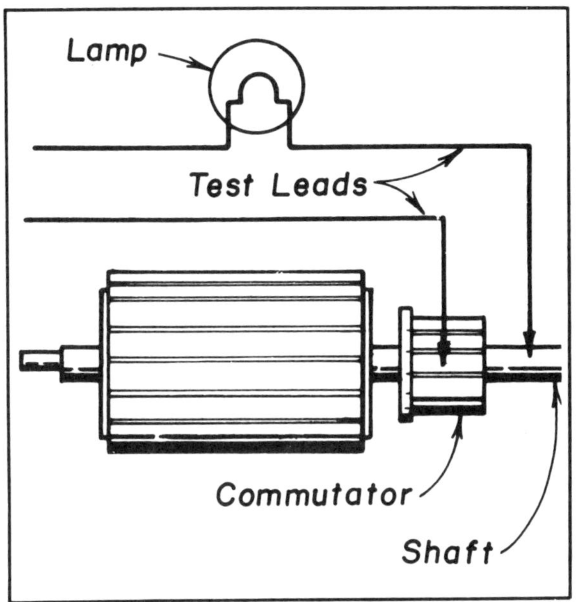

Fig. 3-31. Using a test lamp to test an armature for a grounded circuit.

used to press the bearing on the shaft. The sleeve bearing is generally removed; the new one is installed using a piece of round stock and a press. Sometimes the sleeve bearing is removed from the end bell using the special tool illustrated in Fig. 3-32. Often the inner diameter of a new bearing is made slightly smaller. It is then necessary to ream the bearing to the correct size using a tool called a reamer.

If a repulsion motor fails to start, severely sparks at the brushes, runs intermittently or with low power; the causes may be a dirty or worn commutator, worn brushes, misaligned brushes/brush holder, or defective brush spring.

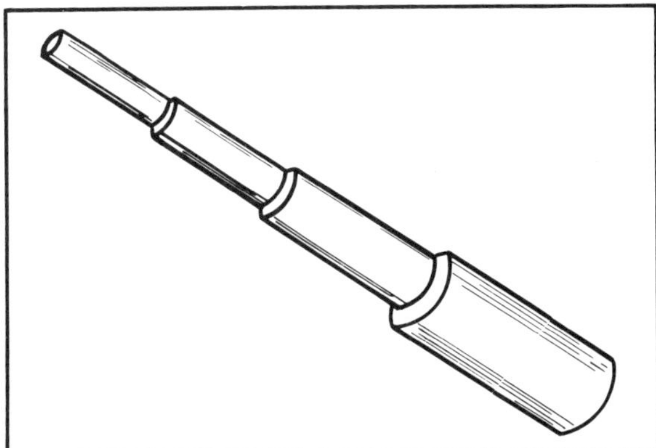

Fig. 3-32. A special tool used to remove sleeve bearings from an end bell.

To correct any of these problems, check the condition of the brushes, brush holder and spring, and commutator.

If the brushes are badly worn, they will need to be replaced. When replacing brushes, always use the correct replacement. The exact curvature and size of the brush is extremely important for satisfactory operation of the motor.

Before inserting the brushes, make sure the brush holder is clean, allowing the brushes to move freely. Also, the spring tension should be sufficient to keep a constant pressure allowing the brush to make good contact with the commutator. Once the brushes have been secured in place, the last step is to match the curvature of the brushes with the commutator. This is called seating and is done using a brush seater stone.

To seat the brushes, operate the motor at normal speed. Place the brush seater stone directly against the rotating commutator. Make sure the brushes are riding firmly against the commutator. Hold the brush seater stone against the commutator for only a few seconds. The stone will give off a powder of granules that ride under the brush, shaping the brush to best match the curvature of the commutator. Never over-seat brushes! This will cause excessive wear to both the commutator and brushes. Lastly, it is a good idea to blow off or vacuum the brush area to remove any powder granules from the commutator. The method used to seat brushes is illustrated in Fig. 3-33.

Fig. 3-33. The method used to hand stone a commutator (courtesy Ideal Industries, Inc.).

If the spring tension or position of the brush holder is wrong, the motor may not operate properly. Check the spring tension. If the spring is not pressing the brushes firmly against the commutator, they should be replaced. Make sure the brush holder is in a position to allow the brushes to make a firm, even ride on the commutator.

If the commutator appears to be out of round, be dirty, or contain high mica, it will need to be resurfaced and undercut. A commutator can be resurfaced with a hand stone or by lathe turning. Depending on the condition of the commutator, hand stone grinding tends to be the more effective and quicker method (Fig. 3-33). This method allows the commutator to be resurfaced with the motor operating at its normal speed. Lathe turning, on the other hand, requires the motor to be taken apart at the armature and placed on a lathe where it is "turned down." Never resurface a commutator more than is needed for a clean, concentric surface. Over-grinding or resurfacing will completely wear out the copper bars of the commutator.

If a commutator has been resurfaced, the mica should be resurfaced. This is called hand slotting or undercutting. Undercutting can be done by hand using a hacksaw blade or by using a direct motor powered mica undercutter. Hand slotting with a hacksaw blade is not done much today since it is more time consuming and less efficient. Undercutting is performed to remove the mica between the commutator bars to a depth approximately equal to the width between the bars. The mica is removed to allow the brushes to ride smoothly on a commutator that is uniform, concentric, free from ridges and burrs. Figure 3-34 shows an illustration of a direct drive universal mica undercutter.

Fig. 3-34. Removing mica from the commutator slots with a direct drive mica undercutter (courtesy Ideal Industries, Inc.).

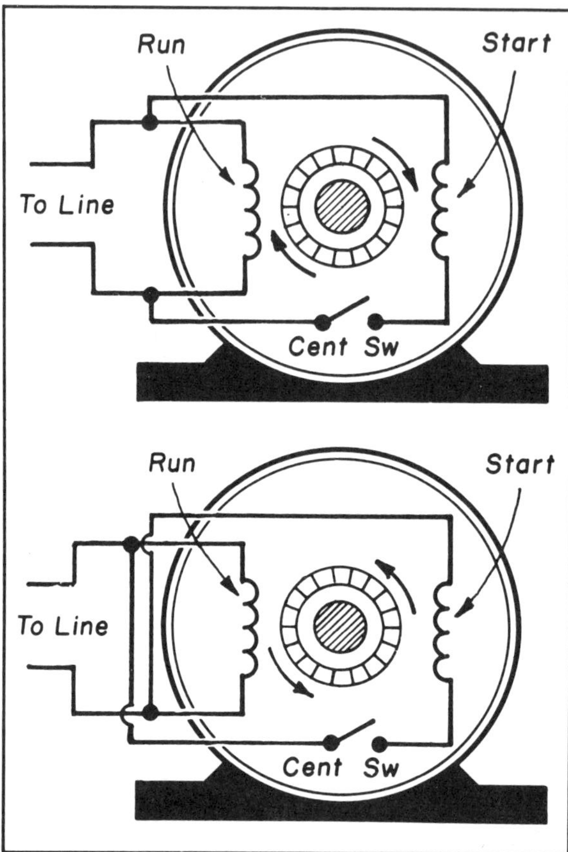

Fig. 3-35. Reversing the rotation of a single-phase motor.

Often a single-phase motor needs its shaft rotation revised. The direction of rotation of most single-phase motors can be changed by reversing the starting or running wires of the motor (Fig. 3-35).

To change the direction of rotation in a shaded-pole motor, it is usually necessary to dismantle the motor and reverse the stator end for end. This is done because the direction of rotation is dependent upon the *shading* coil effect from the main pole to the shaded pole (Fig. 3-36).

To change the direction of rotation of a dc motor, simply reverse the polarity of either the field poles or brushes.

The direction of rotation of a three-phase induction

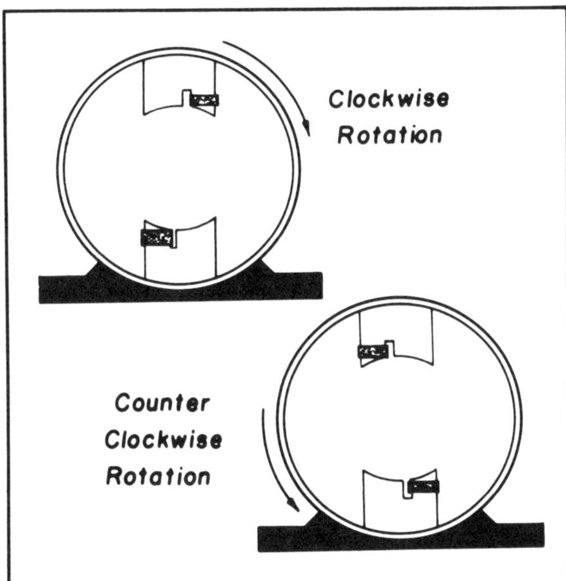

Fig. 3-36. Reversing the rotation of a shaded-pole motor.

motor can easily be changed by switching any two of the three motor leads. Usually the outer two leads are interchanged (Fig. 3-37).

When reassembling a motor, it is important that the wires are not in contact with the iron part of the motor. It is extremely important not to pinch any wires between the end bells and frame. This would result in a ground or short circuit.

When reassembling the motor, it is important to align all the marks punched before the motor was dismantled. It is common practice to slowly tighten all nuts consecutively. It is also common practice to tap the end bells of the motor to prevent misalignment.

Problems that are unique to gearmotors relate to improper lubrication, bad seals, gaskets and gears. Oil is used for lubricating most gearmotors. Oil provides constant lubricating film to the gear teeth. However, smaller gearmotors use grease instead of oil because of sealing problems. Sufficient and clean lubricant must be maintained; otherwise, damage will result to the seals and gears. Also, excessive and abnormal operation (e.g., too low or high ambient temperature) can reduce the life of gears.

Some of the problems associated with stepping motors include bad bearings, shorted windings, defective transmission line, and drive system malfunctions. It is important to first isolate the problem by determining whether it is the stepping motor, transmission line, or drive system. The stepping motor can be checked for shorts, open, and grounds similar to a conventional motor. Substituting a new stepping motor for the suspected defective one is also a good way to check its performance.

PREVENTIVE MAINTENANCE

The life of an electric motor is basically determined by the type of maintenance it receives. A poorly maintained motor can generally be detected by its dirty, corroded appearance. A thorough maintenance program includes the periodic inspection, recording, and servicing of electric motors. A minor adjustment or a simple cleaning of the motor can prevent expensive, time-consuming repairs later on.

The frequency of breakdowns and abuse of electric motors can be reduced through periodic inspections. All

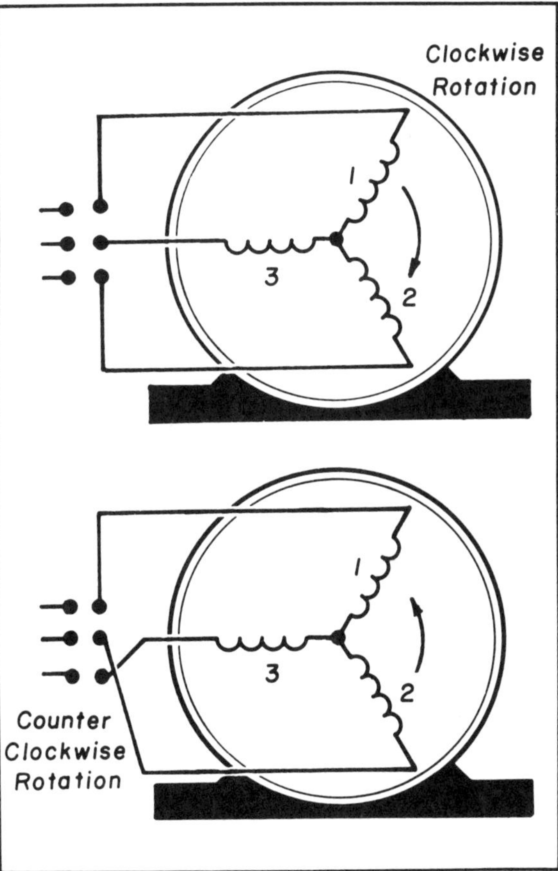

Fig. 3-37. Reversing the rotation of a three-phase motor by switching the outer two leads.

inspections should be accompanied by a log book, where the condition and services administered can be recorded. Some motors need more detailed and regular inspections than others. Motors that are used in very dirty or wet applications, such as those in industry, need to be inspected more regularly.

Some of the obvious, key contaminants to look for when inspecting motors are dirt, grease, water, and chemicals. Any of these key contaminants can cause shorts or grounds and cause the motor to run excessively hot. Also, these contaminants quickly wear down the components of the motor. Eventually, the motor will break down and will need to be repaired.

Dirt and grease and other contaminants have a tendency to block ventilating openings of the motor. They often accumulate on the commutator, preventing proper electrical conduction. They also add to the wearing down of the brushes and commutator. Excessive water accumulation can cause shorts or grounds in the armature or the stator windings, and the motor may break down.

Repulsion-type motors should have their brushes and commutator inspected regularly. Check the brush tension and alignment of the brush and holder. Gently tap on the commutator bars with a small rawhide hammer for presence of loose bars. Bad brushes should be replaced. A dirty commutator should be cleaned using a clean cloth, sandpaper, or cleaning stone. Note, never use *emery* to clean a commutator since the material will short out the bars on the commutator.

The rotor shaft of a motor should be regularly inspected for misalignment. An instrument called a *dial indicator* is often used to check the degree of rotor shaft misalignment.

Switch contacts should be given regular attention by cleaning and reshaping the contacts with sandpaper, a file or cleaning stone whenever they are dirty or disfigured. Most electrical/electronic distributors sell chemical contact cleaner.

All bolts and nuts should be kept tight, and all motor lead wires and windings should be inspected for dirt, breaks, and worn insulation—this prevents major breakdowns later on. It is common practice to clean and reinsulate motor windings using an air-drying epoxy or some other insulating enamel.

Regular inspection for wearing is essential in preventing motor breakdowns. Sleeve bearings should be regularly oiled but should not be over-lubricated. Ball bearings can be lubricated either by oil or by grease, depending upon the manufacturer's specifications. Grease, which is a combination of oil and soap, is generally the lubricant for ball bearings. Do not forget that over-lubrication of ball bearings can cause overheating, resulting in premature ball bearing failure. Squeaky, dirty, or tight bearings should be replaced with new bearings.

Incorrect motor end play should be given immediate attention. The motor end play can be tested by pulling and pushing the motor shaft back and forth. Maximum end play is usually about 1/64 inch. Incorrect end play can be corrected by adding or removing washers, replacing bearings, lubricating bearings, or tightening nuts or bolts.

Any decrease or increase in motor temperature, increase in noise, discoloration, or disfiguration in appearance are common danger signs of motor troubles. These conditions should be given immediate attention in order to establish the reason for their occurrence.

SELF-EXAMINATION

Select the best answer:

1. Basically all electric motors operate on the principle of repulsion or ________.
   A. Magnetism.
   B. Capacitance.
   C. Resistance.
   D. Induction.
   E. Semiconduction.
2. The basic techniques of troubleshooting electric motors consist of amperage, test lamp, growler, and ________.
   A. Substitution.
   B. Heat/freeze.
   C. Bridging.
   D. Megohmmeter.
   E. Oscilloscope.
3. A hot smoky motor is a good indication of:
   A. A ground.
   B. An open.
   C. A short.
   D. A and B.
   E. B and C.
4. The best way to test a capacitor when used in a 115-volt electric motor is by:
   A. Bridging.
   B. Spark test.
   C. Voltmeter.
   D. Ammeter.
   E. All of the above.

5. The direction of a three-phase motor can best be changed by:
   - A. Switching 2 of the 3 leads.
   - B. Dismantling the motor and switching 2 leads.
   - C. Changing the voltage.
   - D. Reversing the stator.
   - E. None of the above.
6. When cleaning a commutator never use ________.
   - A. Sandpaper.
   - B. Clean cloth.
   - C. Both A and B.
   - D. Emery.
   - E. None of the above.
7. An instrument often used to check the degree of motor shaft misalignment is the ________.
   - A. Voltmeter.
   - B. Clamp-on ammeter.
   - C. Growler.
   - D. Megohmmeter.
   - E. Dial indicator.
8. The bearing that should be oiled regularly is the:
   - A. Sleeve bearing.
   - B. Ball bearing.
   - C. Both A and B.
   - D. None of the above.
   - E. All of the above.
9. Grease is basically a combination of:
   - A. Water and soap.
   - B. Vaseline and soap.
   - C. Gas and water.
   - D. Oil and soap.
   - E. None of the above.
10. Maximum end play is usually about:
    - A. 1/64 inch.
    - B. 1/8 inch.
    - C. 1/4 inch.
    - D. 1/2 inch.
    - E. 1 inch.
11. The gear that provides the least amount of noise is the:
    - A. Spur
    - B. Helical
    - C. Worm
    - D. All of the above
12. The gear that offers very high power application is
    - A. Spur
    - B. Helical
    - C. Worm
    - D. All of the above
13. The lubrication used for most gearmotors is:
    - A. Oil
    - B. Grease
    - C. Both A and B
    - D. None of the above
14. The selection of a gearmotor is generally based on:
    - A. Speed
    - B. Torque
    - C. Both A and B
    - D. None of the above
15. A drive system used for the stepping motor is the:
    - A. Microprocessor
    - B. Translator
    - C. Computer
    - D. All of the above

QUESTIONS AND PROBLEMS

1. Explain the basic difference between an electric motor and a generator.
2. Explain the basic principle of the electric motor.
3. Name the basic parts of an electric motor.
4. Explain the difference between a repulsion and induction motor.
5. Name several types of electric motors.
6. What is an open circuit?
7. What is a ground circuit?
8. What is a shorted circuit?
9. Explain how to test a motor for an open, ground, or shorted circuit.
10. Explain the best way to dismantle a motor.
11. Explain the different ways to check a capacitor.
12. Explain the difference between an internal and external growler.
13. Name two types of bearings.
14. Explain how to clean a dirty commutator.
15. List some common problems and causes of electric motor breakdowns.
16. What are the different types of gears used in the gearmotor?
17. Why is grease rather than oil used to lubricate small gearmotors?
18. Explain the operation of a stepping motor.
19. Describe the drive system for the stepping motor.
20. List the applications of the stepping motor.

Chapter 4

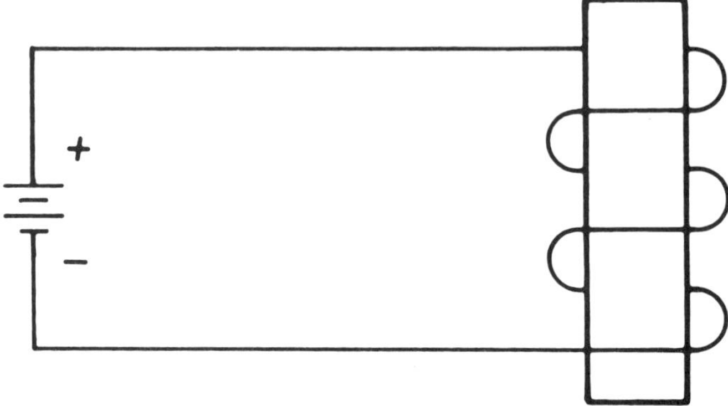

Troubleshooting Motor Controls

Every electric motor must have a control system, whether it be a simple switch to start and stop it or a complex microprocessor for stepping operations. Motor controls provide a wide variety of functions for the motor, such as starting and stopping, reversing, accelerating, decelerating, braking, and time-controlled operations. The motor controller is as important to the motor as the power that drives it.

In the last chapter, a review of the most common types of electric motors and repair of them was presented. The purpose of this chapter is to examine how electric motors can be controlled. The basic theory of motor controls is presented along with types of controllers, testing procedures, and preventive maintenance.

FUNDAMENTALS

The basic functions of a motor control to the motor are in providing such things as starting/stopping, protecting, sequence operations, reversing and speed control. Besides adding protection to the motor, the motor control also helps protect the operator. The simplest form of a motor control device is a simple single pole/single throw switch. This switch, as illustrated in Fig. 4-1, has the responsibility of controlling current to start and stop this ac squirrel cage induction motor.

Figure 4-2 shows how a simple fuse and variable transformer act as control devices. The fuse acts to protect the motor and operator. The variable transformer acts as a speed control for this series wound dc motor.

Many motor controls operate on the principle of electromagnetism. If insulated wire is wrapped around an iron bar and the two ends of the wire are connected to an electrical source, we have what is called an electromagnet (Fig. 4-3).

Reversing the current flow in an electromagnet will reverse the polarity of the iron bar. This principle is illustrated in Fig. 4-4.

A coil of wire connected to a battery has a magnetic flux that surrounds the coil much like that in a permanent magnet. Figure 4-5 illustrates this magnetic flux.

The relay is another motor control device. It is a magnetic device that is used to open or close circuits. Figure 4-6 illustrates a simple relay schematic diagram.

Circuit breakers are a special type of relay often thought of as a manual switch. They are frequently used in homes, business, and industries to protect electrical circuits from excessive currents or overloads. Figure 4-7 illustrates the magnetic portion of a simple schematic diagram of a circuit breaker. A large amount of current will cause the electromagnet to pull the lever down, thus

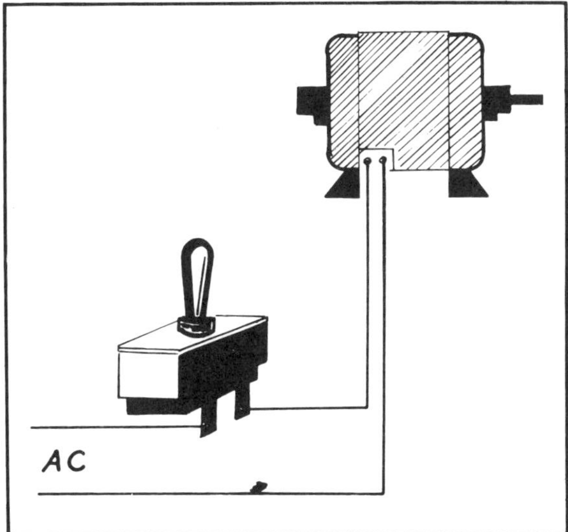

Fig. 4-1. A single pole/single throw switch is a simple motor control device.

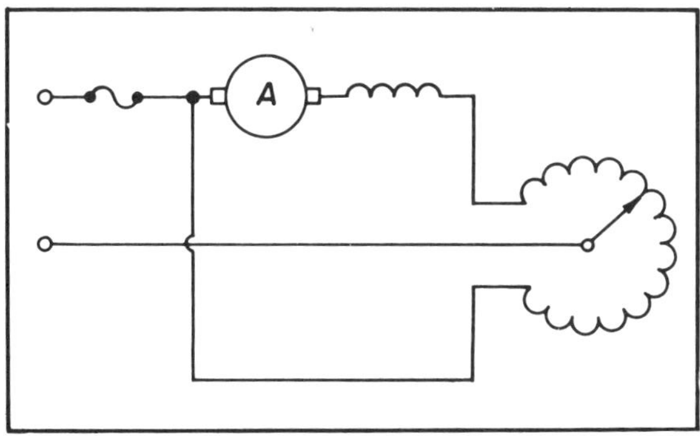

Fig. 4-2. A fuse and a variable transformer act as a control device.

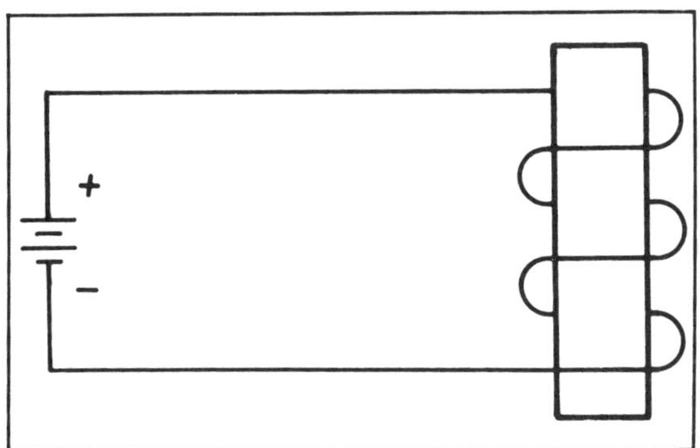

Fig. 4-3. An iron bar wrapped with wire and supplied with voltage forms an electromagnet.

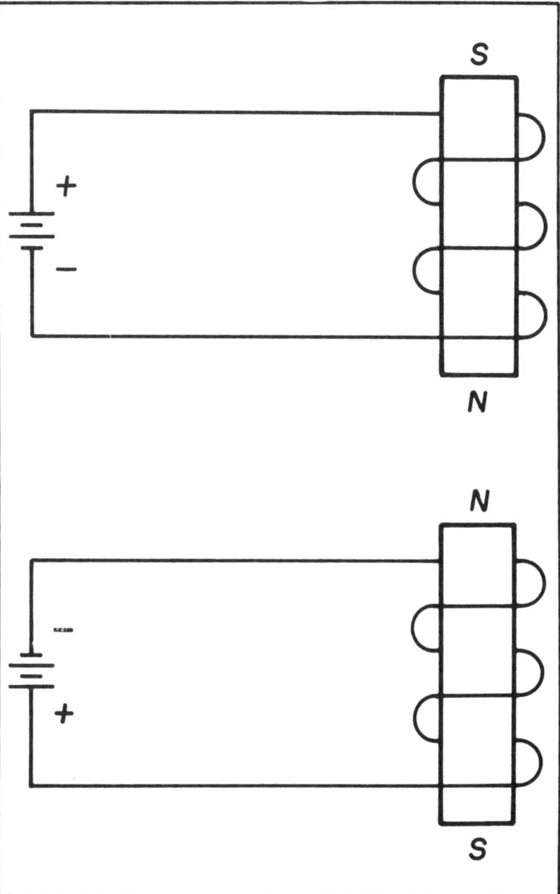

Fig. 4-4. The polarity of an electromagnet is reversed by reversing the current flow.

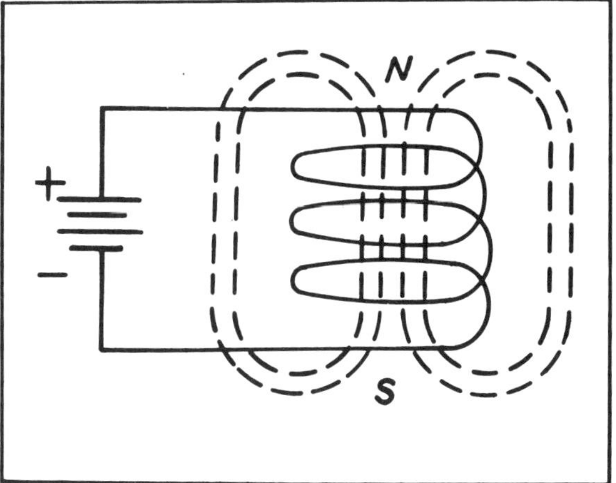

Fig. 4-5. A magnetic flux surrounds an electromagnet.

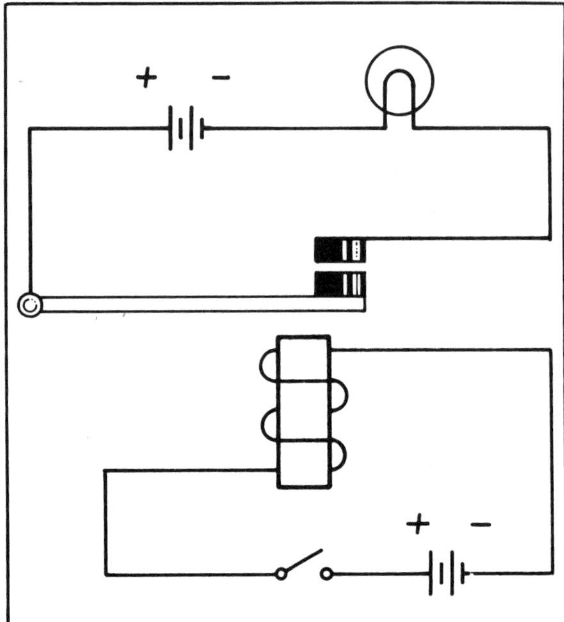

Fig. 4-6. A simple relay.

separating the contacts and breaking the circuit.

There is a wide variety of motor controllers and devices. Each has its own particular characteristics. A number of different types of motor controllers and devices are listed as follows:

1. Manual starters.
2. Magnetic starters.
3. Reversing starters.
4. Timing relays.
5. Drum switches.
6. Push-button stations.
7. Limit switches.
8. Snap switches.
9. Foot switches.
10. Float switches.
11. Thermostats.

Starters. The motor controllers, manual starter, magnetic starter and reversing starter mentioned above have some type of overload protection. One type of overload device is the melting alloy thermal overload relay (Fig. 4-8). Basically, when too much current is drawn in an overload situation, the "solder pot," which is made of brass, contains an eutectic alloy that melts. This solder pot which held the ratchet from turning can no longer do so. The melting alloy allows the ratchet to turn. The turning action opens the control contacts and the motor stops running. When the solder pot cools down, the relay must be manually reset.

The bimetallic thermal overload relay works on the principle of a bimetal strip that expands under heat (Fig. 4-9), opening the contacts. These relays come in automatic reset or manual.

Another type of overload relay is the magnetic overload relay. It works on the principle of electromagnetism (Fig. 4-10). If excessive current is drawn, the coil pulls in the core which opens the contacts.

The manual starting switch is a switch manually operated by a toggle handle. It is generally used to turn on or off fractional horsepower motors of 1 horsepower or less. Some single phase manual starting switches provide overload protection through use of a thermal unit that trips when too much current is being drawn. Once the

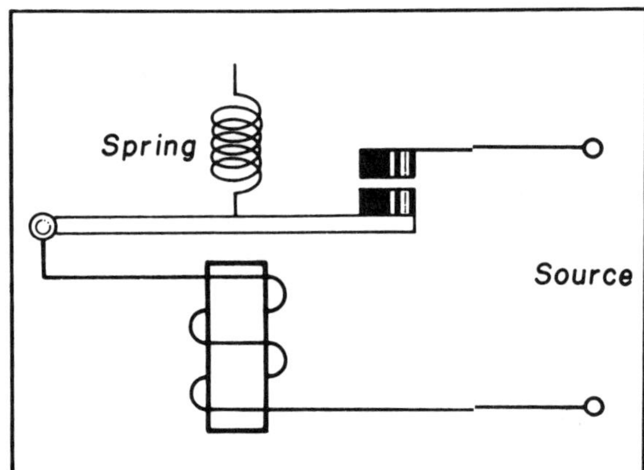

Fig. 4-7. A simple circuit breaker.

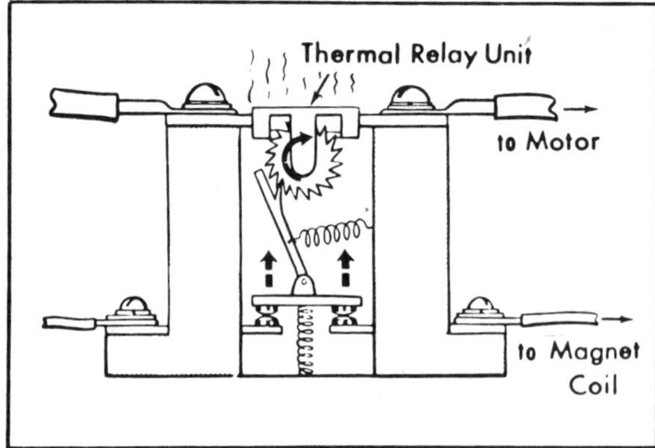

Fig. 4-8. Melting alloy thermal overload relay (courtesy Square D. Co.).

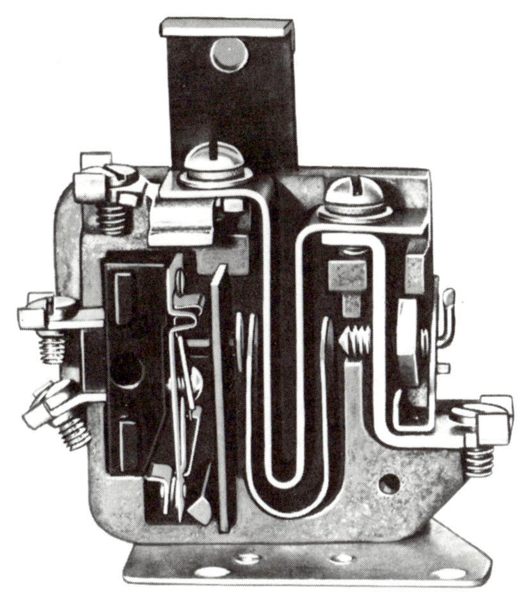

Fig. 4-9. Bimetallic thermal overload relay (courtesy Square D. Co.).

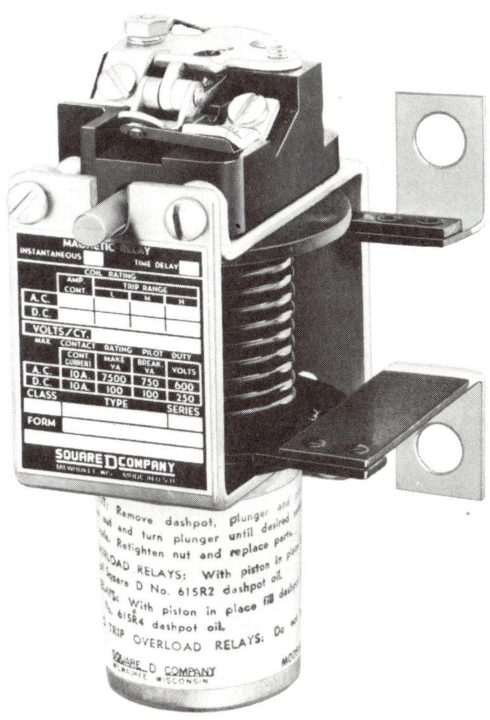

Fig. 4-10. Magnetic overload relays (courtesy Square D. Co.).

thermal unit has tripped, it must be allowed to *cool off* before operation can be restored. Thermal units do not provide low voltage protection. Also, if a power failure would result, the contacts of the switch will remain closed and operation of the motor will resume when the power returns. The motor could unexpectedly turn on which could be dangerous to the operator.

Three-phase manual-starting switches are sometimes used to turn on or off three-phase motors less than 5 horsepower that do not require overload protection. They are generally used on small machinery that has separate overload protection. Figure 4-11 illustrates a manual motor starting switch with no overload protection.

Another type of manual operated starter switch often used to control single-phase motors up to 5 horsepower and polyphase motors up to 10 horsepower is a line voltage manual starter. This starter provides overload protection through use of a thermal relay. When *tripped* the thermal relay requires resetting after a sufficient cooling down period. Figure 4-12 illustrates this type of starter.

The magnetic controller is generally used for controlling a motor from a distant location. The magnetic starter or magnetic contactor basically means the same

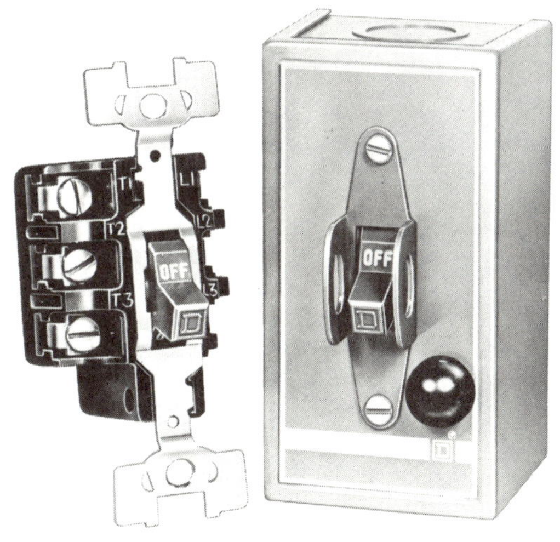

Fig. 4-11. Manual motor starting switch without overload protection (courtesy Square D. Co.).

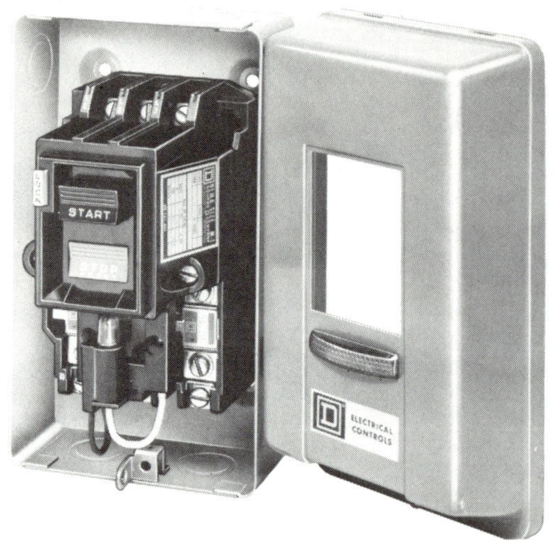

Fig. 4-12. A typical manual starter (courtesy Square D. Co.).

thing. Both have the capability to handle high currents. The main difference between the two is that a magnetic starter contains overload protection and the contactor does not. The contactor requires separate overload protection. Figure 4-13 illustrates a magnetic starter and circuit diagram.

The primary principle of the magnetic starter is when current is supplied to the magnetic coil, the coil becomes energized and pulls in the armature, which closes the contacts of the starter, and starts the motor. To prevent chatter between the magnet and the armature, due to the variation of the magnetic field sinusoidally with time, a *shading coil* is added. This shading coil helps seal in the armature by offsetting the phase of the magnetic coil. Also, a small air gap is left in the laminated iron in order to prevent the residual magnetism (magnetism left in the coil after current has been removed) from locking in the armature when the magnetic coil is de-energized (Fig. 4-14).

In addition to overload protection, the magnetic starter also contains a holding interlock. This normally open interlock holds in the circuit for the magnetic coil that would otherwise be broken when the push button was released. Figure 4-15 shows a picture of this holding interlock.

The reversing magnetic starter is used to operate the motor in the forward or reverse direction. Actually, two magnetic contactors are interwired. This starter consists of two magnetic contactors with the motor terminals T1, T2, and T3 connected to L1, L2, and L3 on one contactor. Terminals T1 and T3 are reversed on the other contactor (Fig. 4-16). Also, it presents a picture of the magnetic reversing starter. Neither the forward or re-

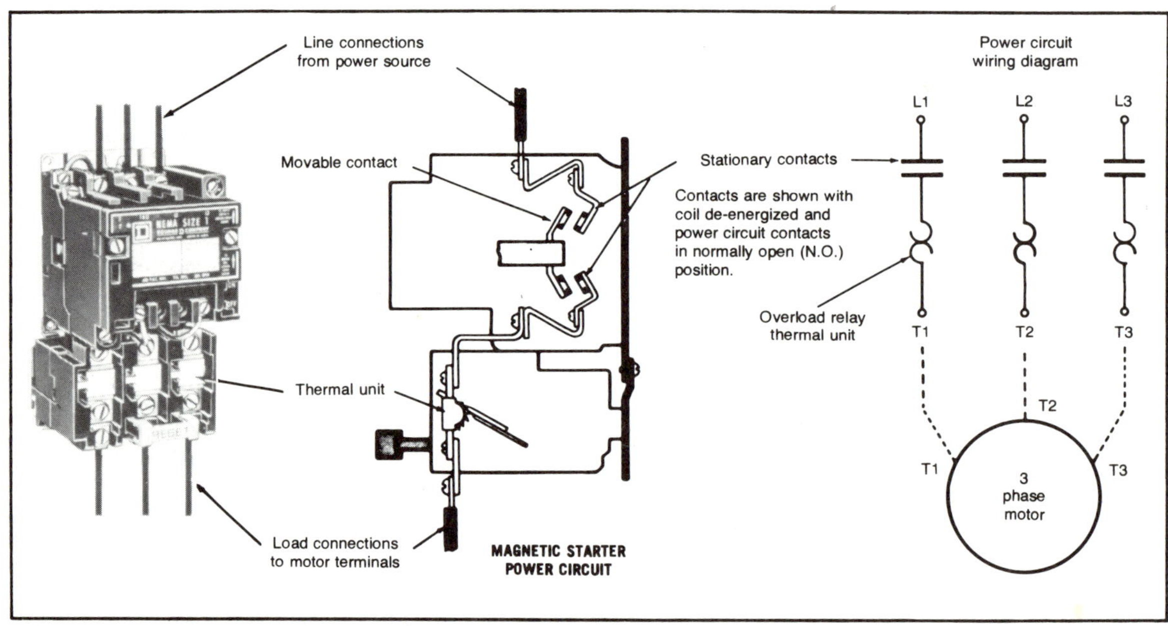

Fig. 4-13. A magnetic starter with circuit diagram (courtesy Square D. Co.).

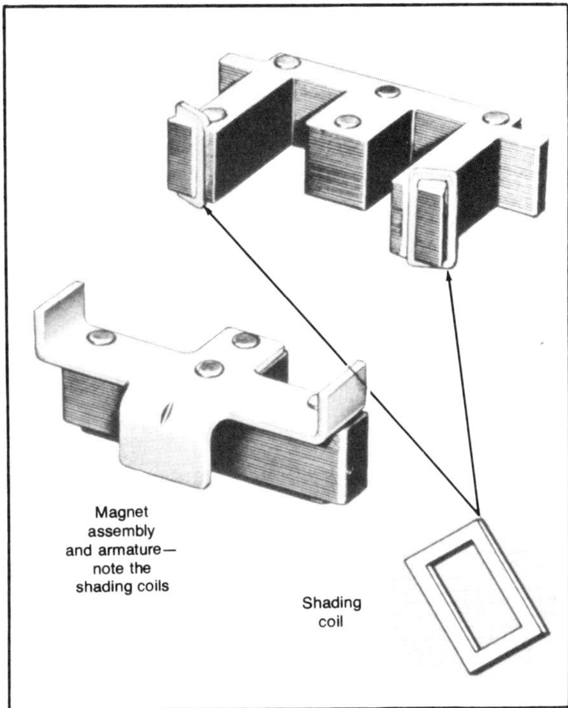

Fig. 4-14. A magnetic armature and shading coils (courtesy Square D. Co.).

Fig. 4-15. A magnetic starter with two magnetic contactors (courtesy Square D. Co.).

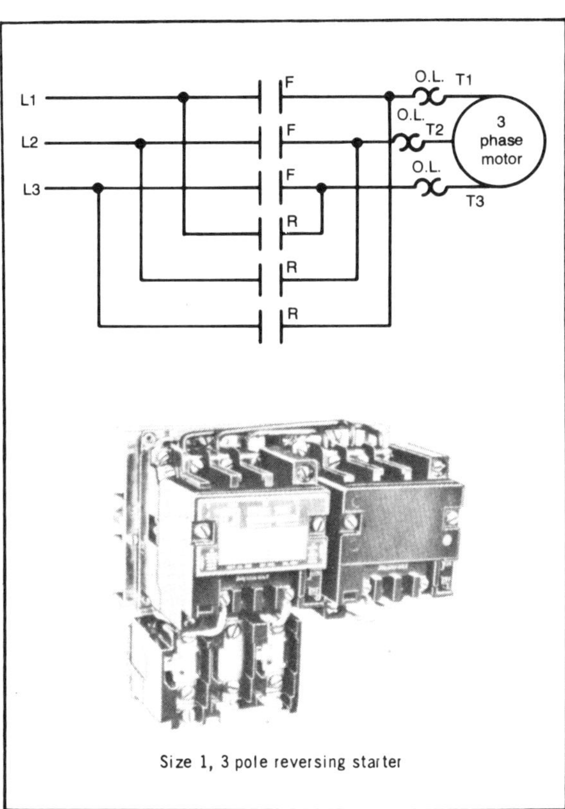

Fig. 4-16. A magnetic reversing starter with circuit diagram (courtesy Square D. Co.).

verse contactor can be energized with the other contactor already energized. This is accomplished by mechanical and electrical interlocks that prevent this action. If the forward contacts are closed, the mechanical and electrical interlocks will prevent the energization of the reverse contactor. The forward contactor must be first de-energized in order for the reverse contacts to be closed.

Relays. Figure 4-17 shows three different types of relays commonly used in motor control. The relay is used with other control devices to open or close the current to the motor.

Push-Button Stations. There are several types of push buttons used in motor control operation. Generally, a push-button station contains two sets of contacts. One is normally open, the other is normally closed. This means that when one set of contacts closes, the other set must open and vice versa. Push buttons are used in conjunction with magnetic controllers. They are not required to be located close to the controller. Push buttons help to control the functions of starting, stopping, jog-

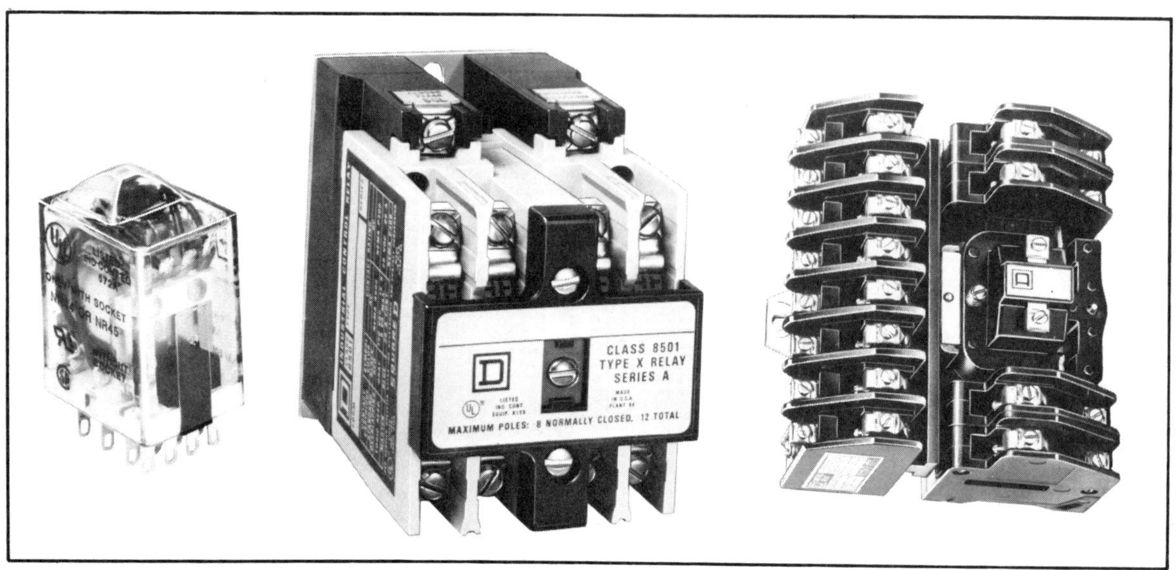

Fig. 4-17. Typical relays (courtesy Square D. Co.).

ging, reversing, and other operations. They also are designed for various operating conditions and may include indicator lights, key operation, or padlocking provisions (Fig. 4-18).

Limit Switches. Limit switches like the pushbutton switch are commonly used in conjunction with magnetic starters. One of the main differences of the limit switch is that it is often used to convert motion of machinery into electrical control signals (Fig. 4-19).

Drum Switch. The drum switch is a three pole manually operated switch used for reversing single or three-phase motors. Figure 4-20 shows the interal switching positioning of the switch as well as an overall picture.

Timers. There is a large variety of timers used in motor control, such as interval, pulse, percentage, and revolution. One type, illustrated in Fig. 4-21, is a pneumatic time relay that works on an air/diaphragm principle. The needle valve controls the flow of air back into a partially evacuated chamber. When the diaphragm returns to its initial position, with equal air pressure on both sides, the contacts operate.

Once these contacts operate the circuit will turn on or off depending upon whether the normally open or normally closed contacts are being used.

There are many other types of motor controls commonly used today. The actual introduction of all these controls are beyond the scope of this book, but the basic theory and repair of them is somewhat similar.

REPAIR AND TESTING PROCEDURES

Common techniques used in troubleshooting motor control circuits include substitution of controls and parts, amperage and voltage readings, etc.

Basic manual switches can be checked with an ohmmeter. Infinite resistance when the switch is in both positions would indicate a defective switch. Also, if the thermal device fails to reset or continues to trip with normal current operation, the thermal overload may need to be replaced. Keep in mind that an incorrect thermal unit selection may have been made in the first place. Also, don't disregard the fact that there may be parts broken inside the overload relay while absolutely nothing is wrong with the thermal unit itself or its selection.

Relays can also be checked using an ohmmeter. First begin with a visual inspection. Look for burnt contacts or charred coils. When the relay is manually *pushed in* the ohmmeter should indicate either continuity or no continuity depending on whether the contacts are normally open or closed. The ohmmeter will show continuity one way and not the other. Also, the total resistance of the windings can be checked. Depending on the coil size, a relay coil should not read zero ohms. Zero ohms might very well indicate a shorted coil. On the other hand, infinite coil resistance would indicate an open circuit. Likewise, if power is supplied to the coil, but it fails to pick up the armature, the coil is probably open.

A manual starter can easily be checked by visually inspecting the contacts (Fig. 4-22). When contacts appear

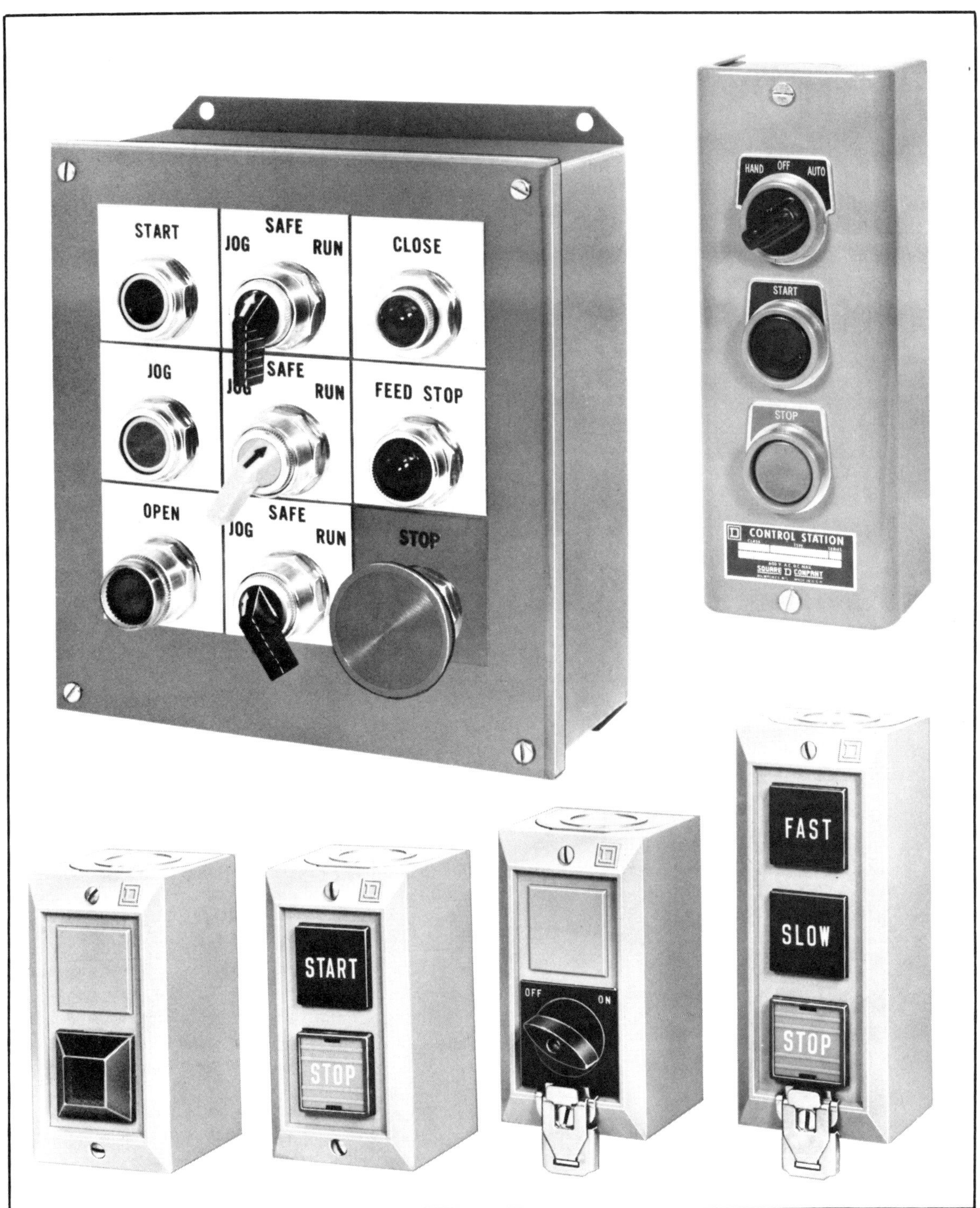

Fig. 4-18. Push-button control stations (courtesy Square D. Co.).

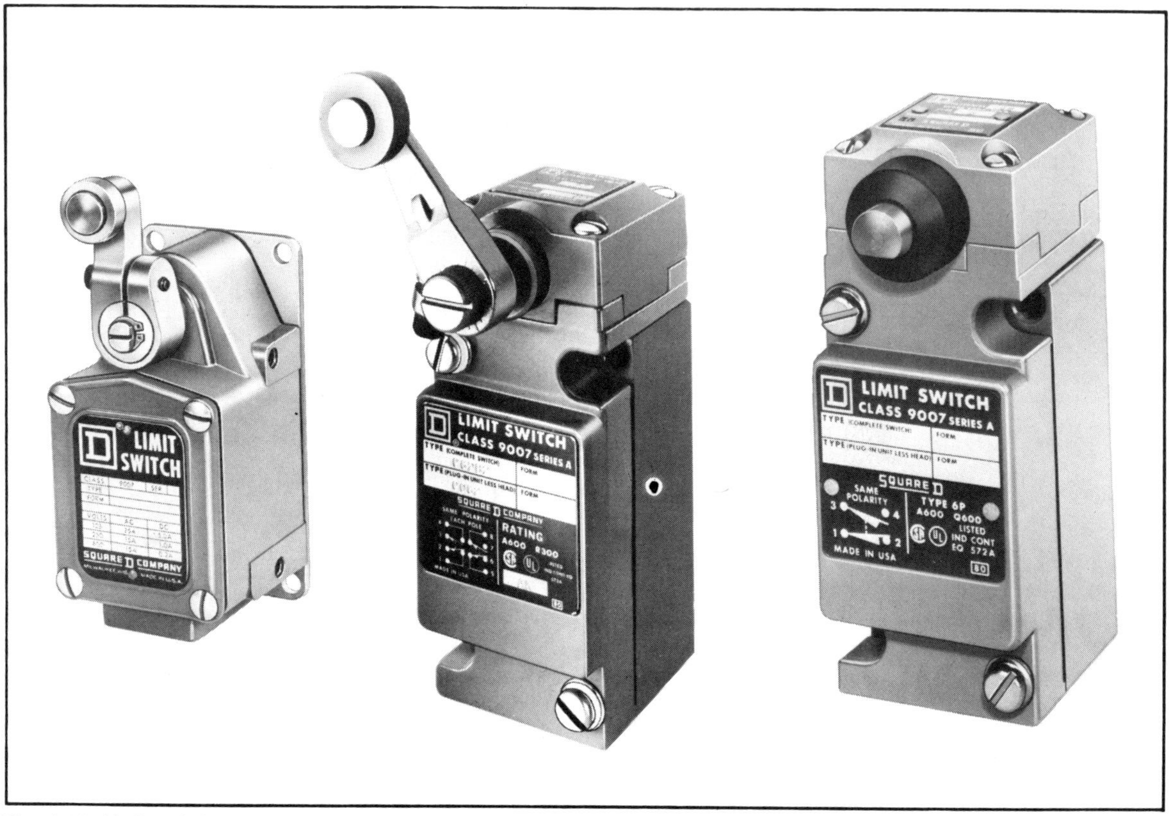

Fig. 4-19. Limit switches (courtesy Square D. Co.).

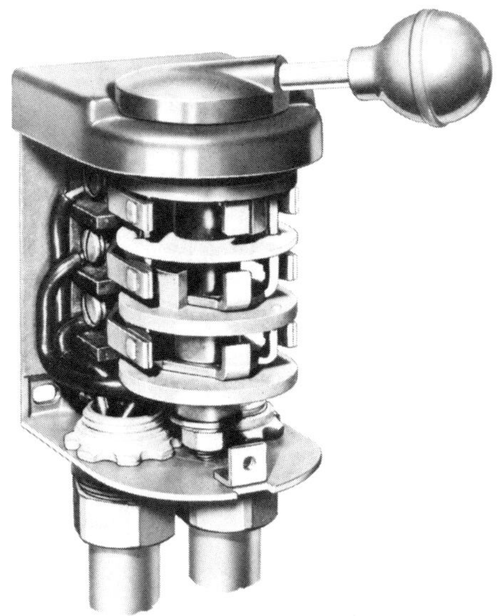

Fig. 4-20. Drum switch shows internal switching position and overall view (courtesy Square D. Co.).

to be dirty, deformed, or pitted, they should be replaced. Most contacts can be cleaned with acceptable cleaning solvents and dressed up with a file. Note: Silver contacts should never be filed! Dirty, gummy starters, coils, and other devices may need to be enclosed in a more appropriate enclosure. Although hot soapy water sometimes is used to clean control equipment, it is not recommended by the author. Control equipment should never intentionally be exposed to water. There are many excellent cleaning solvents available by motor control distributors that would be more appropriate to use.

If the thermal overload relays fail to hold under normal current or fail to reset, they should be replaced. Also, make sure that there are no broken parts inside the overload relay that might be causing the problem.

Always inspect the starter for loose connections and check for any type of physically damaged parts. Replace any broken wires or terminals with the same replacements.

When troubleshooting overload relays, check to see if the coil continues to trip. Also, check the ambient

Fig. 4-21. A pneumatic timing relay (courtesy Square D. Co.).

temperature. It may be necessary to replace the coil with one of the next larger size. Remember, never take anything for granted. Other factors to consider when an overload continues to trip are as follows:

1. Service factor of the motor.
2. Jogging for plugging duty.
3. Long acceleration time.
4. Overloaded motor.
5. Slow-speed motors.

Any of these factors may cause the motor to draw more current than is normal for a given horsepower.

When troubleshooting contacts, observe whether the contacts fail to hold. If they do, the voltage may be too low. Also, if the contacts tend to wear quickly, freeze, or weld together, it may be due to a short circuit, low voltage, or an abnormal voltage surge. One should never disregard the possibility of physical abuse of the starter by its operator! Excessive abuse will quickly wear down the contacts.

When troubleshooting magnetic starters, common problems include dirty gummy parts, worn contacts, loose connections, defective heater overloads, damaged or worn mechanical parts, etc., (Fig. 4-23). For example, failure of a magnetic coil to *pick up* could indicate a defective coil, wrong coil, low voltage, incomplete control circuit, a break in wiring continuity, or some type of mechanical problem. On the other hand, if a magnetic coil appears to be overheated, check for incorrect voltage, incorrect amperage, wrong coil, or starter abuse by the operator. If a coil fails to drop out, check the physical position of the coil and inspect the starter for dirt or gum, wrong voltage/amperage, welded contacts, or mechanical damage. A noisy magnetic starter can often be attributed to a broken shading coil, low voltage, or any type of dirt or corrosion.

If a magnetic starter continues to trip the overloads, check for any possible shorts or grounds in the circuit. Also, the overloads may need to be replaced. If the contacts chatter, check for low voltage or defective contacts. Remember, any type of starter abuse by the operator or dirty conditions can cause premature breakdown of the magnetic starter. Figure 4-24 shows a servicer measuring current with a clamp-on volt/amp/ohmmeter.

Figure 4-25 represents a typical three-phase 240 volt motor with a 120 volt (low voltage) control circuit. Here, the fused transformer provides low voltage for safer operation. If the motor is completely dead when the start push button is pushed, first make sure that the overload relay is reset. If the relay check is good, begin by checking the transformer line fuse for continuity. Check each fuse in the disconnect box, control overloads, and/or main panel box. If a blown fuse is found, it is advisable to determine what caused the fuse to blow. To do this, check the motor for possible grounds or shorts. Examine the motor control for possible loose connections, overheated coils, and dirt. Figure 4-26 shows a servicer using a clamp-on volt/amp/ohmmeter as an ohmmeter to check fuses.

If the motor fails to run when the start push button is pushed, manually push in the magnetic control armature. If the motor now operates, check for continuity in the push-button station, transfer line fuse, magnetic coil, holding contacts, wire to and from transfer line to magnetic control, etc. A defective component or loose connection must be the problem. Isolating and testing

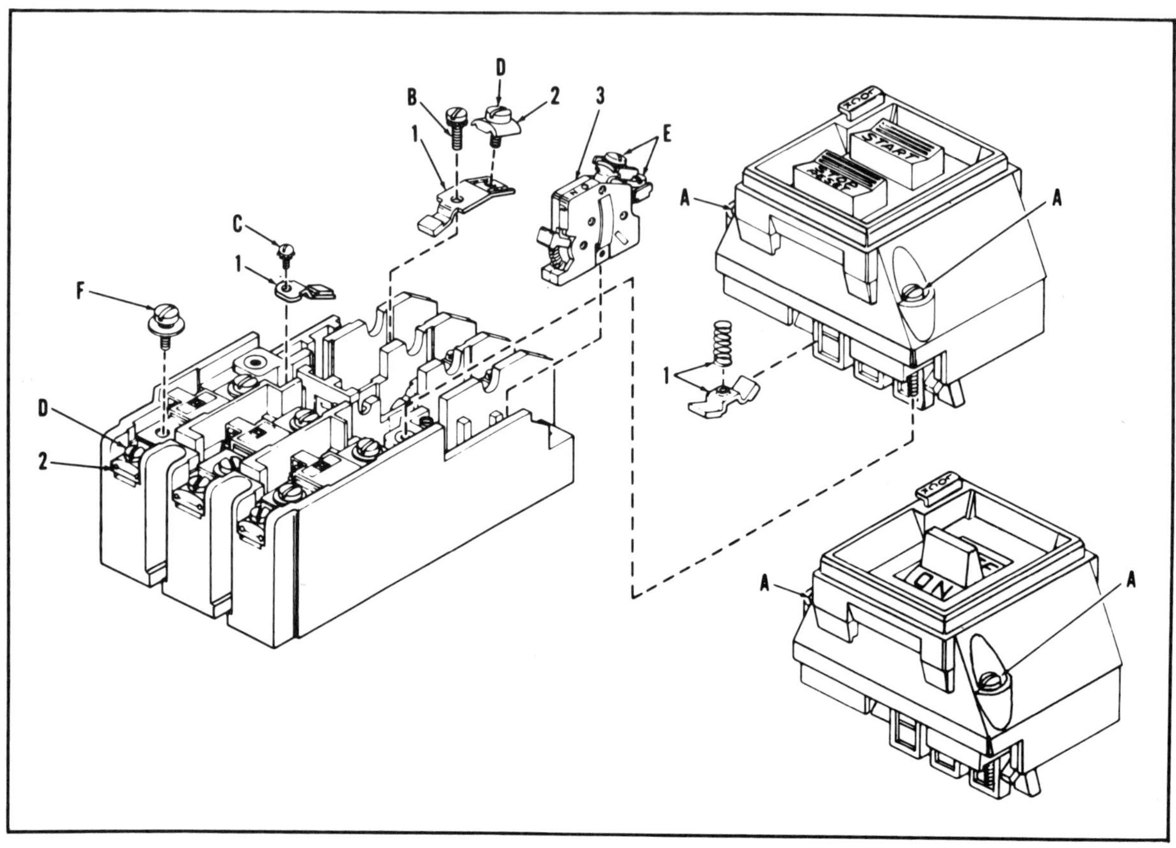

Fig. 4-22. An exploded view of a manual starter (courtesy Square D. Co.).

each stage will determine the cause. If an open fuse is found, replace the fuse only with one of the same rating. If necessary, use a fuse below the rated ampere value, but never one above the rated value.

If the motor runs, but fails to stop consistently when the stop button is pushed, look for a defective or miswired switch or welded holding circuit interlock. Examine and check the push-button station. The stop contacts may be badly worn and corroded, thus preventing proper conduction. Last, check all wires for possible breaks, by both a visual inspection and a continuity check.

If the motor runs at half speed or "sits and hums," look for one of the phases to be open. Begin by taking a voltage measurement across each phase by the motor terminals and determine which phase is open. After locating the open phase, trace back in the circuit with the voltmeter until voltage is restored. A loose connection or broken wire may be the cause of the problem. Note: If all line phases have the correct voltage check the motor with a *Megger* for ground or shorts. A shorted motor may act in the same manner as an open line phase.

If the operator complains of noisy operation, try to isolate the noise by determining if it is being caused by the motor or the control. A chattering or humming in the control could indicate a broken shading coil, dirty or worn contacts, dirty control, mechanically worn or misaligned armature, or low voltage. If the noise comes from the motor, this could indicate worn bearings, a bent shaft, or excessive vibration due to loose motor base connections.

If the motor continues to trip the magnetic control overloads, check the amperage at the motor terminals to determine if excessive current is being drawn. The motor may be shorted, the overloads may be faulty, partially shorted/grounded, or loose wires may be causing the problem. Remember, don't overlook the obvious. Excessive moisture, dirt, heat, or corrosion can all contribute to tripped overloads or blown fuses.

There are other factors that may cause an increase of heat and cause motor problems. They are sustained jogging (repeated starting/stopping) and plugging (rapidly

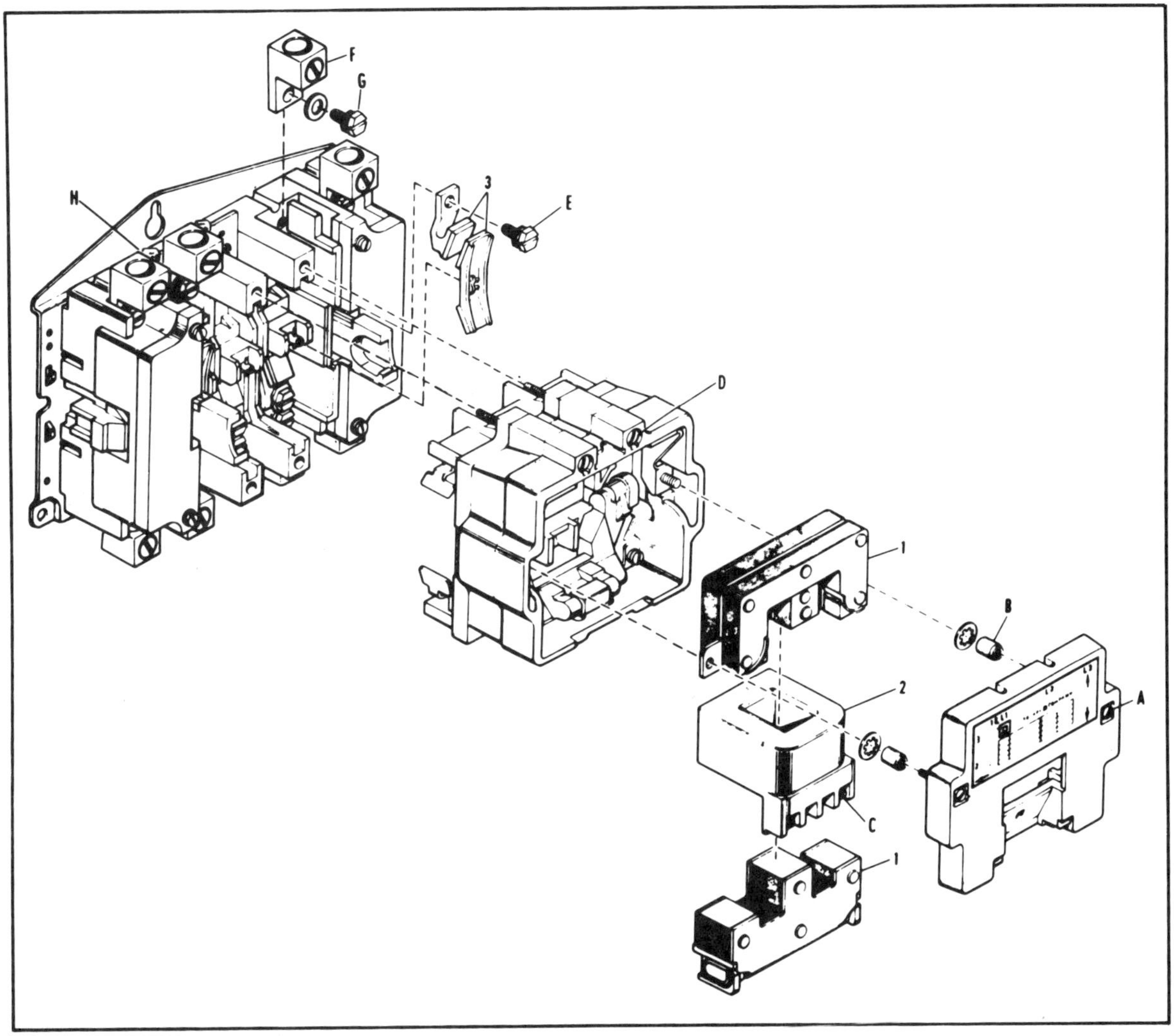

Fig. 4-23. An exploded view of a magnetic starter (courtesy Square D. Co.).

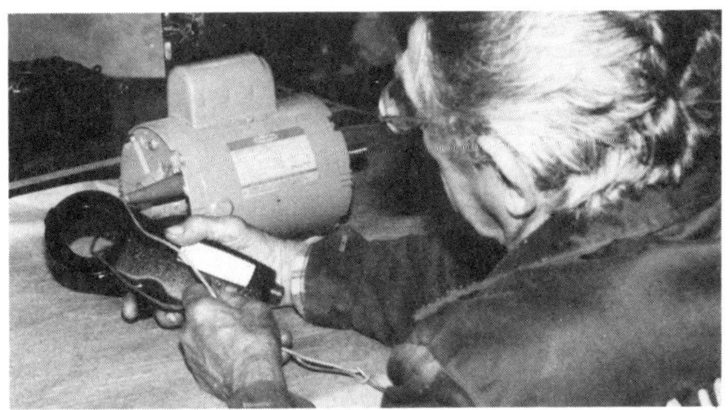

Fig. 4-24. Measuring current with a clamp-on volt/amp/ohmmeter.

stopping of a motor by momentarily connecting the motor in reverse direction). Operator abuse is also a factor that should be considered.

If the motor fails to run and the transformer is suspected, various methods of determining a defective transformer can be used. Many transformers can be visually checked by looking for charred or burnt insulation and windings. A shorted transformer will feel hot when touched, and will have a bad odor. The windings can be checked with an ohmmeter—zero resistance probably indicates a short. An open winding will read infinite resistance. A voltmeter can also be used to check the performance of the transformer. If correct voltage is read

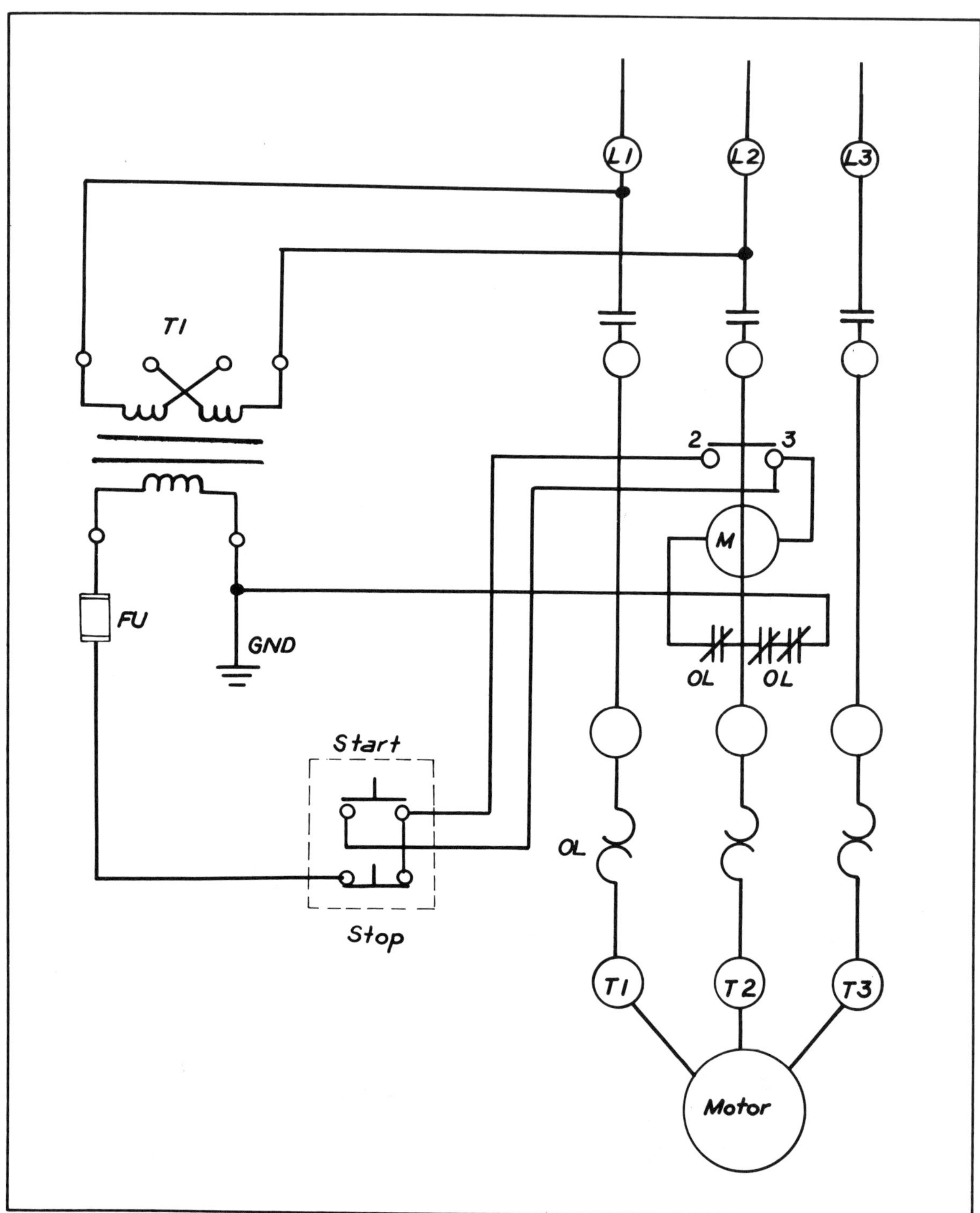

Fig. 4-25. A schematic diagram of a three-phase motor control circuit.

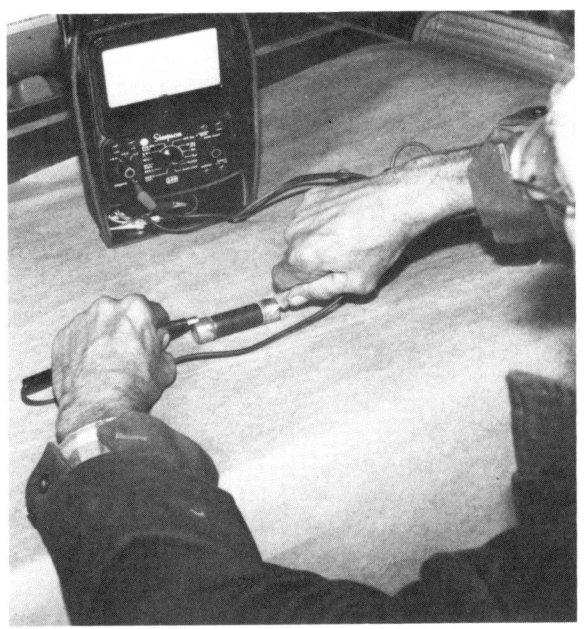

Fig. 4-26. Using a vom meter to check fuses.

at the primary side, but not at the secondary side, the transformer is probably *open*. Incidentally, check to make sure the transformer lead connections are secured. A loose connection would indicate an open transformer circuit.

PREVENTIVE MAINTENANCE

It has been generally accepted that the motor controls successful operation is dependent upon the type of maintenance and operating conditions of the control. Dirty, gummy controls should be scrubbed with a stiff brush and cleaning solvent. If control equipment should get wet, it is important to completely dry out the control by heating it or drying it in air or sun. Finally, check the coils, with an ohmmeter or megohmmeter for correct insulation resistance.

Controllers should be periodically checked for worn parts, loose connections, or weak spring tension. Dirty, corroded copper or cadmium-plated contacts should be kept clean and refiled to maintain correct shape. Silver contacts should *not* be filed.

Lubrication of motors and controls is important in

Fig. 4-27. Using a heat gun to shrink an insulating sleeve (courtesy Ideal Industries, Inc.).

order to avoid breakdown due to excessive friction. Usually, controllers do not require lubrication because this attracts dirt and other contaminants that wear down the controller. (But most motors should be periodically cleaned and bearings regreased or replaced.)

Most water jackets used in cooling operations should be flushed out periodically and replaced to eliminate rust and corrosion. Also, proper oil level and oil/air pressure should be constantly checked for possible leaks.

Using common sense and a little care can prevent many accidents as well as control breakdowns. Keep a lookout for loose guards, worn belts, worn controls, and cracked, charred, or brittle wire. If a section of wire is brittle, cracked, or frayed, the wire should be replaced. If the entire wire cannot be conveniently replaced, a section of it can be spliced in. Use the exact replacement wire and make sure the connections are both mechanically and electrically secured by solder or splicing terminals. Lastly, make sure the connection is well insulated by electrical tape or sleeving. Figure 4-27 shows a heat gun being used to shrink sleeving on a cable of wire.

SELF-EXAMINATION

Select the best answer:

1. Reversing the flow in a coil:
   A. Does not change the magnetic polarity.
   B. Destroys the coil.
   C. Reverses the magnetic polarity.
   D. None of the above.
2. The simplest form of a motor controller is the:
   A. Toggle switch.
   B. Magnetic switch.
   C. Drum switch.
   D. Relay.
3. Another name for a magnetic starter is a:
   A. Manual switch.
   B. Contactor.
   C. Manual starter.
   D. Magnetic control.
4. What type of overload contains a solder pot?
   A. Melting alloy.
   B. Bimetallic.
   C. Magnetic.
   D. Fuse.
5. What type of overload operates on the principle of electromagnetism?
   A. Melting alloy.
   B. Bimetallic.
   C. Magnetic.
   D. Fuse.
6. A 15-amp fuse is found open, but the servicer does not have a replacement and has a choice between a 10-amp and 30-amp fuse. Use the:
   A. 10-amp fuse.
   B. 30-amp fuse.
   C. Either one.
7. Low voltage often causes the magnetic control to:
   A. Gum up.
   B. Run more efficiently.
   C. Chatter.
   D. None of the above.
8. Rapid stopping of a motor by momentarily connecting the motor in reverse direction is called:
   A. Jogging.
   B. Inching.
   C. Plugging.
   D. Sequence operation.
9. A hot, smoky transformer indicates a/an:
   A. Open.
   B. Short.
   C. Both A and B.
   D. None of the above.
10. A shorted motor will draw ________ current.
    A. Low.
    B. High.
    C. Both A and B.
    D. None of the above.
11. Three-phase manual-starting switches are sometimes used to turn on or off motors up to ________ horsepowers.
    A. 5.
    B. 10.
    C. 15.
    D. 20.
12. The device used to control a motor from a distant location is the:
    A. Manual starter.
    B. Drum switch.
    C. Magnetic controller.
    D. Snap switch.
13. Which of the following devices is a type of timer?
    A. Interval.
    B. Pulse.

C. Percentage.
D. All of the above.

14. The device that controls the flow of air back into chamber of the pneumatic time relay is the:
    A. Diaphragm.
    B. Needle valve.
    C. Shading coil.
    D. All of the above.
15. If a motor runs but fails to stop, the problem may be:
    A. Open fuse.
    B. Welded holding circuit interlock.
    C. Jogging or plugging duty overload.
    D. None of the above.
16. If the motor runs at half speed or "sits and hums," look for __________ in the magnetic controller.
    A. An open phase.
    B. Welded holding circuit interlock.
    C. Both A and B.
    D. None of the above.
17. Repeated starting and stopping is referred to as:
    A. Stepping.
    B. Jogging.
    C. Phasing.
    D. Long acceleration time.
18. A chattering or humming in a magnetic contractor may be due to:
    A. A broken shading coil.
    B. Dirty/worn contacts.
    C. Both A and B.
    D. None of the above.
19. Which of the following contacts should not be filed?
    A. Cadmium-plated.
    B. Silver-plated.
    C. Both A and B.
    D. None of the above.
20. An increase in heat in a motor can be caused by:
    A. Sustained jogging.
    B. Repeated plugging.
    C. Both A and B.
    D. None of the above.

QUESTIONS AND PROBLEMS

1. State the basic functions of a motor controller.
2. Explain the basic theory of a circuit breaker.
3. What is a toggle switch?
4. What is an advantage of a magnetic starter over a manual starter?
5. State the characteristics of a magnetic starter.
6. What is a limit switch?
7. Explain the theory of operation of a melting alloy thermal overload relay.
8. Explain the theory of operation of a bimetallic thermal overload relay.
9. Explain the theory of operation of a magnetic overload relay.
10. What is a pneumatic timer?
11. How do you check a relay?
12. Sate the various methods to test a transformer.
13. What is the proper technique to clean a coil?
14. What might be wrong with a motor that runs at half speed?
15. State different types of noise from a motor and motor control. What are the causes of this noise?
16. What is a shading coil?
17. Define the term jogging.
18. Define the term plugging.
19. Why is low control voltage often used in motor control operations?
20. Explain the procedure in maintaining or servicing motor control contacts.

Chapter 5

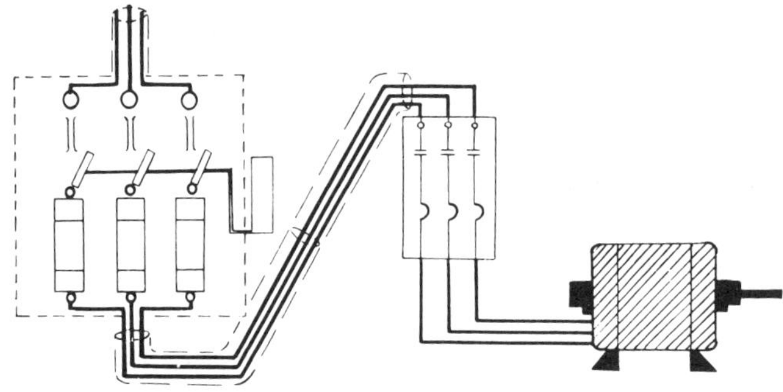

Troubleshooting Residential/Industrial Wiring

Electrical wiring provides the medium by which power is transmitted to electric motors, controls, and other electrical/electronic apparatus. In this chapter, fundamentals of residential and industrial wiring are presented, along with troubleshooting techniques. In addition, basics and testing procedures are presented for television distribution systems and residential/industrial lighting systems.

FUNDAMENTALS

Basically, most wiring circuits consist of four basic sections (Fig. 5-1) as follows:

1. Power source.
2. Transmission line.
3. Control device.
4. Operating device.

The single phase *power source* usually consists of a distribution panel box that supplies 120/240 volts. It has a total ampere capacity that generally ranges from 60 amps to 200 amps—100 amperes is minimum today. In the panel box, each individual circuit is connected to a *hot* side and a *neutral* ground side. Figure 5-2 illustrates an example of a 120- and 240-volt transformer supply.

In the panel box the hot leg that is often called the power leg is usually a black wire and/or red if used for 240 volts. The neutral ground wire is a white wire, and the green bonding wire acts as a safety appliance/equipment ground. Fuses or circuit breakers protect the hot side from overloads. The neutral side is almost always grounded to earth by connecting this wire to a cold water pipe or ground rod. Figure 5-3 shows a typical distribution panel box layout.

An appliance/equipment ground consists of a small bonding wire that is connected to the neutral (common ground) terminal block in the panel box. It is then connected to the metal receptacle box and third odd hole of the receptacle. When an appliance, such as a drill, is plugged into this receptacle, it will also have a third prong and wire that is connected to the metal frame of the drill (Fig. 5-4).

If a grounded circuit should arise while a person is operating the drill, the current will not go through the operator. Instead, it will return through this bond wire and back to the source. If there is no protective appliance ground, the current from the hot side of the line may go through the grounded drill through the operator and back to the earth ground (Fig. 5-5).

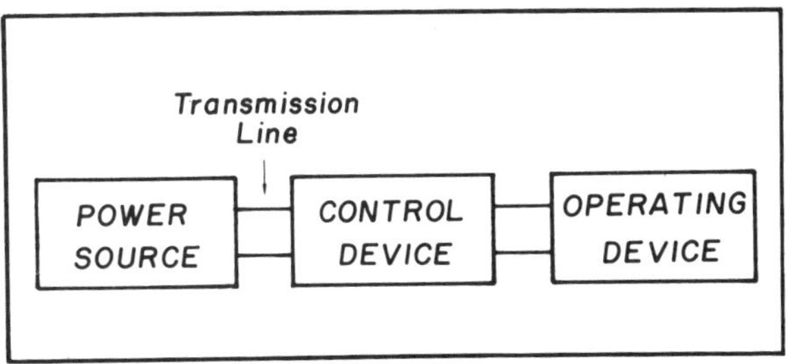

Fig. 5-1. The four sections of a wiring circuit.

Most circuits found in the home usually use No. 14 wire for 15 amps, No. 12 wire for 20 amps, No. 10 wire for 30 amps, No. 8 wire for 40 amps, and No. 6 wire for 55 amps.

The maximum capacity of a 20 ampere general-purpose appliance circuit is approximately 2400 watts and for a 15 ampere circuit, about 1800 watts. The total number of appliances, if used at one time, should not exceed the rated capacity. In calculating residential wiring circuits, a 15 ampere circuit is commonly used for every 375 square feet of floor space and a 20 ampere circuit for every 500 square feet. A convenience outlet should be accessible every 12 feet of running wall space. Typically, an electric range (8,000 to 18,000 watts) should have a separate 240 volt (40 amp) circuit. An electric water heater (2,000 to 4,000 watts) requires a separate 240 volt (40 amp) circuit. Kitchens usually have two to three separate 20 ampere appliance circuits. One general-purpose circuit (15 to 20 amperes) often can

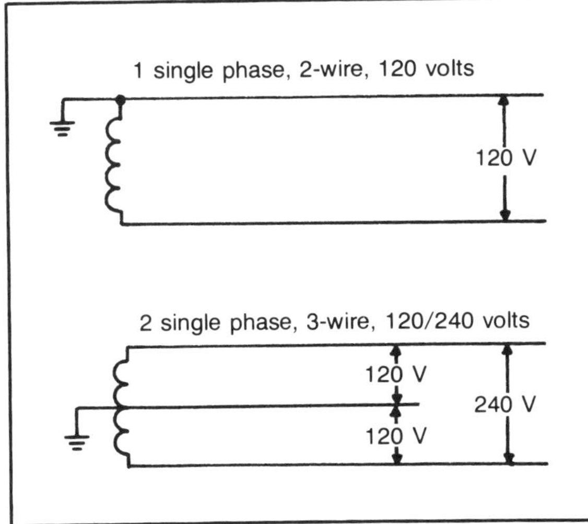

Fig. 5-2. Example of 120 volt and 240 volt transformer supply.

supply two bedrooms and a bathroom.

The three basic kinds of cable used are nonmetallic sheathed cable, flexible armored cable, and conduit emt (electrical metallic tubing) or "pipe." Conduit usually comes in 10-foot lengths and comes in sizes: ½ inch carrying 4 No. 14 wires, or 3 No. 12 wires, ¾ inch carries 4 No. 10 wires or 5 No. 12 wires, etc. Different types of conduit include *thin wall, rigid, plastic,* etc. Nonmetallic cable basically consists of indoor or outdoor types.

Switching Circuits. Some of the more common switching circuits used are single pole, three-way, and four-way. Figure 5-6 illustrates a typical single-pole switch controlling a lamp. The current travels through the switch to the lamp and returns back to the power source.

A three-way switch controls one lamp or more from two locations (Fig. 5-7). In the position shown the lamp is off; turning either switch the lamp will turn on.

A four-way switch controls one lamp or more from three different locations (Fig. 5-8).

A three-phase four-wire system consists of three phases having equal voltage and a fourth grounded wire that is used for protection and reduced voltage. Figure 5-9 illustrates a typical wye and delta system commonly used in schools, businesses, and industry.

REPAIR AND TESTING PROCEDURES

Like the electric motor, common problems that occur in residential/industrial wiring circuits include opens, grounds, and shorts. Besides these problems, a typical problem found in three-phase industrial systems is "low power factor." Any type of ac "inductive reactant" circuit or devices like heaters, lamps, motors, controls, etc., will cause a lagging of current that is expressed in percentage. Power factor percentages often range somewhere between 30 and 90 percent—90 percent being excellent and 30 percent very poor. Power factor actually is the cosine of the phase angle between voltage and current. A low power factor rating uses excessive current, is wasteful and expensive, and increases heat. A power factor rating is usually corrected in a circuit by increasing the size of the cable or by adding capacitance in parallel with the circuit (Fig. 5-10).

Power factor correction is usually expressed in a term called var (volt-ampere reactive) or kilovars. The correction is accomplished by the addition of capacitors that decrease the phase difference between the voltage and amperage used in the circuit. Therefore, a motor with corrected power factor simply draws less current which saves power, money, and maintenance. In Fig. 5-10, note

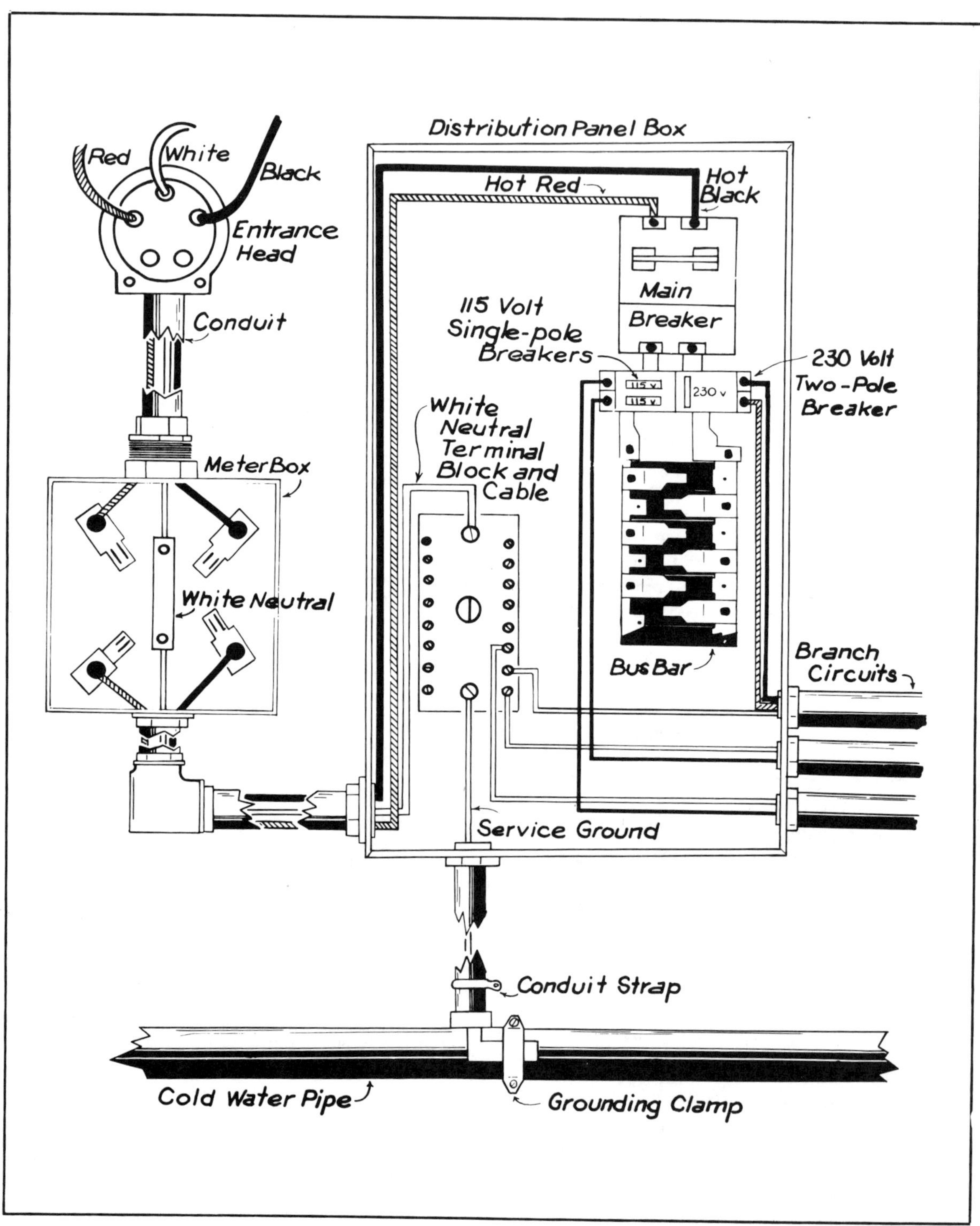

Fig. 5-3. A typical distribution panel box layout.

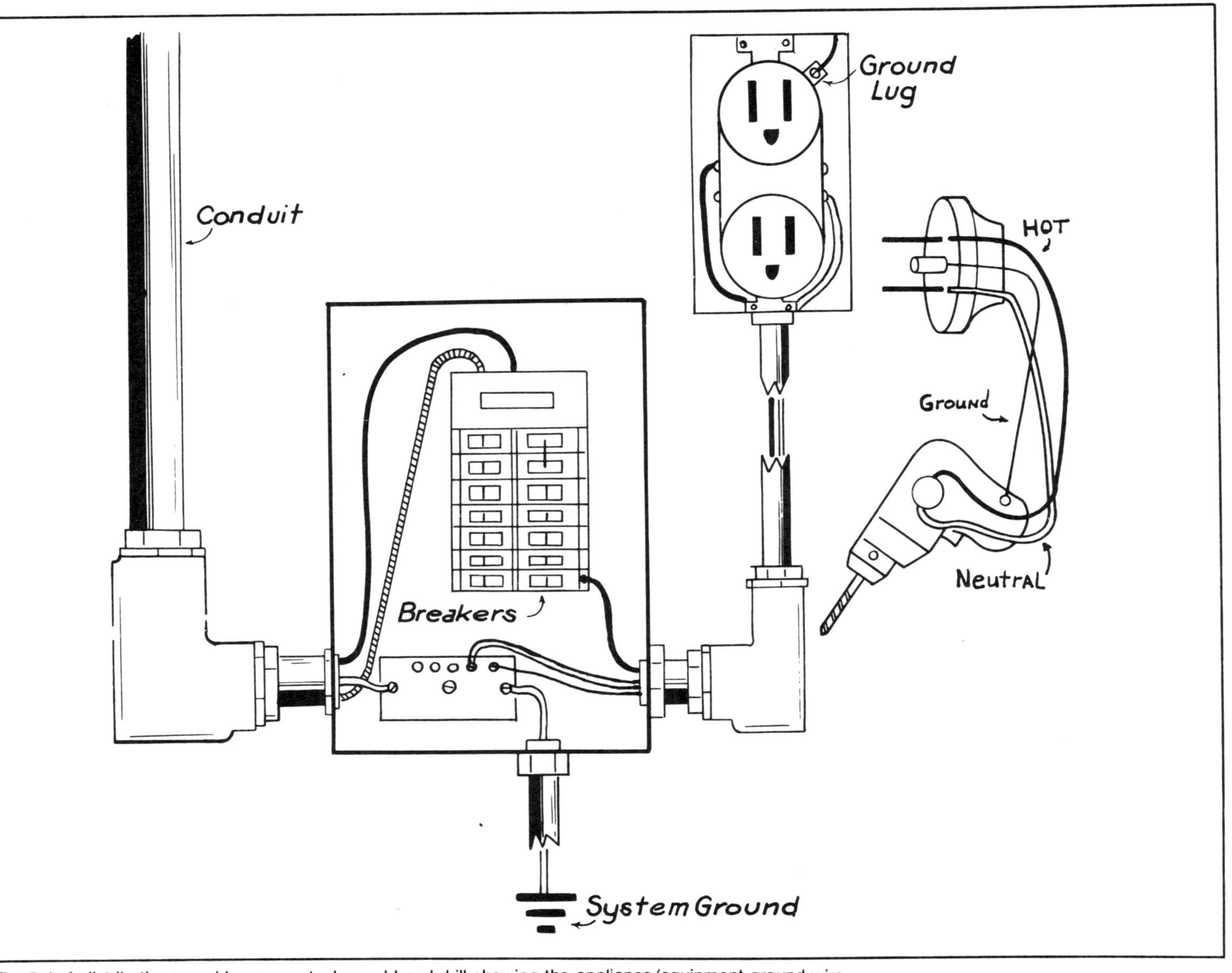

Fig. 5-4. A distribution panel box, receptacle, and hand drill showing the appliance/equipment ground wire.

Fig. 5-5. Example of a grounded circuit and a circuit that is not grounded.

the capacitor box has a fuse in each line for added protection in case one of the capacitors should "short circuit." Also, bleeder resistors are shunted across the capacitors to decrease shock hazards and increase the life of the capacitor. Notice that actual physical placement of the "capacitor banks" can be located at the power source entrance, across the motor, and/or ahead of the motor loads.

In troubleshooting wiring systems, use your senses. If a circuit breaker has just tripped, instead of resetting the breaker, stop and think what might have caused the breaker to trip. Did someone just plug in a device over-

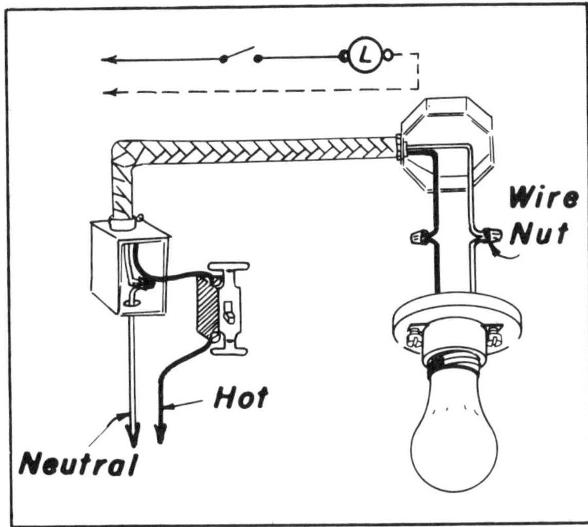

Fig. 5-6. A typical single-pole switch controlling a lamp.

loading the circuit? Did you smell any funny odors?

What are the steps used to analyze an operating device failure? First, examine the system for cracks, charring, odors, mechanical problems, etc. Next, check the main fuses using a voltmeter. Figure 5-11 illustrates the technique of locating a blown fuse in a single-phase system.

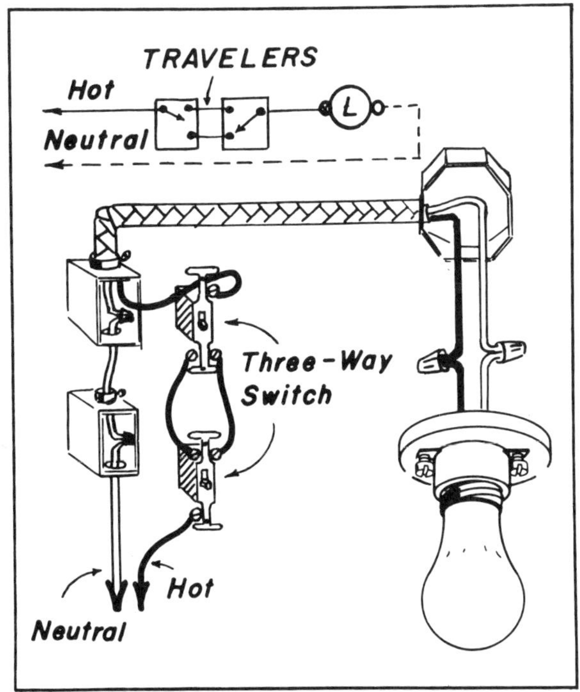

Fig. 5-7. A typical three-way switch controlling a lamp.

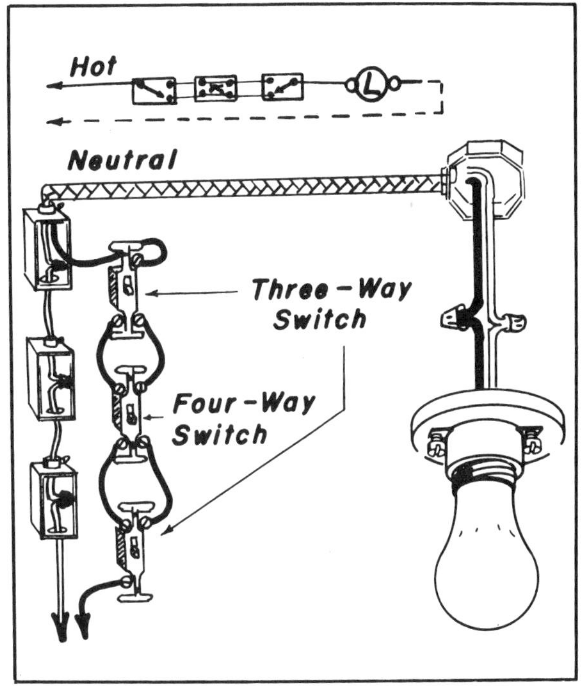

Fig. 5-8. A typical four-way switch controlling a lamp.

In a three-phase system, fuses can be checked with a voltmeter or ohmmeter in the same manner as they are checked in a single-phase system (Fig. 5-12).

Another common device used to check the final installation of a housewiring circuit is a receptacle tester. It indicates typical circuit faults such as open hot, open neutral, hot and neutral reversed, and reversed polarity. A typical receptacle tester is shown in Fig. 5-13.

Shorted wires, open wires, and wires grounded to conduit are common problems in industry. Figure 5-14 shows a typical industrial circuit. Troubleshooting tips are as follows:

1. Check the distribution line coming from the main panel box.
2. Check for dirty/loose connections or blown fuses.
3. Check the motor control and wires leading to and from the control. Check overloads, loose connections, dirt, oil, shorted windings, bad contacts, etc.
4. Check for shorted or grounded wires in conduit.
5. Check inside the junction box for poor pigtail

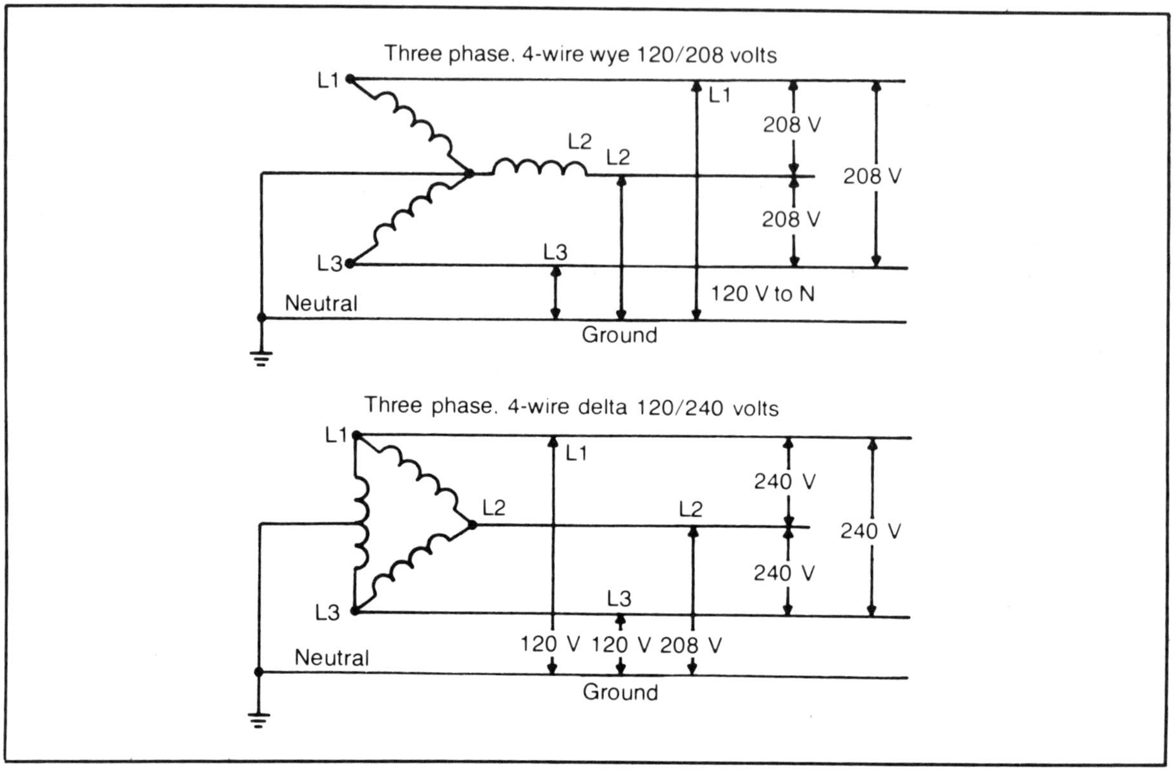

Fig. 5-9. Common wye and delta circuits.

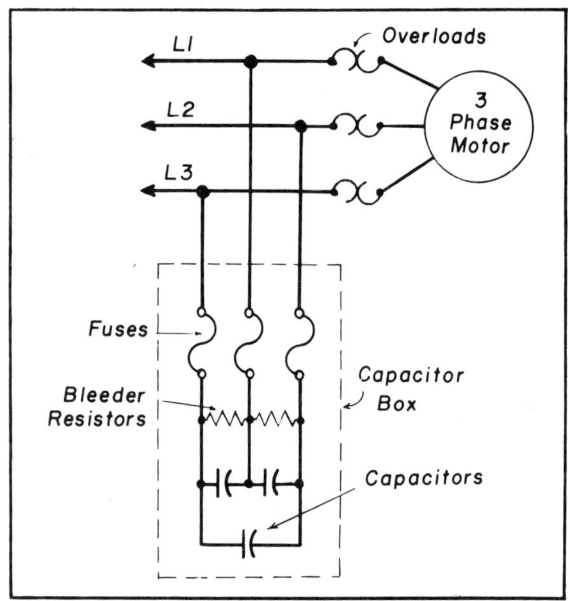

Fig. 5-10. Low power factor corrected by adding capacitor.

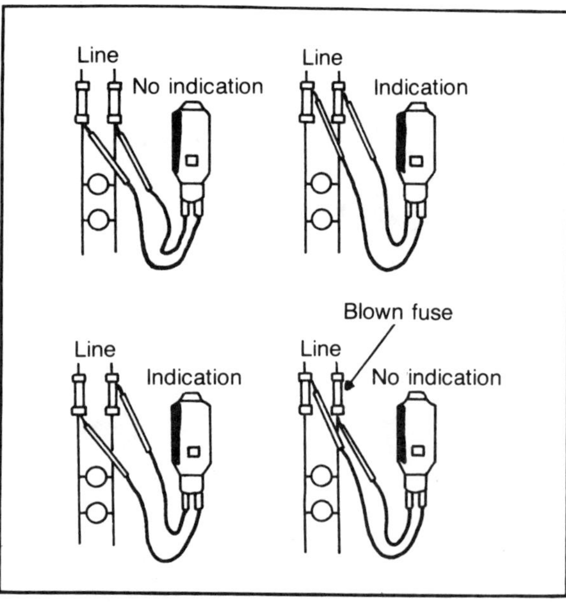

Fig. 5-11. Checking for blown fuse in a single-phase system (courtesy Square D. Co.).

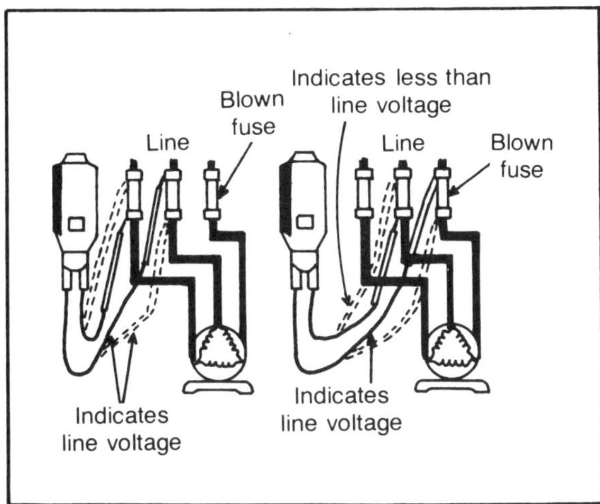

Fig. 5-12. Checking for a blown fuse in a three-phase wiring system (courtesy Square D. Co.).

connections, wire nuts, wires pinched inside the box, etc. Also check for the presence of moisture, dirt, and oil.

Keep in mind, if a cable blows fuses or trips a breaker without a load attached to it, and all control devices have been checked, you can assume that the cable is shorted. Here is where conduit has its advantage over nonmetallic cable. The old *shorted* wires can easily be pulled out of conduit and new ones *fished through* using a steel flexible coil of wire called *fish tape*. If nonmetallic cable needs to be replaced it may be difficult to replace it without cutting into the walls. If at all possible, when pulling out the old nonmetallic cable, attach the new cable

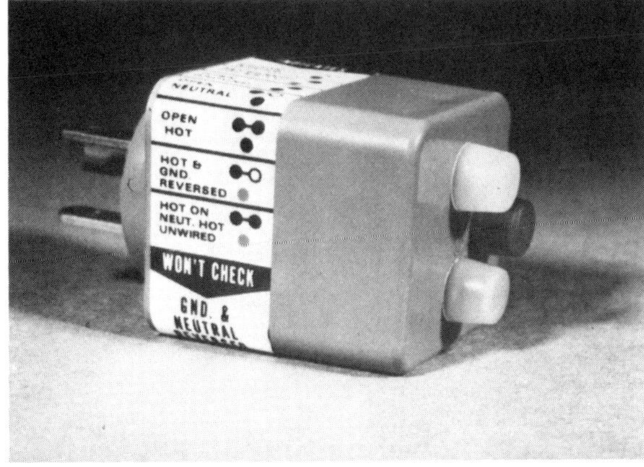

Fig. 5-13. A typical receptacle tester (courtesy ETCON Corp.).

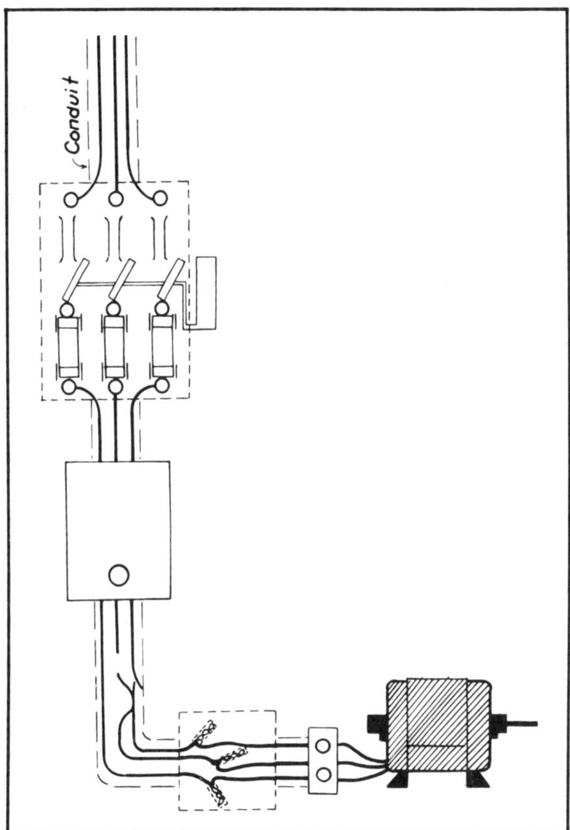

Fig. 5-14. A typical industrial circuit.

to it and pull it through as you pull the old cable out. Conduit is also mechanically stronger and protects the copper wire better. With this in mind, all exposed wiring should be installed using conduit. Figure 5-15 illustrates the proper method to bend conduit using a conduit bender.

If a GFI (ground fault interrupter) is used in a circuit, it will be necessary to trace the entire branch circuit looking for moisture and/or shorted/grounded wires. The GFI is actually a magnetic switch that *opens* or trips whenever the hot line in any way "grounds to neutral" drawing a few milliamperes or more of current. The main advantage of this GFI is that it can save lives by automatically shutting off the power. One of the disadvantages of the GFI is that any type of moisture can trip it. For example, many times on a hot, rainy day or during a rain storm, the GFI will trip although there is no immediate threat to anyone. Another disadvantage is that even though the GFI trips, a surge of voltage (even a microsecond in duration) can pass through the GFI and enter

Fig. 5-15. Using a conduit bender to bend conduit (courtesy Ideal Industries, Inc.).

a person causing him to lose his balance and fall. This could be dangerous to a person working in high places. Other protective devices when used with an *isolation transformer* could eliminate this disadvantage of the GFI. Figure 5-16 illustrates a typical household ground fault interrupter.

A megohmmeter can be used to check the insulation resistance in three phase distribution cables. Test one conductor at a time by connecting the megohmmeter to one line and tieing the other two lines together and grounding them (Fig. 5-17). Compare the ohmmeter reading with that of the specified standard of the cable, which depends on its size, length, and operating conditions.

Besides the actual line cable, broken or charred insulators may need to be replaced or rebuilt. Many times a service technician may need to design and build new insulators. This situation most often occurs when this particular insulator is no longer available.

Another method of troubleshooting wiring systems is through the process of *elimination*. Using a megohmmeter or voltmeter, trace a system from one point to another disconnecting one apparatus at a time until the short, open, or ground is found (Fig. 5-18). For instance, in order to locate a ground first disconnect the motor at point C and check it out using a megohmmeter, and check the line voltages. Next, disconnect the motor control and check the voltages. Then proceed to the panel box.

Loose wiring connections, shorted wires, poor pigtail connections, due to moisture, dirt, oil, or physical abuse are typical problems in industrial wiring. When making cable connection in electrical boxes, allow 6 inches of wire for the pigtail connections. These pigtail connections are usually made by using plastic wire connectors. Figure 5-19 illustrates the proper method of making pigtail connections using plastic wire connectors.

LIGHTING SYSTEMS

Residential and industrial lighting systems often

Fig. 5-16. A typical household ground fault interrupter (courtesy Harvey Hubbell, Inc.).

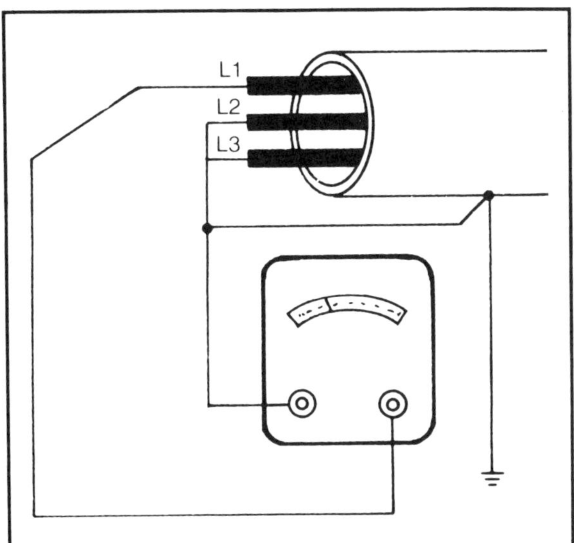

Fig. 5-17. Checking insulation resistance with a megohmmeter.

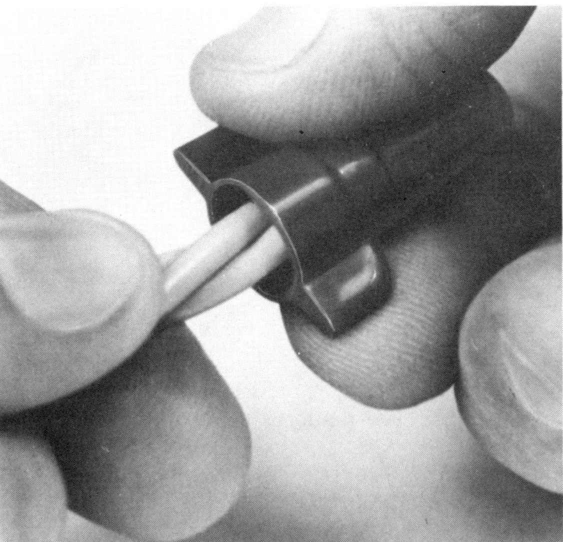

Fig. 5-19. Making a pigtail connection using plastic wire connectors (courtesy Ideal Industries, Inc.).

need servicing. Some of the common lamps include incandescent, fluorescent, mercury vapor, and high-intensity discharge. The incandescent lamp (i.e., light bulb) is generally used in homes and employs a tungsten filament. Typical problems with this lamp are cracked lamp sockets, low voltage, and vibration. Low voltage greatly reduces the efficiency (i.e., brightness) of the lamp. Vibration is a major problem to this lamp, because it causes the tungsten filament to break. Measures should always be taken to safeguard against vibration.

There are several types of fluorescent lighting systems. Some of them include the preheat, rapid start, and instant start. Preheat lamps require an auxiliary starting switch. The starting switch, when activated, allows the ballast to generate current for preheating the electrodes of the fluorescent lamp. After a few seconds, the starter opens and the lamp ignites.

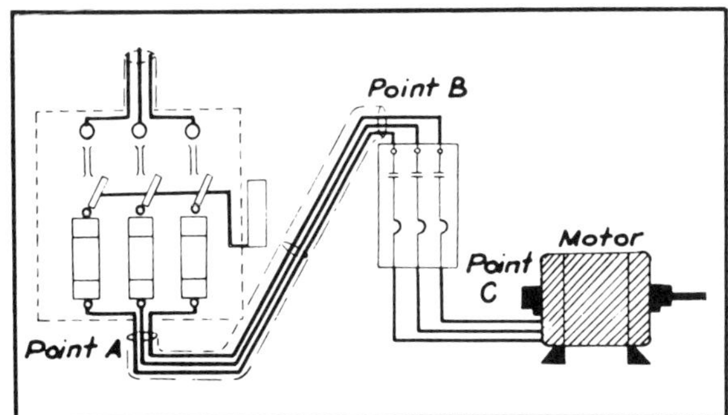

Fig. 5-18. Troubleshooting procedure in a typical industrial wiring circuit.

Some common problems of preheat lamps include flickering and failure of the lamp to ignite. These problems may be due to a broken socket, improper lamp seating, low voltage, dirty lamp, improper wiring, or a defective lamp, ballast or starting switch. The most common problems are low voltage, dirty lamp, or a defective starting switch.

Rapid start and instant lamps differ from the preheat lamp, because they do not require a starting switch. The ballast incorporates a built-in winding, which provides the necessary current to ignite the lamp. A typical problem with these lamps is that when the lamp becomes dirty, especially under high humidity, the lamp does not ignite but rather flickers. When this flickering occurs, the lamp needs to be cleaned or replaced. Other problems include defective sockets, low voltage, improper wiring, or a defective ballast.

Mercury vapor lamps are generally used for industrial applications. These lamps offer high brilliance but require a constant and reliable voltage source. If the voltage varies, the lamp will not operate. Common problems include low voltage, power interruptions, and defective transformers. If the voltage varies, the lamp will shut down and it will require five to ten minutes to cool before it will return to full brilliance.

The high-intensity lamp operates similarly to the mercury vapor lamp. These lamps also require a constant, reliable voltage source and sufficient cooling time

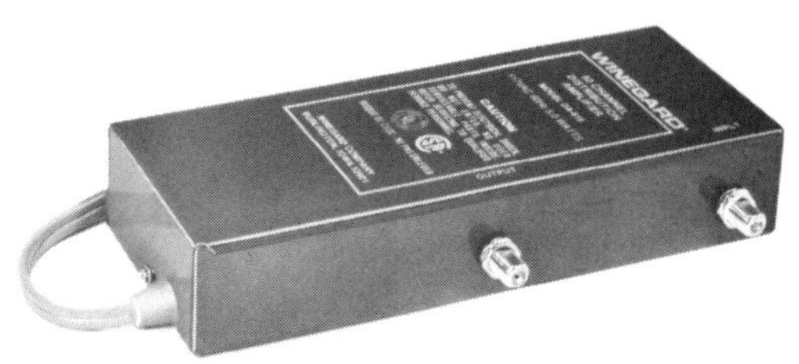

Fig. 5-20. A solid-state distribution amplifier designed to amplify VHF-UHF signals (courtesy Winegard).

for restarting. Common problems include inadequate voltage, power interruptions, or defective lamps or ballasts.

TV DISTRIBUTION SYSTEMS

There are many devices used in TV distribution systems. The preamplifier mounted in the housing of the antenna increases the signal level at the point of reception. The distribution amplifier (Fig. 5-20) is used to boost the signal to provide a signal sufficient for good reception. The splitter divides or couples signals to a transmission line. The most common types of splitters are the 2-way and 4-way splitter. Figure 5-21 shows a 2-way splitter.

Taps are used to provide a portion of the signal from the trunk line to the receiver. The two main types of taps are the line tapoff and the drop tap. Line tapoffs are mounted in the wall near the TV set. The line tapoff provides isolation so that receivers are unaffected by each other, preventing such problems as smears and ghosts. The drop tap is used when there are several outlets (receivers) a long distance from the main trunk. Drop taps provide lead signals directly to TV sets which simplify installation. Figure 5-22 shows an example of a single outlet directional line drop tap.

Couplers are designed to match the signal(s) from the antenna(s) to the TV set(s). There are several different types of couplers. For example, if separate VHF and UHF antennas are used, a coupler designed to combine these separate signals into a single 75- or 300-ohm output eliminates the need to run two separate lines to the TV set. Couplers are often mounted on the antenna or in the attic.

Traps are used to attenuate specific signals (i.e., frequencies), eliminating interference. The CB and FM traps are common ones used to eliminate citizen band and FM radio interference. Most FM traps are designed to attenuate FM signals from 88 to 108 MHz. Attenuators are also used to filter out undesirable frequencies.

An antenna is a conductor that picks up electromagnetic waves and induces current. The length of the antenna and its resonance effect is related to the frequency and wavelength of the electromagnetic wave. Frequency and wavelength are inversely proportional; the lower the frequency, the longer the wavelength. Also, the lower the frequency, the longer (larger) the antenna. The decibel is used to measure the gain, loss, or isolation in antenna and cable distribution. The decibel is a ratio of two voltages (i.e., input and output voltages in an antenna/cable system). One thousand microvolts is equivalent to 0 decibels (dBmV), when the reference voltage is 1 millivolt (mV).

The two basic kinds of antenna wire commonly used are 75-ohm coaxial cable and 300-ohm twinlead. Coaxial cable is generally used for commercial wiring and is the

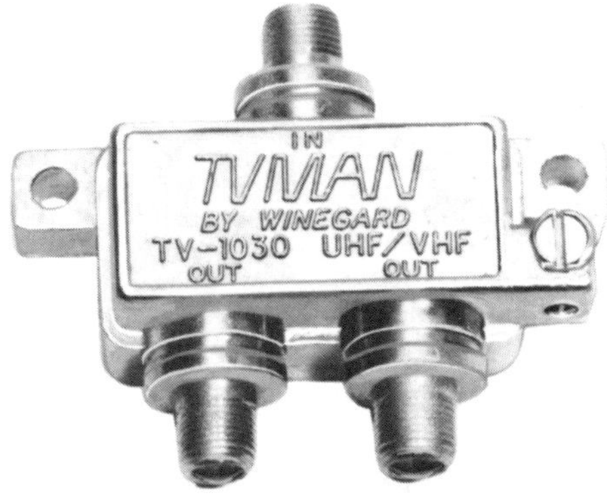

Fig. 5-21. A two-way hybrid splitter for coupling or dividing signals on 75-ohm coaxial line (courtesy Winegard).

Fig. 5-22. A single outlet directional line drop tap (courtesy Winegard).

most popular wire. Coaxial cable is round and is smaller and more expensive than twinlead. It is ideal for running through pipes, has excellent noise rejection, high durability, and is weatherproof. It also is a shielded wire that protects against noise and interference.

There are basically two types of twinlead—the standard flat twinlead and the foam-filled tubular twinlead. The standard flat twinlead is the cheaper but does not stand up well in poor weather conditions. The foam-filled twinlead is more weather resistant than the flat twinlead. Neither twinlead gives significant shielding against noise and interference.

Figure 5-23 shows an 82-channel, 96-outlet distribution system. This system illustrates the use of splitters, line taps, amplifier, power supply, preamplifier, and antenna. Terminating resistors are also used to prevent signals from feeding back up the transmission line.

There are many factors that contribute to faulty antenna systems. The antenna must be of good quality, free from rust and corrosion. It must be mounted securely to maintain proper antenna orientation. If the antenna is not pointing in the proper direction, weak reception or ghosts may result. Ghosts are duplicate images seen on a television receiver. The ghosts are caused by the television picking up multiple signals from the transmitting station caused by the wave being reflected by an object such as a large building. The reflected signal is actually being delayed and arrives shortly after the direct signal and is displaced slightly to the right on the television screen. To remedy this problem, reposition the antenna or install a better antenna system.

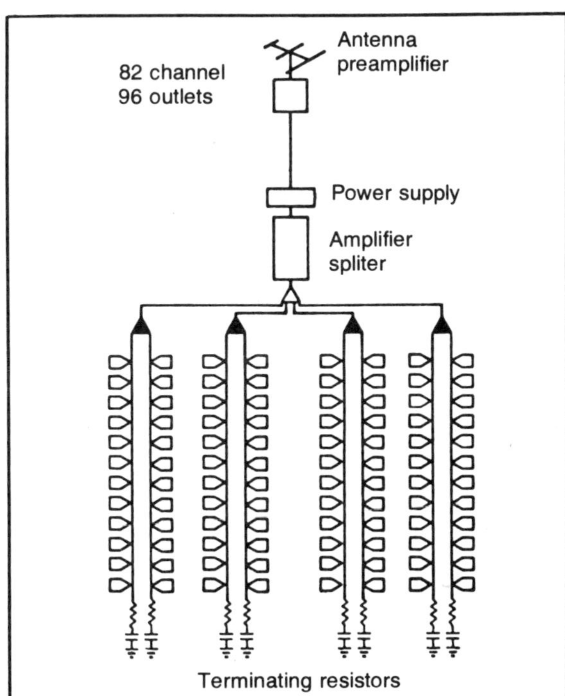

Fig. 5-23. Master antenna TV system for an apartment complex (courtesy Winegard).

The antenna must be mounted away from trees, metal, or large objects. The major cause of antenna faults is corrosion from weather conditions. When the antenna is corroded, and signal loss is evident, the antenna should be replaced.

An antenna can be mounted in the attic of a home, but aluminum backing of insulation and other metals may dampen the signal strength.

The major cause of distribution line faults is insulation breakdown. This is most often caused by prolonged exposure to the elements of the weather, such as rain, ice, and snow. The transmission line can be checked by an ohmmeter. Also, by using a small television or monitor, the antenna system can be traced step-by-step. In this way the transmission line can be diagnosed and, if needed, replaced.

The distribution line can also be checked by substituting a known good cable. Keep in mind that a short in coaxial cable does not always create a short circuit. The shorted cable often acts as a trap, apparently due to the dc resistance and self-inductance developed by the short circuit. Unusual effects can be produced on the television receiver such as color missing on one channel and not the others or only some receivers having weak signals. An open circuit in coaxial cable can also produce strange results, such as smearing, ghosts, and poor reception or absence of signal.

Sloppy installation of the distribution system can cause poor reception, ghosts, or smears. Some of the problems are: failing to install a terminating resistor at the end of a run; pinching the cable with staples; making sharp bends; creating impedance mismatches; and making loose or poor connections.

Actually, whenever a good television has a *snowy* picture the reason must be due to a weak signal. A signal of 0 dBmV, which is equivalent to 1000 microvolts, is usually needed for good, clear, sharp reception. A common problem of many antenna systems is that too many television receivers are connected to the antenna without proper signal amplification. Each television connected to the antenna will decrease the signal strength. Also, if the distribution cable is too long the signal will be decreased. For example, 100 feet of common coaxial cable will attenuate a loss of approximately 6 dB at Channel 14. This is a decrease of approximately one half of the 1000 microvolts needed for a good picture.

When servicing or installing antenna systems, it is essential to calculate the signal strength and use the most suitable amplifier. A strong signal could over-drive one amplifier and not another. Also, it is important to design a system where all dB losses have been taken into account in order to have adequate signal strength.

In areas where undesired noise or reception occur it may be necessary to use a filter to attenuate this undesired frequency. In areas of exceptionally strong FM signals an FM trap may need to be inserted. If necessary, the service technician may need to connect several FM traps together in cascade.

Lightning is a major threat to antennas. Lightning often strikes an antenna causing any number of problems to occur, such as open transmission line, open preamp, open amplifier, open splitter or other component. Usually the preamp that is located in the antenna housing has a protective diode that lessens the chance of more serious problems when struck by lightning. Also, lightning arresters and grounding rods can reduce the chance of more serious problems when struck by lightning.

PREVENTIVE MAINTENANCE

An ideal preventive maintenance program should consist of routine inspections of all distribution systems, panel boxes, and operating devices. Log books should be kept in order to record all breakdowns.

Chattering, noisy controls should be given immediate attention and corrected. Dirty, gummy wet lines and motor controls should be periodically cleaned.

Line voltage should be periodically checked for abnormal low or high readings. Line voltages should never vary more than 10 percent of the specified voltage. During hot summer days when low line voltage tends to be a problem, the maintenance department should try to shut down any nonessential operating lights, motors, and other equipment. Low line voltages cause motors to draw more current which in turn produces more heat. This eventually ruins cables, motors, and other equipment.

The ambient temperature (the temperature close to the area of the operating device) should be routinely checked for abnormal temperatures. Abnormally high temperatures can cause breakdowns and can indirectly be a fire hazard since high heat increases current that causes insulation breakdown. Also, extremely cold temperatures can cause cracking.

Semiconductors are very sensitive to temperature changes. Any change in temperature can destroy diodes, transistors, SCRs, and other semiconductors.

Any unnecessary physical or mechanical abuse should be avoided since it causes a safety hazard to the operator. Abuse also shortens the life of the operating device.

SELF-EXAMINATION

Select the best answer:

1. The minimum ampere capacity panel box used today is:
   A. 30.
   B. 60.
   C. 100.
   D. 150.
   E. 200.
2. The common neutral wire is usually ________ in color and connected to a:
   A. Black, cold water pipe.
   B. White, circuit breaker or fuse.
   C. Black, circuit breaker or fuse.
   D. White, cold water pipe.
   E. Red, circuit breaker or fuse.
3. Number 8 wire is generally used for ______ amps.
   A. 15.
   B. 20.
   C. 30.
   D. 40.
   E. 60.
4. Another name for emt is:
   A. Nonmetallic cable.
   B. Flexible metallic cable.
   C. Conduit.
   D. Armored clad cable.
   E. None of the above.
5. Another name for flexible steel cable is:
   A. Conduit.
   B. Emt.
   C. Synthetic cable.
   D. Nonmetallic cable.
   E. None of the above.
6. A 4-way switch controls a lamp from ______ different locations.
   A. 1.

B. 2.
C. 3.
D. 4.
E. 5.

7. The power factor rating of an inductive reactive circuit can be increased by adding:

A. Capacitors.
B. Inductors.
C. Coils.
D. Fuses.
E. None of the above.

8. The device used to pull wire through conduit is called:

A. Wire tongs.
B. Straps.
C. Connectors.
D. Fish tape.
E. None of the above.

9. The best way to check insulation resistance is with a:

A. Voltmeter.
B. Ohmmeter.
C. Ammeter.
D. Megohmmeter.
E. All of the above.

10. To decrease signal strength in strong signal areas use a:

A. Splitter.
B. Amplifier.
C. Coupler.
D. New antenna.
E. Attenuator.

11. Which of the following lamps uses a starting switch?

A. Preheat.
B. Rapid start.
C. Instant start.
D. All of the above.

12. Which of the following lamps requires a cooling period prior to restarting?

A. Incandescent.
B. Fluorescent.
C. Mercury vapor.
D. None of the above.

13. The device used to attenuate specific signals is the:

A. Drop tap.
B. Line tapoff.
C. Splitter.
D. Trap.

14. The wire that is shielded is:

A. Flat twinlead.
B. Foam-filled twinlead.
C. Coaxial.
D. All of the above.

15. Which of the following wires has 75-ohm impedance?

A. Flat twinlead.
B. Foam-filled twinlead.
C. Coaxial.
D. None of the above.

16. The device that is used to prevent signals from feeding back up the transmission line is the:

A. Trap.
B. Line tap.
C. Preamplifier.
D. Terminating resistor.

17. A cause for ghosts and smears is:

A. Open transmission line.
B. Improperly positioned antenna.
C. Multi-path reception.
D. All of the above.

18. The preamplifier is generally mounted in the:

A. TV receiver.
B. Antenna.
C. Attic or wall.
D. None of the above.

19. The device used to boost a signal is the:

A. Line drop tap.
B. Amplifier.
C. Attenuator.
D. Trap.

20. The maximum capacity for a general 20-ampere general-purpose appliance circuit is approximately:

A. 1500 watts.
B. 1800 watts.
C. 2400 watts.
D. 2800 watts.

QUESTIONS AND PROBLEMS

1. Explain the different types of cable commonly used in residential/industrial wiring.
2. Explain the operation of a three-way and four-way switch.
3. Explain the appliance/equipment ground system.

4. What are the differences in voltages between a delta and wye transformer power supply system?
5. Define and explain a GFI.
6. What are some disadvantages of a GFI?
7. Explain some techniques used in troubleshooting wiring systems.
8. State some reasons why a television has a *snowy* picture.
9. Explain the effect of ambient temperature on an operating device.
10. State some preventive maintenance procedures one should use in industry.
11. Explain how ghosts are developed in a television receiver.
12. What can be done to attenuate strong FM signals?
13. State some common problems in antenna/cable systems.
14. What is the difference between coaxial cable and twinlead?
15. How many microvolts are generally required for a good clear television picture?
16. Describe what may happen when coaxial cable develops a short circuit.
17. What are some possible results of an open circuit in coaxial cable?
18. Explain the difference between a line tapoff and the drop tap.
19. What is an attenuator?
20. What is the purpose of a splitter?
21. Describe the benefits of coaxial cable.
22. Describe the difference between the standard flat twinlead and the foam-filled tubular twinlead.
23. What is a decibel?
24. What is the difference between an amplifier and a preamplifier?
25. Why are terminating resistors used?

Chapter 6

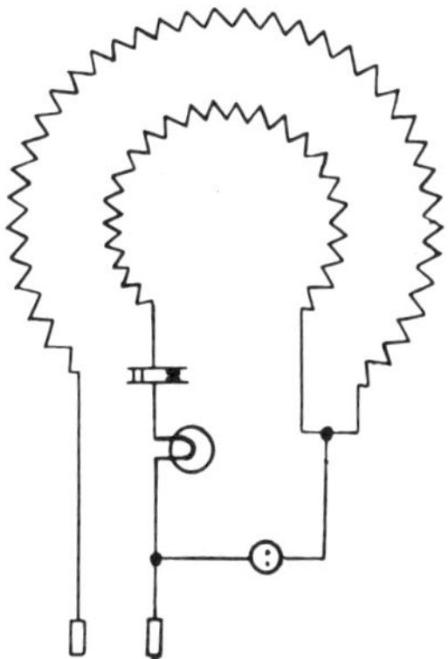

Troubleshooting Appliances

Today, more than ever, our homes are being filled with numerous simple and complex appliances. Microwave ovens, refrigerators, hot water heaters, vacuum cleaners, hair dryers, shavers, and garbage disposals are but a few of the appliances that have found a place in our homes. The complete discussion and repair procedure for every product is beyond the scope of this book, but the discussion of a few common products should provide the necessary carry-over knowledge needed to repair many of them.

APPLIANCES WITH HEATING ELEMENTS

Some of the more common appliances that use heating elements include irons, toasters, waffle irons, ovens, rotisseries, skillets, hair dryers, roasters, and coffee makers. Most appliances use a heating element made of nichrome wire or quartz rod. These are low resistance materials that provide heat when current passes through it. The bimetallic thermostat is another device common to all heating appliances. The thermostat consists of two different metals that expand when heated. The bimetallic thermostat acts as a switch to automatically turn on or off electrical circuits. Figure 6-1 shows the operation of a bimetallic thermostat controlling an appliance.

One of the more simple types of appliances is the *coffee percolator.* Many coffee percolators provide automatic high/low heating positions. The high heating element is used to heat the coffee and the low heating element is used to keep the coffee warm. Figure 6-2 shows how the magnetic switch maintains contact to supply current to the high heating element. At this time, the low heating element and thermostat will be off, due to the high heat. As soon as the water moves to the upper bowl, the magnetic switch releases and the low heating thermostat kicks in due to the lower heat.

Figure 6-3 shows a typical diagram of a coffee maker. Note the appearance and location of the thermostat (8), lead wire assembly (5), keep warm and lead assembly (7), and body with element (1).

The actual servicing of coffee percolators is quite simple. If the percolator does not heat, check for a defective cord, switch, or thermostat. Often, the thermostat contacts become dirty and need cleaning. Use fine sandpaper to touch up the contacts. Most heating appliances use cords made of thick rubber or braided wire covered with asbestos. We recommend replacing a defective cord rather than trying to repair it. To test a cord for continuity use an ohmmeter or test lamp. If there is no

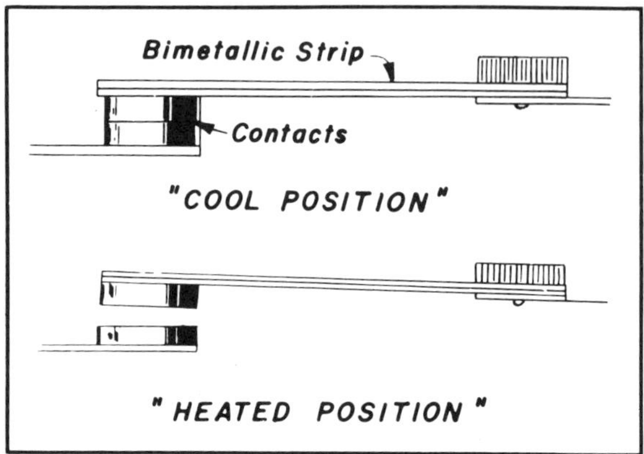

Fig. 6-1. Principal operation of a bimetallic thermostat.

continuity through the cord, try squeezing and bending the cord at various places. This will often locate the *open* spot. Keep in mind that most cords will open close to the plug due to excessive use or abuse by the owner.

If the percolator fails to shut off after the coffee has perked, check the thermostat or magnetic switch—the contacts may have frozen. If the cord tends to overheat, check the terminal posts for corrosion. Replace the cord if necessary. Remember, never touch an exposed wire

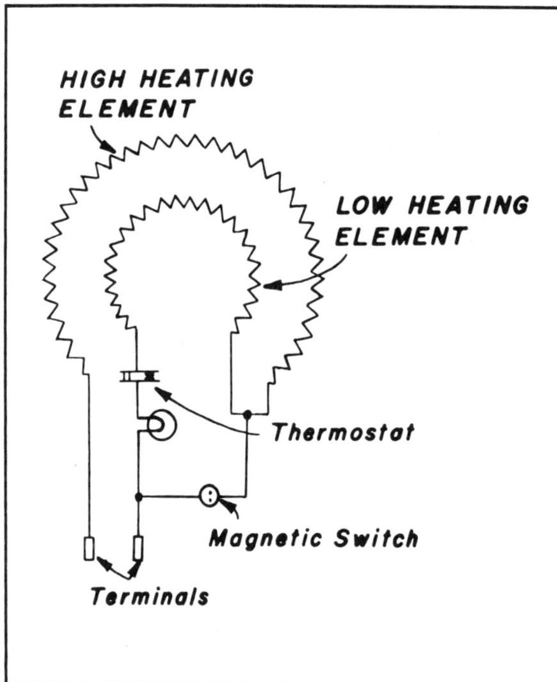

Fig. 6-2. High/low heating circuit of a coffee percolator.

when working on any type of appliance—severe shocks can result.

If the percolator fails to heat, the heating element may have broken. Check this element for continuity (visually or by using an ohmmeter). Also, many percolators use a heat sensitive fuse which, although simple to replace, is often quite expensive.

Frypans, roasters, and skillets rarely need maintenance. The most common problems are dirty terminals, thermostat, or heating element. Often, the thermostat can be repaired by cleaning or adjusting. The temperature can often be adjusted by using a jeweler's (small) screwdriver and turning the temperature adjusting screw.

Often, heating elements are "sealed in" or "built in" the appliance. If they become defective, the entire appliance will need to be replaced. Keep in mind that a service technician's time is very expensive and unless he/she can quickly repair any type of appliance, the repair bill will exceed the cost of the appliance.

Remember, after any type of appliance has been repaired always check the appliance for grounds, shorts, or opens before returning the appliance to the customer. A common mistake for beginning technicians is that of pinching a wire to the frame of the appliance when reassembling it. Although the appliance may operate perfectly, severe shocks can result. The testing procedures for appliances are about the same as for motors, as explained in Chapter 3.

The repair procedure for *steam irons* is very common to any other type of appliance except steam irons have a tendency to leak after extended use. Determine where the leak is and if necessary replace the gaskets used to seal the unit. Figure 6-4 shows an exploded view of a typical steam iron. If an iron tends to steam when it is not in the steam position, check the valve assembly (Nos. 29, 30, 31, and 50). A bent pin, weak spring, or improperly sealed valve often will cause the problem.

Like other types of heating appliances, if the steam iron overheats, check the thermostat (43). It may need to be readjusted or replaced—look for dirty or fused contacts. Insulators may also be broken or missing. Insulators are often made of mica, ceramic, porcelain, or plastic.

If the iron fails to spray properly, check the nozzle assembly (17)—often it becomes dirty and clogs up. A typical problem common to other appliances as well as the steam iron is a broken line cord. If the cord fails to operate, the line cord should be one of the first considerations. The line cord can be easily checked for continuity by using an ohmmeter. Lastly, if the iron has difficulty

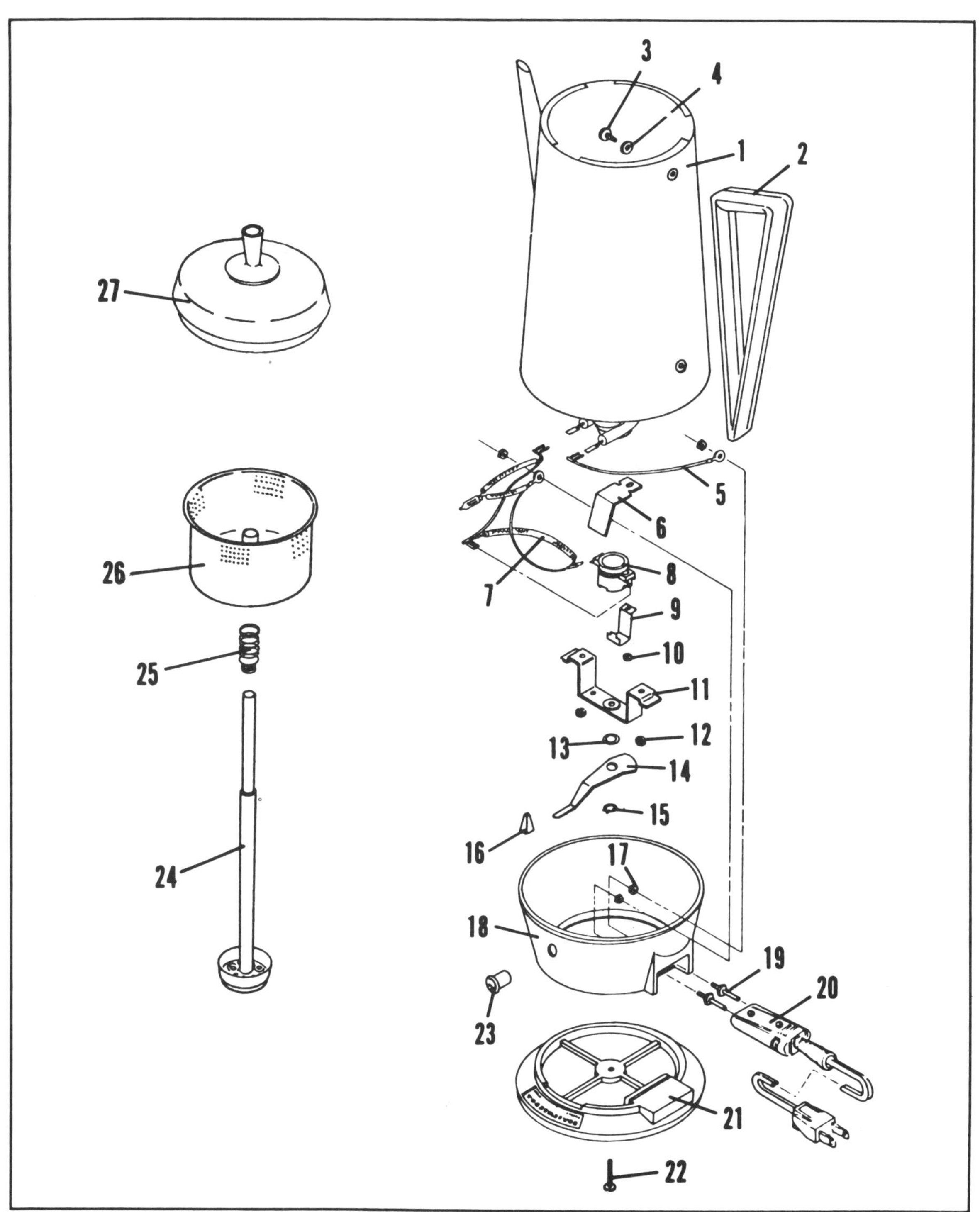

Fig. 6-3. An exploded view of a typical coffee maker (courtesy McGraw-Edison Co.).

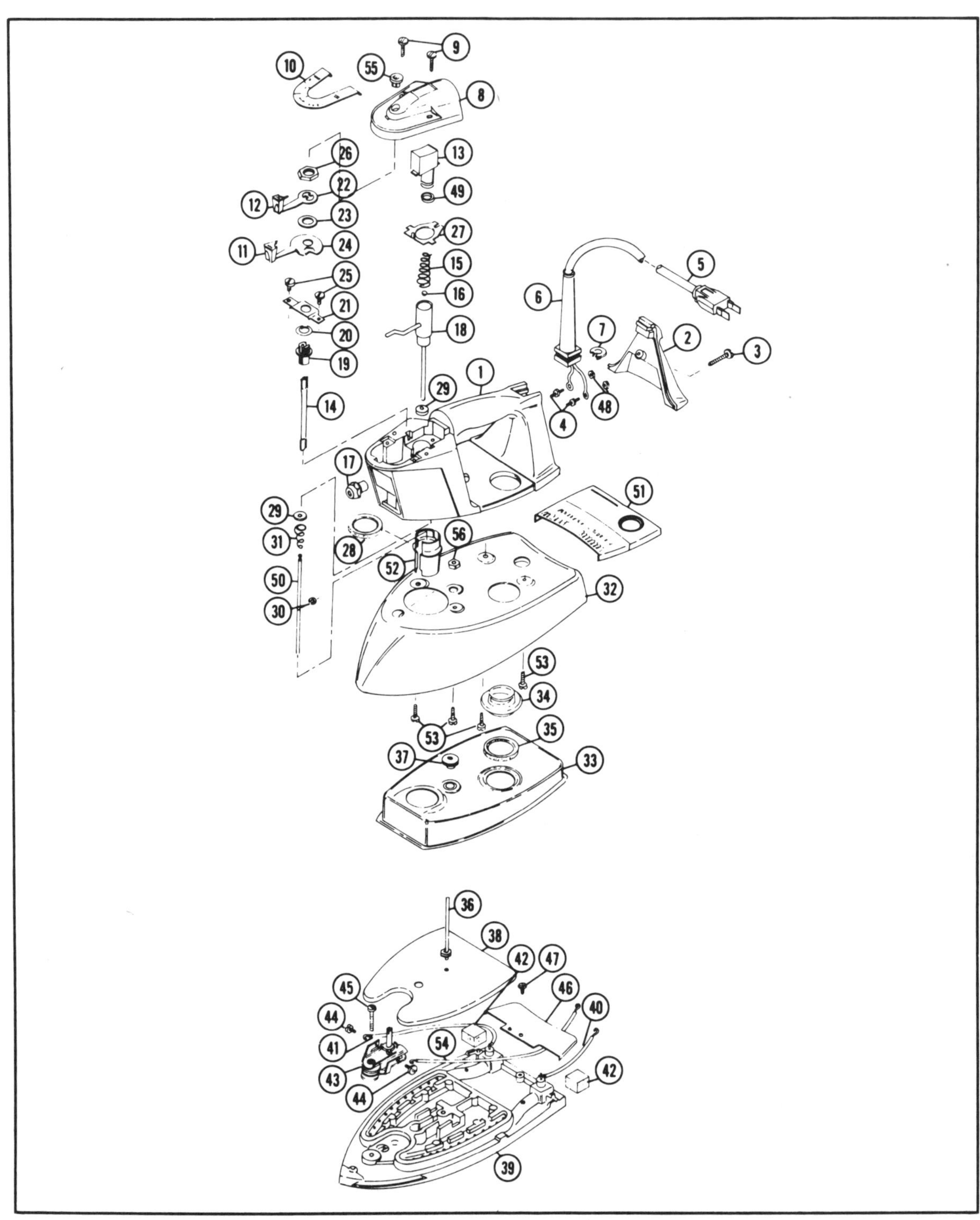

Fig. 6-4. An exploded view of a steam, spray, and dry iron (courtesy McGraw-Edison Co.).

steaming, the steam chamber and passages may be clogged. This happens because of the accumulation of mineral deposits from the water. The mineral deposits develop a hard whitish coating. The deposits can be cleaned by flushing the unit with vinegar or other chemical solutions. Also, deposits in passages can be removed by using a small round file or other similar object.

Toasters are probably one of the most mechanically complex appliances. Although, electrically quite simple, most problems with toasters tend to be mechanical in nature. For this reason many appliance servicers tend to stay away from toaster repair. Figure 6-5 shows an

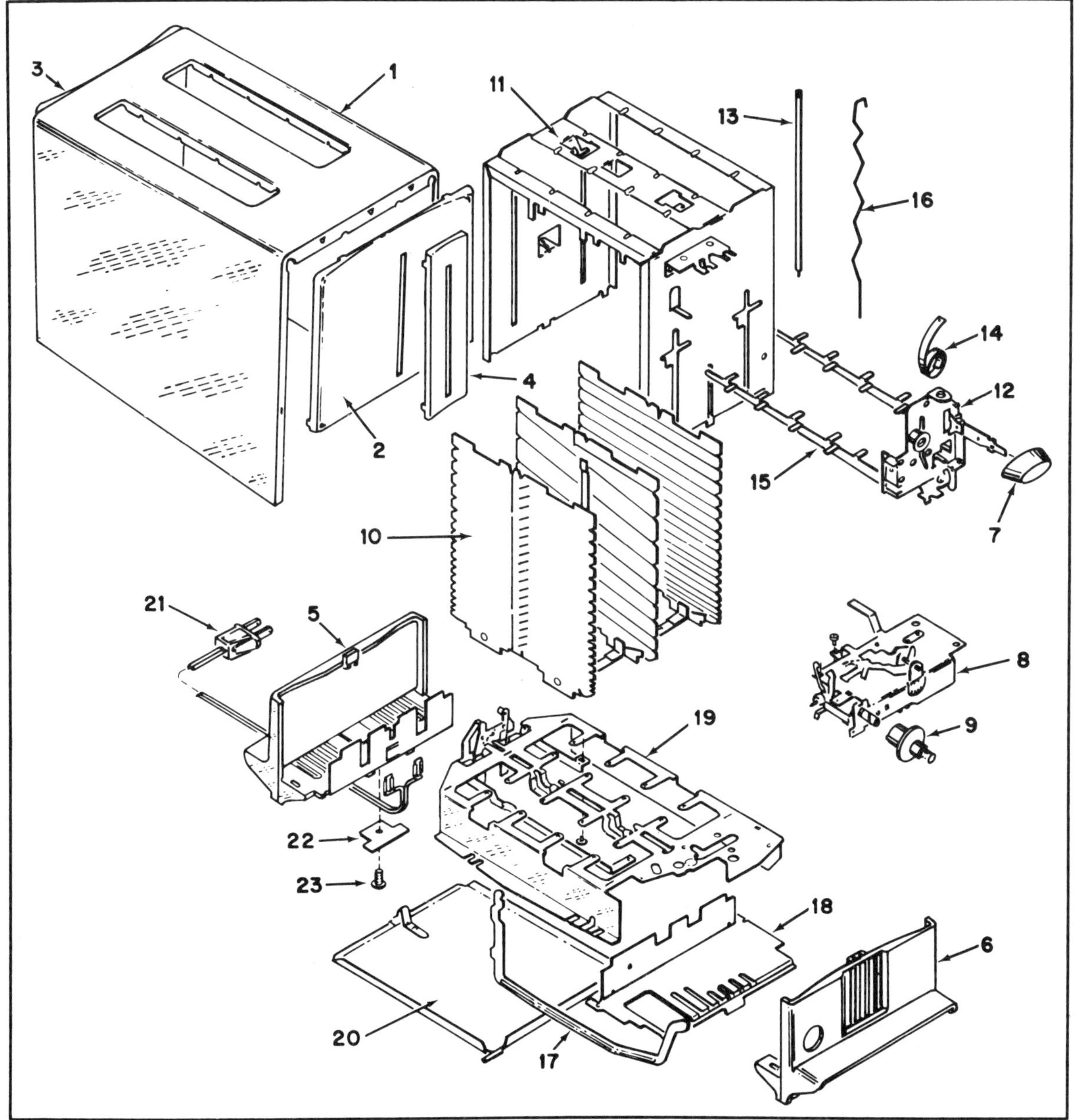

Fig. 6-5. An exploded view of a 2-slice toaster (courtesy McGraw-Edison Co.).

exploded view of a typical toaster.

One of the most common causes of toaster problems is bread crumbs and butter. They tend to coat or clog up the latches, hinges, guides, and especially switches. A typical problem with a toaster is failure to *pop up* the bread, which is usually directly caused by bread crumbs clogging up the thermostat switch. A simple cleaning will often repair the defective toaster—use a small brush (toothbrush) and cleaning solvent. Many appliance repair shops and electrical supply stores carry solvents. Remember, never lubricate any of the mechanical parts of a toaster. Usually this can do more harm than good.

Another problem common to toasters is lack of toasting uniformity that is often caused by a defective thermostat, timer assembly (8), or a misplaced mechanical part.

Remember that any time a heating element needs to be replaced in an appliance, it is important that the exact replacement be purchased (10). Match the rating (watts) and element length. Also, remember that many spring elements may easily pull out, but may be very difficult to put back in. Try not to stretch or bend these elements. When replacing the element attach one end of the element to the insulating support, and slowly follow the path to the other insulating support. Many broken elements may be obsolete and can be repaired by a "nickel/silver" tube by crimping the tube over the two broken ends (Fig. 6-6).

APPLIANCES WITH ELECTRIC MOTORS

Appliances with electric motors include mixers, vacuum cleaners, blenders, hair dryers, shavers, carving knives, electric toothbrushes, sewing machines, shoe polishers, can openers, etc. Some of the appliances discussed here are the slicing knife, can opener, mixer, and hair blower/styler.

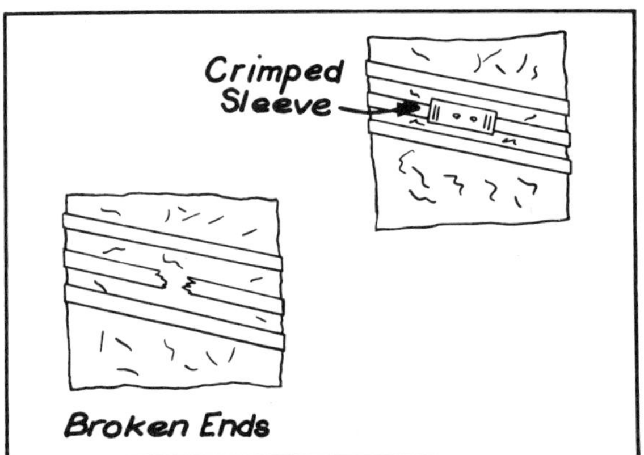

Fig. 6-6. Broken heating elements can be repaired with a nickel/silver tube. This tube is crimped over the two broken ends.

Figure 6-7 shows an exploded view of the parts of a common *slicing knife*. Some common parts of all slicing knives are the brushholder (9), armature assembly (12), field assembly (13), gear (16), switch blades (21, 22), and housing (2, 3).

Most electric slicing knives work on the principle of a small electric motor and gear box attached to a pinion that turns and moves the knife blade back and forth. With this in mind, the most common problems are usually associated with the motor, switch, gear box for binding or lack of lubrication. The motor, which is a repulsion type, may need new brushes. Visually check the brushes and commutator for wear. The commutator can be cleaned with sandpaper and/or a commutator stone.

The cordless models use a battery pack and a charger unit. The batteries can be easily checked by using a voltmeter. A low voltage indicates the batteries need to be recharged or replaced. Keep in mind that discharged batteries may be caused by a defective charger unit. The charger unit is usually sealed in a plastic block and cannot be repaired. Since this charger rectifies ac to dc for the charging of the batteries, a common problem is a defective rectifier. If the charging fails to produce a dc voltage or light a small 6.3 volt lamp, replace the unit.

The *electric can opener* is a common appliance that uses an electric motor. Figure 6-8 shows a picture of a typical electric can opener.

A complete pictorial view of an electric can opener is shown in Fig. 6-9.

Most electric can openers are reliable and don't require a lot of servicing. Typical problems with a can opener may be poor cutting action, failure of the can opener to properly hold the can in position, improper latching, or an inadequate cutter wheel. An adjustment of the cutter wheel or latch may be necessary to help secure the can in position. Also, the cutter wheel may need to be replaced.

If the can opener fails to rotate the can, check the motor, switch, or drive mechanisms.

Food mixers are usually fairly easy items to service. Inefficient motor operation can be due to a defective governor switch assembly, open resistor or capacitor, or any other problem common to electric motors which are discussed in Chapter 3. Check for obvious problems, such as a defective cord, switch, connection, worn bear-

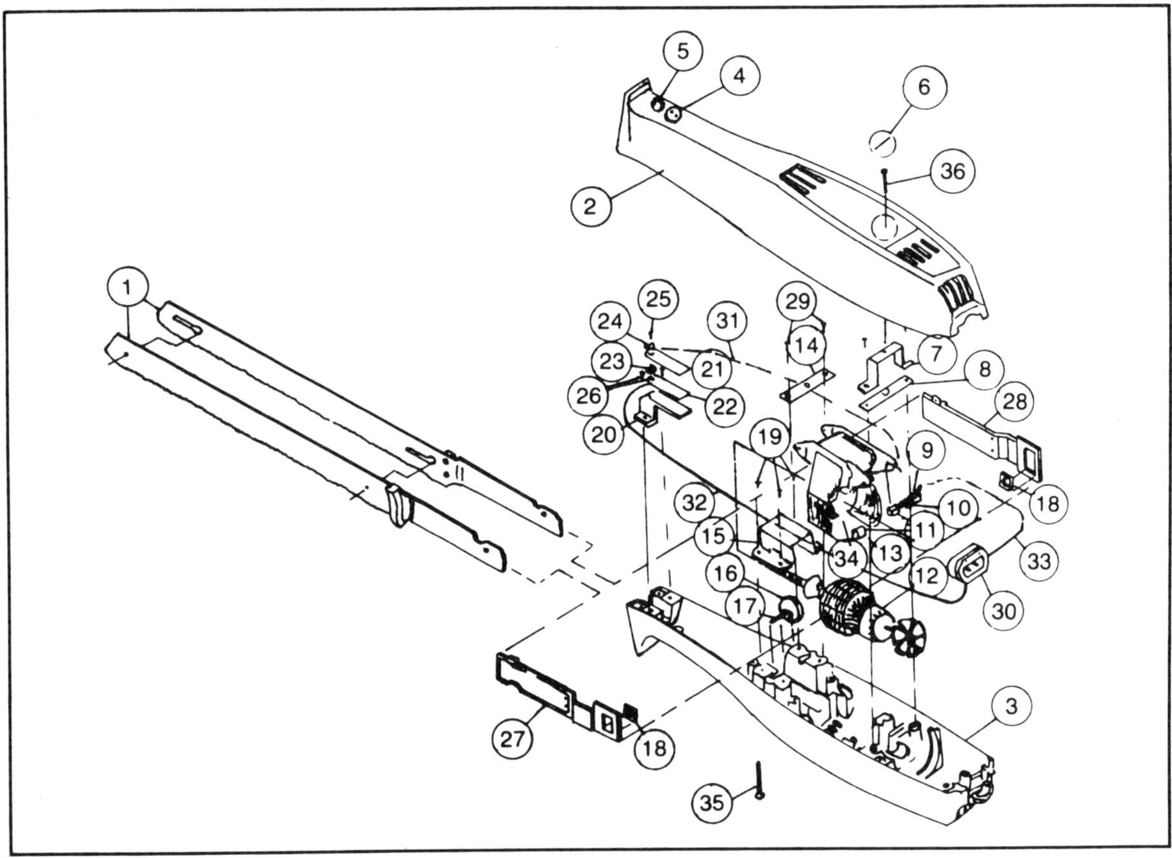

Fig. 6-7. An exploded view of a typical electric slicing knife (courtesy McGraw-Edison Co.).

ings, shorted field coil, armature, worn brushes/commutator, or worn drive mechanisms.

The servicing of *hair blowers* is usually quite simple. Basically, the only thing that can go wrong is one of the following:

1. Cord.
2. Switch.
3. Rectifier.
4. Heating element.
5. Motor.
6. Fuse.

Undoubtedly, the most common problem of the hair blower is the cord that becomes *open* close to the handle. Usually this is caused by the constant twisting and bending by the operator. Often an open cord can be located by bending the cord at different points. When the open point is bent the hair blower will run intermittently. A defective cord should be replaced.

Another common problem with the hair blower is an open fuse which is often caused by a wearing down of the heating element, overuse, misuse, low voltage, etc. Always replace the fuse with an exact replacement. Never jump a fuse since this could cause overheating, short circuits, etc.

Keep in mind that the rectifier, heating element, switch, and cord can be checked with an ohmmeter.

The *garbage disposal* is an appliance that employs a high-power electric motor to drive a fly cutter which grinds kitchen garbage. Some of the common problems with disposals include failure of the motor to start, poor grinding action, noisy operation, and leaky pipes and fittings. If the motor fails to start, the cause may be a jammed flywheel caused by an obstruction such as a piece of silverware, bone, or dishcloth. When this problem occurs, the overload relay trips, which shuts off the unit. After removing the obstruction, the overload relay, gen-

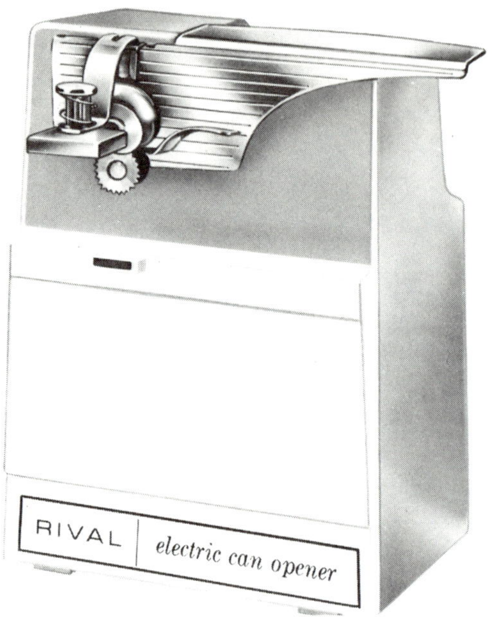

Fig. 6-8. A typical electric can opener (courtesy Rival Manufacturing Co.).

erally located on the bottom of the disposal, can be reset by pushing in its button. If the unit still fails to start, check to see if the power-supply circuit breaker tripped. The problem may also be due to a shorted motor, etc.

If the disposal grinds slowly/poorly, the impeller may be stuck or the flywheel may need replacement. If the disposal is noisy, it may be due to a bad bearing or impeller. Another common problem is leaky drain pipes and fittings that have rusted. In this case, replacement of the pipes or fittings is necessary.

Lastly, some disposals include a water-flow interlock that prevents the disposal from operating unless water is flowing through the system. If the motor fails to start, this interlock may need to be adjusted or replaced.

The *electric water heater* is a common appliance used in homes and businesses. The principle of operation is simple. Basically, it consists of an insulated tank of water which is heated by one or two heating elements (Fig. 6-10). The insulation used in most water heaters is fiberglass or rock wool. The heating element, which is manually or thermostatically controlled, is generally shielded by copper. A relief valve is used for protection against excessive temperature and pressure. The water heater also has a cold water inlet valve and a hot water supply valve. Temperature is controlled by the thermostat. For safety, most electric water heaters have a non-adjustable high temperature limit control which will open the electrical circuit when the water temperature is excessive.

Before repairing the water heater, it may be necessary to shut the system down by turning off the electrical power and draining the water. If shut-down is necessary, turn off the cold water inlet valve. The water can then be drained through the heater drain valve. The drain valve should be left open during the shut-down period. Also, a nearby hot water faucet should be left open during shutdown to relieve water and air pressure. When filling the heater with water, close the drain valve, open the cold water inlet valve, and close the hot water faucet when water starts to flow. Then turn on the electrical power.

If excessive water temperatures are reached, the high temperture limit control will open. Some models have a limit control which can be manually reset; others use a fuse which must be replaced. Excessive temperatures may be caused by a defective thermostat or by abrupt changes in temperature.

When there is insufficient hot water, check for defective heating elements, leaky hot water faucet, a blown fuse or tripped circuit breaker, a leaky tank, or a low thermostat setting. Keep in mind that the demand for hot water may be exceeding the capacity of the tank. On the other hand, if the water is too hot, see if the thermostat setting is too high and reset to desired level.

Another problem with water heaters is an excessive hissing sound. This is caused by lime or scale buildup in the tank and heating elements. In this case, remove the heating elements and clean them with a wire brush. Also, consult a local dealer for a chemical to help remove the deposits.

If there is water leakage around the tank, check for a leaky relief valve. The leak may be caused by excessive temperature, pressure or a faulty valve. Water leakage may also be caused by condensation or an open drain valve.

If the water heater fails to operate, check for an open fuse, tripped circuit breaker, open limit control, or inadequate water supply. A common problem unique to electric water heaters is that, in time, the heating elements get corroded and eventually become defective. In this case, remove the heating elements and visually inspect them. A defective element generally will be corroded, deformed, and/or broken. An ohmmeter can also be used to check for continuity. An open element will not have continuity. If in doubt, replace the heaters. Keep in mind to always replace both heating elements even if only

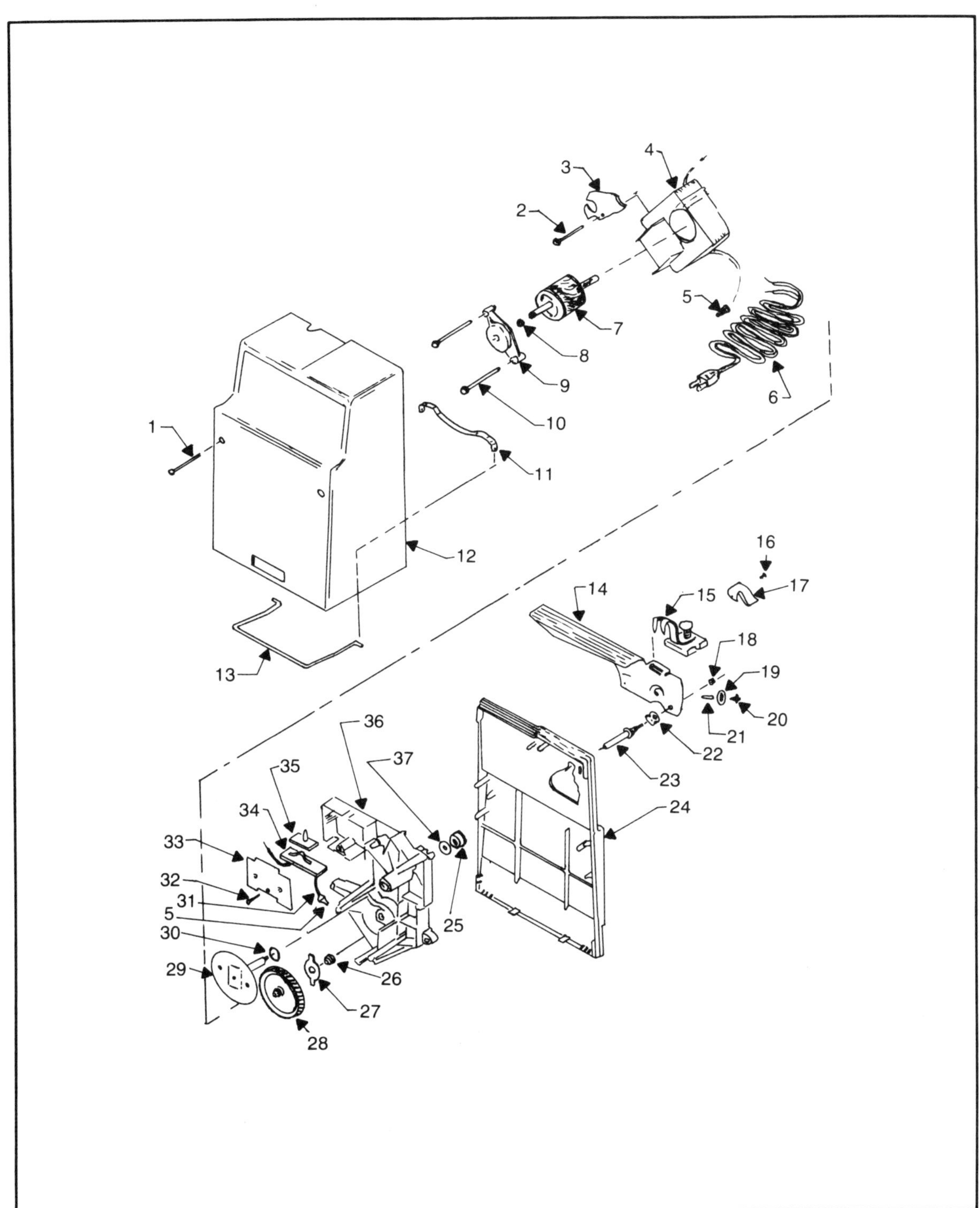

Fig. 6-9. Pictorial parts diagram of an electric can opener (courtesy Rival Manufacturing Co.).

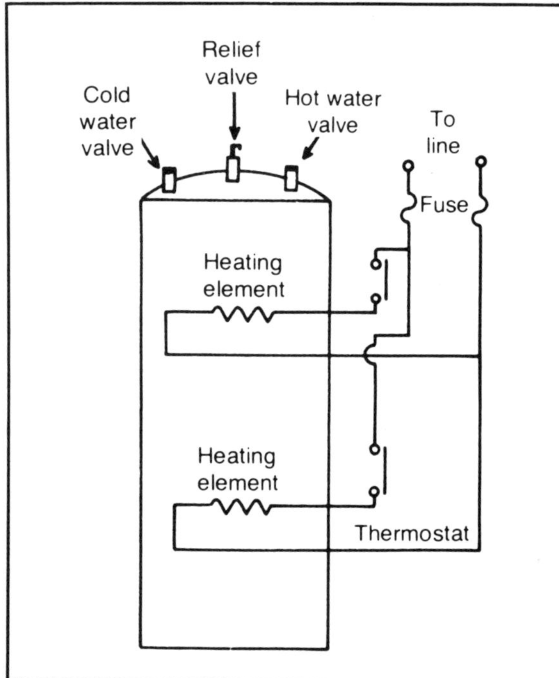

Fig. 6-10. A simplified electric water heater diagram.

one element is defective. This ensures uniform heating of the water.

Another problem with water heaters is that the tank rusts, especially when located in old/damp basements, and eventually leaks. When this problem occurs, it is time for a new water heater.

Lastly, there are a few preventive maintenance measures that can be taken to ensure long life of the water heater. Periodically, open the drain valve and allow some of the accumulated dirty/corroded water to run out. This will help to reduce lime and scale buildup. Check to make sure the relief valve operates by lifting up on its handle (be careful, since hot water will flow). The desired water temperature should be adjusted. Also, the tank and heating elements can be cleaned.

SELF-EXAMINATION

Select the best answer:

1. The type of thermostat most commonly used in appliances with heating elements is the:
   A. Magnetic relay.
   B. Circuit breaker.
   C. Bimetallic.
   D. All of the above.
2. The appliance that is one of the most mechanically complex is the:
   A. Fry pan.
   B. Hot blower.
   C. Slicing knife.
   D. Toaster.
3. Most heating appliances use a cord made of:
   A. Rubber.
   B. Asbestos cover.
   C. Both A and B.
   D. None of the above.
4. Insulators are commonly made of:
   A. Ceramic.
   B. Mica.
   C. Porcelain.
   D. All of the above.
5. One of the most common causes of toaster problems is:
   A. Bread crumbs.
   B. Low voltage.
   C. High amperage.
   D. Poor installation.
6. Most cordless electric slicing knives use a charger unit and:
   A. Rectifier.
   B. Battery pack.
   C. Diode.
   D. None of the above.
7. Heating elements can be repaired by a ________ tube which crimps the two broken elements together.
   A. Nickel/silver.
   B. Aluminum.
   C. Wire.
   D. Aluminum/nickel.
8. Most heating elements can be checked by using a/an:
   A. Ammeter.
   B. Wattmeter.
   C. Ohmmeter.
   D. Growler.
9. One of the most common problems of a hair blower/styler is a defective:
   A. Cord.
   B. Transistor.
   C. Thermostat.

D. SCR.

10. An overheated cord often indicates:

    A. Corroded terminals.
    B. Defective cord.
    C. Both A and B.
    D. None of the above.

11. If the motor of the garbage disposal fails to start, check for:

    A. Tripped overload relay.
    B. Leaky pipes.
    C. Loose fittings.
    D. All of the above.

12. The insulation used in most water heaters is:

    A. Fiberglass.
    B. Rock wool.
    C. Both A and B.
    D. None of the above.

13. Excessive temperatures in the water heater may be caused by:

    A. Closed drain valve.
    B. Defective thermostat.
    C. Defective heating element.
    D. Tripped circuit breaker.

14. The hissing sound in water heaters is generally caused by:

    A. Excessive temperature.
    B. Leaky relief valve.
    C. Lime or scale buildup.
    D. Inadequate water supply.

15. Insufficient hot water may be caused by:

    A. Defective heating element.
    B. Tripped circuit breaker.
    C. Low thermostat setting.
    D. All of the above.

QUESTIONS AND PROBLEMS

1. Explain the principle operation of an automatic high/low percolator.
2. State the most common problems of percolators.
3. State the most common problems of irons.
4. Explain a couple of ways to test an appliance cord.
5. What are common problems with toasters?
6. What important factors should one keep in mind when replacing a heating element?
7. Explain how to repair a heating element.
8. How can a battery pack from an electric slicing knife be checked?
9. How can a charger unit from an electric slicing knife be checked?
10. List the six basic things that generally go wrong with a hair blower.
11. Explain some of the possible causes if the garbage disposal fails to start.
12. Describe the operation of an electric hot water heater.
13. Explain the procedure for shutting down a water heater prior to servicing the unit.
14. Explain the possible causes of insufficient hot water in an electric water heater.
15. Explain the troubleshooting procedure if the hot water heater fails to operate.

Chapter 7

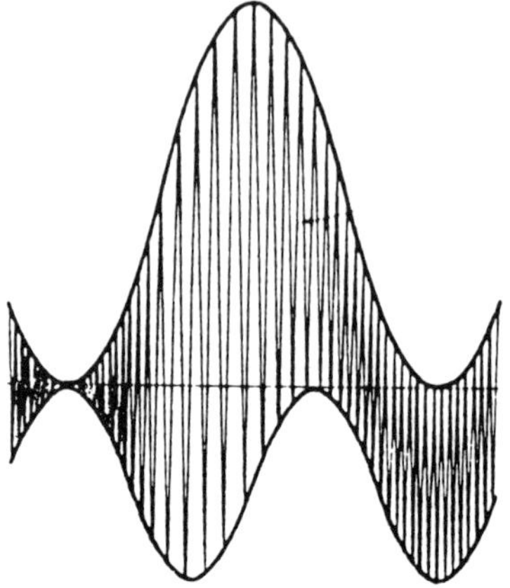

Troubleshooting Radio and Audio Equipment

A radio receiver is a piece of electronic equipment which picks up electromagnetic signals from the air, amplifies them, extracts the intelligence, and reproduces the sound which enters the microphone at the transmitter. There is a type of receiver to match each type of transmitter. The type of receiver is specified by the type of modulation the receiver is designed to handle; amplitude modulation (AM), frequency modulation (FM), frequency multiplexing (FM MUX) or stereo. We will examine each of these receivers and other audio equipment with basic troubleshooting techniques and tips on preventive maintenance.

AM FUNDAMENTALS

A radio wave is basically an electromagnetic energy vibration. This wave travels at a speed of 300 million meters per second or 186 thousand miles per second with each wave having a certain length. The lower the frequency, the longer the wave; the higher the frequency, the shorter the wave. An audio wave is approximately between the frequency of 20 Hz and 20 kHz (20 thousand Hz) and can be heard by most people. Waves above this frequency, which are called radio frequency (rf), cannot be heard by people.

Since the physical size of an antenna is proportional to the length of the wave, using an antenna would be impractical to transmit and receive audio waves. Therefore, a high frequency continuous wave produced by an oscillator is mixed with the low frequency audio wave. This produces a modulated wave used to transmit the intelligence from the transmitter to the receiver. Figure 7-1 illustrates the amplitude modulation (AM) principle in which the intelligence or low audio frequencies, varies the amplitude of the carrier. The AM wave or signal is a composite of the carrier and the upper and lower side bands.

Most AM receivers are called superheterodyne receivers because of the mixer stage. A block diagram of an AM receiver is shown in Fig. 7-2 with the signals that are processed at each stage. The antenna receives many radio frequencies within its frequency band and the tuner—which consists of a variable capacitor and a coil—selects a desired band of frequencies. The broadband rf stage amplifies the radio frequencies selected and passes them on to the mixer stage.

At the mixer stage, the incoming rf signal is combined with a constant amplitude continuous wave which oscillates at a frequency of (generally) 455 kHz (the usual

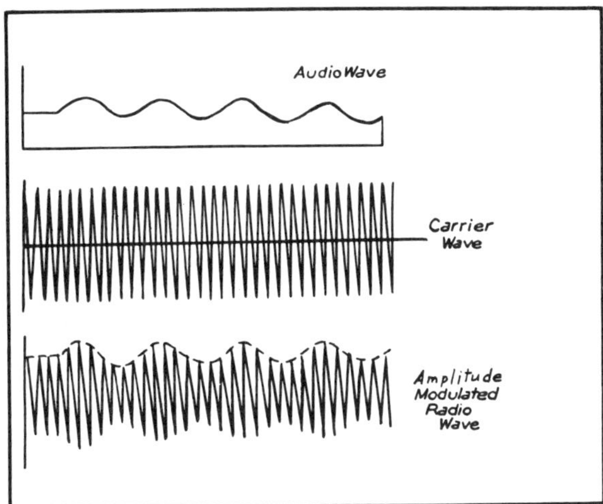

Fig. 7-1. The two waves that form a modulated rf wave.

i-f frequency), above the incoming rf signal. The beating of the signals together produces many frequencies composed of the sums and differences. The output of the mixer passes through a tank circuit tuned to 455 kHz.

Suppose the rf tuning circuit is set to receive a signal at 1000 kHz and the i-f is set at 455 kHz. Then the local oscillator will be set at 1455 kHz above the incoming signal so that the new intermediate difference frequency of 455 kHz will be created. The 455 kHz i-f is the difference between the incoming frequency of 1000 kHz and 1455 kHz. It should be noted that 455 kHz is the most common intermediate frequency; many receivers are designed to operate at other intermediate frequencies.

The next stage of signal processing is a single stage or multiple stages (up to three) of i-f amplification. Each i-f stage is tuned to the receiver's specific intermediate frequency for improving selectivity.

The function of the detector stage is to separate the audio intelligence from the i-f carrier. It does it in two steps. First, by rectifying the composite signal which leaves the upper envelope of the composite AM signal. Then by filtering the i-f carrier component through a capacitor to ground and passing only the low frequency audio signal. The automatic gain control circuit (AGC) feeds back a little of the signal to provide control for constant volume.

The audio signal from the detector is passed through an audio frequency (af) amplifier stage. Here, enough power is given to the audio signal to drive the speaker.

Figure 7-3 shows a schematic diagram of an AM receiver. Notice the dotted lines between variable capacitors. It indicates they are *ganged*. Ganged means that the variable capacitor for the rf stage and the one for the local oscillator are mounted on the same rotating shaft. The dotted line around the coil indicates a metal cover over the oscillator coil or transformer coils. The diode at point 18 on the schematic does the rectifying while the capacitor C17 filters the rf leaving only the audio signal at point 20.

FM FUNDAMENTALS

The frequency modulation (FM) technique of transmission begins with an rf carrier wave and an audio frequency called the modulating signal. When the rf carrier is modulated with the audio signal, the frequency of the rf carrier varies with the amplitude of the modulating

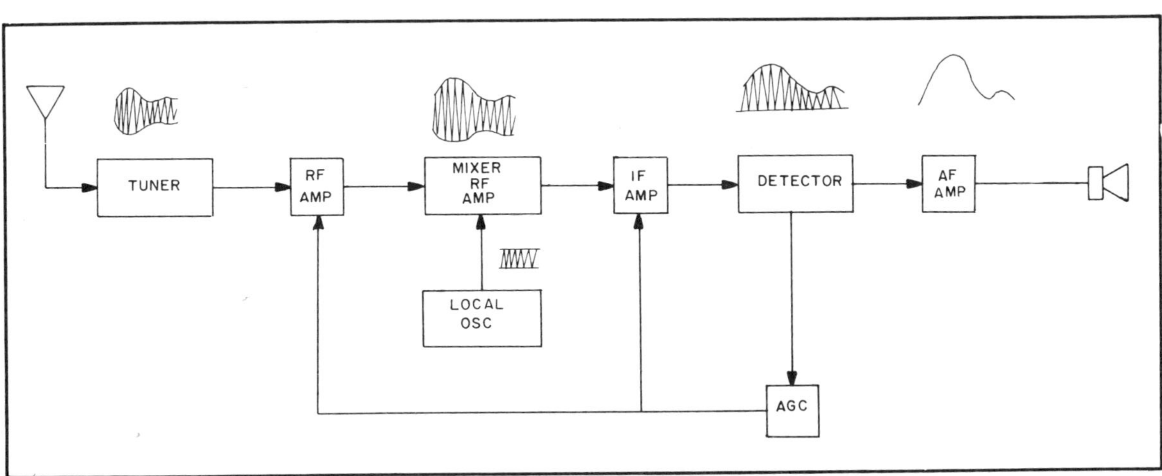

Fig. 7-2. AM receiver block diagram.

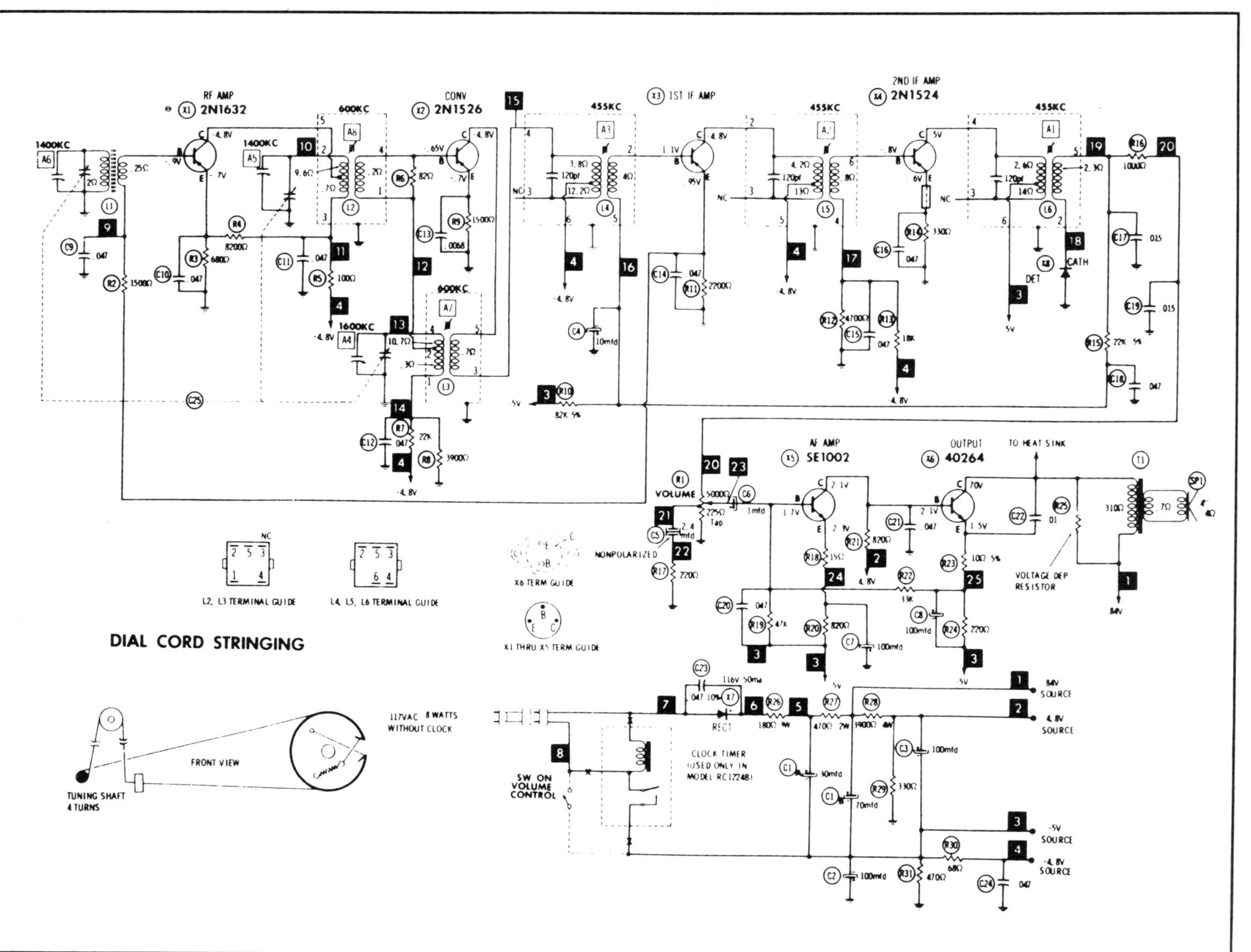

Fig. 7-3. Schematic diagram of a superheterodyne receiver (courtesy Howard W. Sams & Co., Inc.).

signal. Figure 7-4 illustrates the principles of frequency modulation. A block diagram of an FM receiver is shown in Fig. 7-5. The antenna receives FM signals within its band and the tuner stage selects a specific band of frequencies. The rf amplifier strengthens the FM signal. The local oscillator generates a constant amplitude rf signal which is mixed with the FM signal, producing an intermediate frequency. The i-f stages are generally tuned to 455 kHz. The i-f stage or stages pass as well as amplify the i-f carrier.

The FM detector stage is different from the AM detector stage. The FM detector must convert the frequency variations into its audio representation. Then the detected audio signal is fed through a de-emphasis network. The de-emphasis network restores the relative amplitudes of the signal's frequency components. At the transmitter, the high frequencies are further amplified—called pre-emphasis—to improve the signal-to-noise ratio for transmission. Therefore, at the receiver a reverse process must be done.

After de-emphasis processing, the audio signal is amplified to drive the speaker at the audio amplifier stage. Notice that there is an automatic frequency control (AFC) which keeps the receiver oscillator properly tuned.

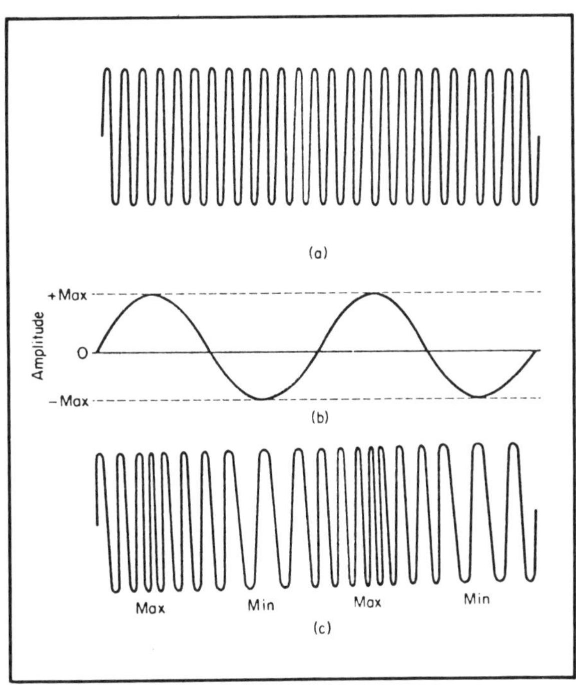

Fig. 7-4. Frequency modulation principle—an audio wave mixed with a carrier wave producing frequency variations.

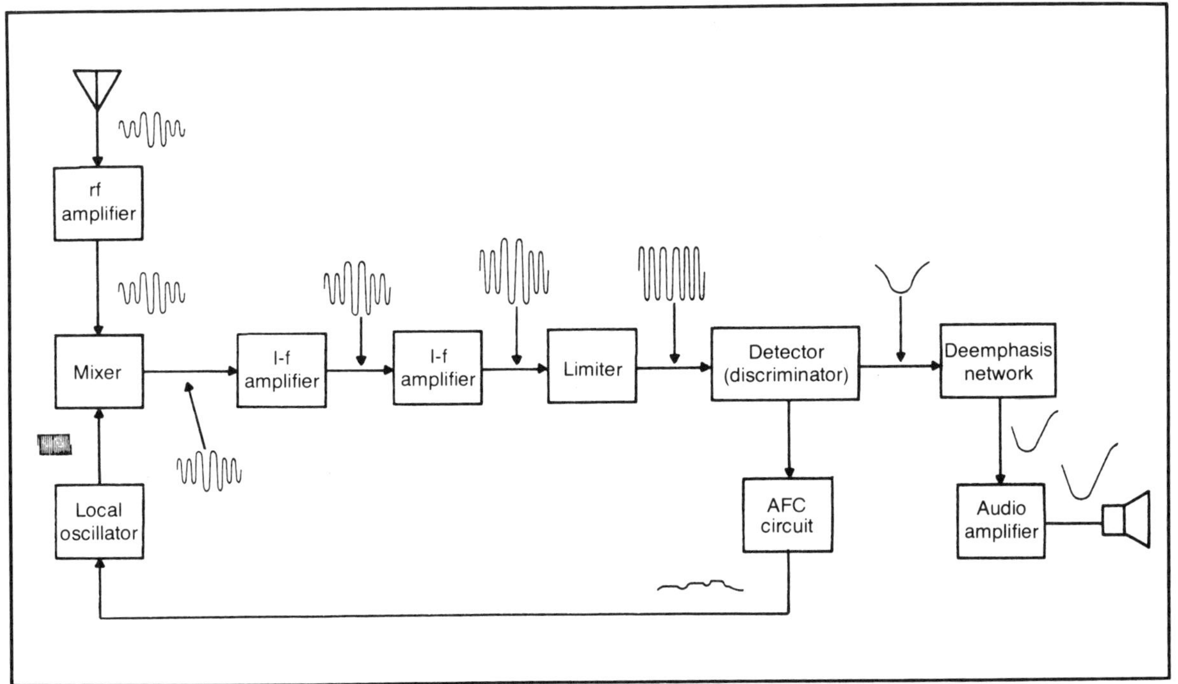

Fig. 7-5. FM receiver block diagram.

Some of the AM circuit can be used by an FM receiver. Figure 7-6 illustrates a block diagram of a combination AM and FM receiver. When you switch from AM to FM reception the unique circuits are all switched in simultaneously. Figure 7-7 is a schematic diagram of an AM/FM receiver.

FM MULTIPLEX FUNDAMENTALS

When you listen to an FM stereo station, it is very apparent that there are two separate audio channels coming from the two speakers. Back at the radio station the FM stereo multiplex signals begin with two separated microphones picking up the audio signals. The signals are designated L and R based on the relative positions of the microphones, left (L) and right (R). The L and R signal are sent to a circuit called a matrix network which produces two new signals. Figure 7-8 shows all the components of an FM stereo multiplexed transmitted signal. One of the new signals is the sum of L and R called L+R signal while the other is L minus R called L−R signal. The L−R signal is used to amplitude modulate (AM) a 38 kilohertz subcarrier which produces side bands above and below the 38 kHz. The 38 kHz subcarrier is suppressed after the modulation. Next, both the L+R signal and the AM side bands of L−R are used to frequency modulate the radio frequency carrier for transmission. Figure 7-9 shows a block diagram of a stereo FM transmitter. Notice the matrix network.

A frequency spectrum analysis of a modulated FM stereo multiplex carrier is shown in Fig. 7-10. Notice the low end of the frequency spectrum contains the L+R signal for monophonic receivers (30Hz-15kHz). The L−R side bands with suppressed carrier (23kHz-53kHz) occupy the high end. A 19 kilohertz pilot carrier is also transmitted as part of the FM multiplexed carrier to be used by the receiver for synchronization and recreating the 38 kHz for demodulation. The use of the suppressed subcarrier is to reduce the energy content of the composite stereo signal for optimum signal-to-noise ratio.

A receiver works "backward" from the transmitter

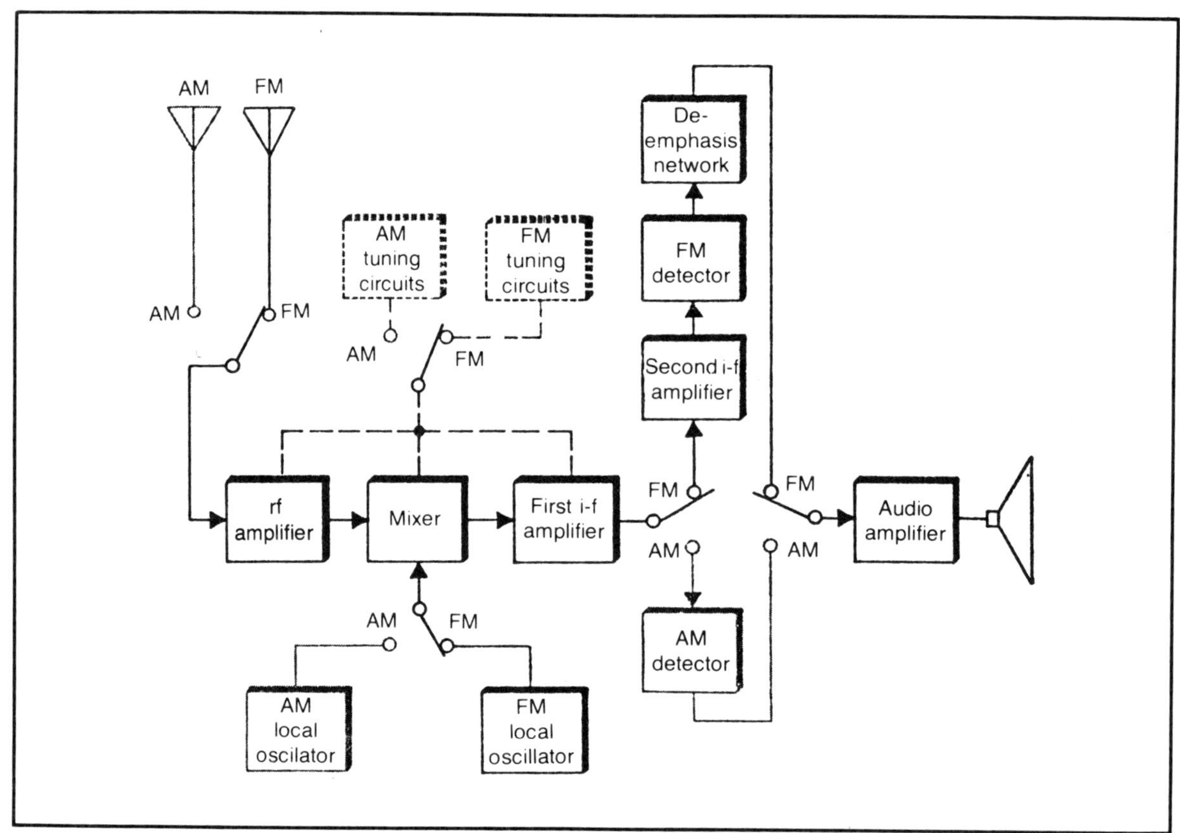

Fig. 7-6. AM/FM receiver block diagram.

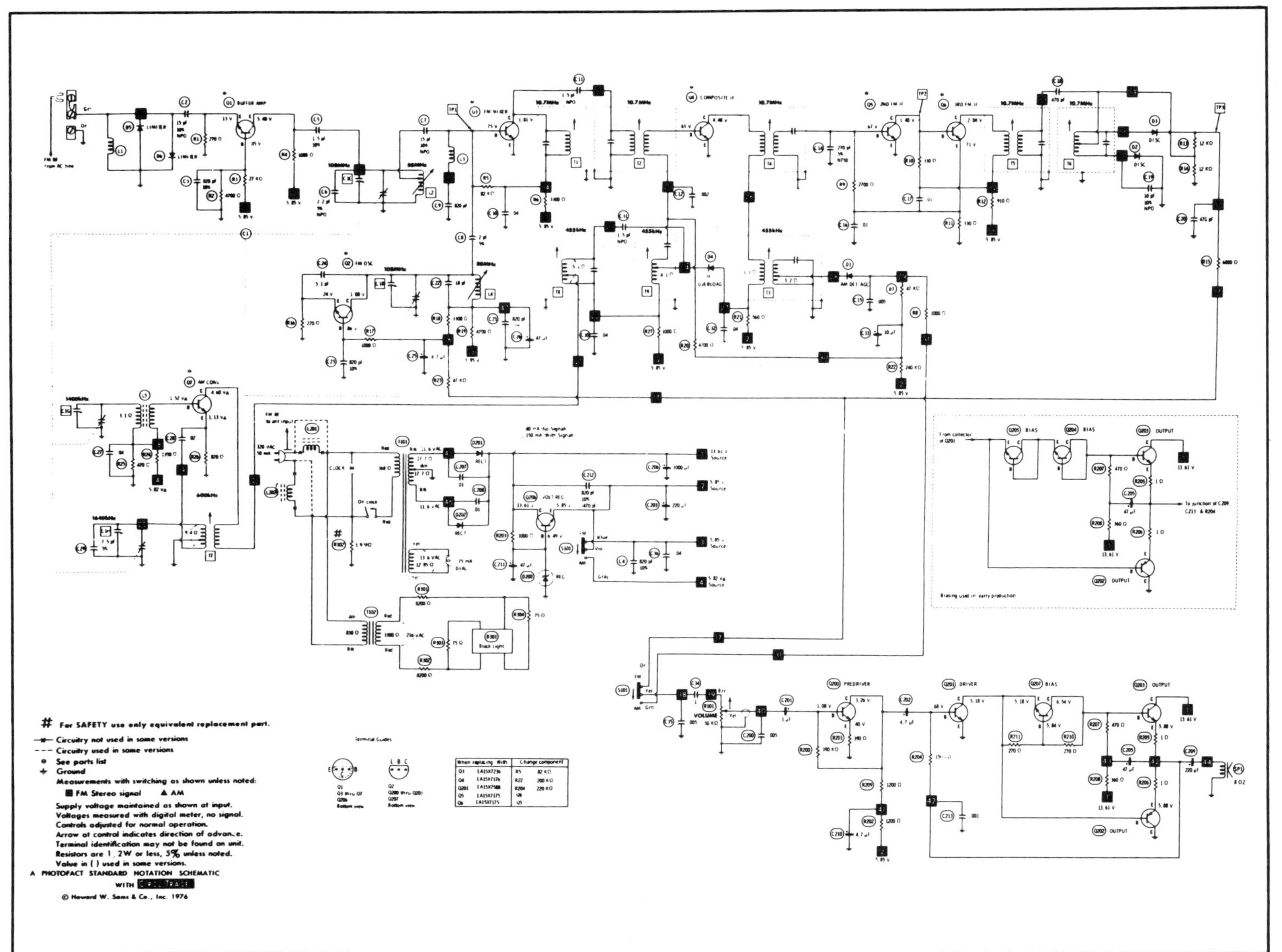

Fig. 7-7. A schematic diagram of a complete AM/FM receiver (courtesy Howard W. Sams & Co., Inc.).

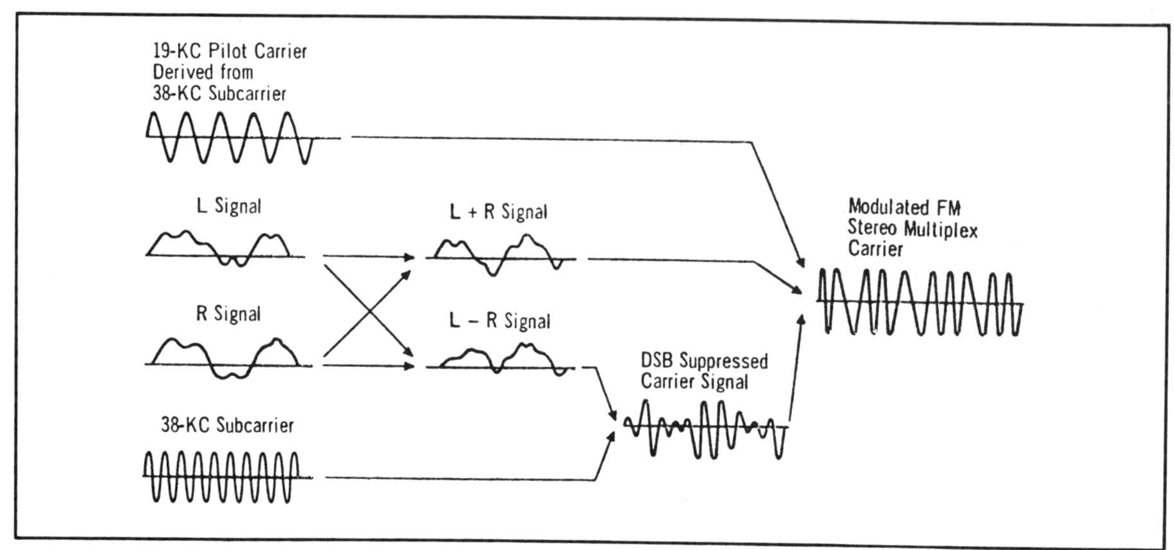

Fig. 7-8. Signal components of FM stereo transmission.

in demodulating the composite stereo signal. When the composite stereo signal is received, it consists of the L+R signal, the side bands of the L−R signal and the 19 kHz pilot carrier. If a receiver is not equipped for FM stereo, it responds only to the L+R signal and processes this as a monophonic signal.

However, when the receiver has FM stereo, the L−R signal is recovered by mixing the L−R modulation side bands with a 38 kHz carrier and then extracting the original L−R signal. Remember that the 38 kHz is generated in the receiver and uses the 19 kHz pilot carrier for synchronization. Then the L−R signal and the L+R signal are processed in a matrix network circuit similar to the one used at the transmitter.

In the matrix network, the L+R and the L−R signals are added together which produces the original L signal. Also, in the same matrix the L+R and L−R signals are subtracted which produces only the original R signal. Then the original L and R signals are amplified and sent to their respective speakers.

The bandpass and matrix method of demodulation is shown in Fig. 7-11. After the discriminator, the three signal components are separated by filter circuits. The L+R signal is obtained through a low-pass filter (30Hz to 15kHz) and then is passed through a delay circuit so that it reaches the matrix network at the same time as the L−R signal. A high bandpass filter (23-53 kHz) "selects" the L−R double side band signal. A narrow bandpass filter of 19 kHz "selects" the pilot carrier which is then amplified and frequency doubled (multiplied by 2) to yield 38 kHz. The 38 kHz is the precise carrier frequency of the double side band suppressed carrier.

Passing the L−R signal and the 38 kHz signal through the nonlinear circuit of an AM demodulator produces the sums and differences of the signals of which one is the 30 Hz to 15 kHz L−R signal that is selected by a low-pass filter. The L−R audio signal and the L+R audio signal are sent to the matrix network and processed as shown in Fig. 7-12. Notice that one channel adds L+R and L−R signals which produces only the L signal. The L−R signal is passed through a phase inverter which changes the signs of L−R to −L+R. The −L+R signal and the L+R signal are then added which produces only the R signal. The L and R signals are sent to respective audio amplifiers and then to their speakers.

Another method of demodulation uses electronic switching. A block diagram is shown in Fig. 7-13. The signal coming from the discriminator is passed through a subcarrier regeneration circuit and is then fed into an electronic switch. An illustration of the signal that is fed into the electronic switch is also shown. The electronic switch samples the positive and negative halves of the composite audio stereo waveform and in effect demodulates the stereo signal. Figure 7-14 shows a schematic of an electronic switching multiplex unit. The composite audio signal is comprised of the L+R signal, L−R sidebands and the 19 kHz pilot carrier.

The 19 kHz pilot carrier is amplified and sent to a frequency doubler (CR1 and CR2) which regenerates the 38 kHz subcarrier. The subcarrier signal is then mixed

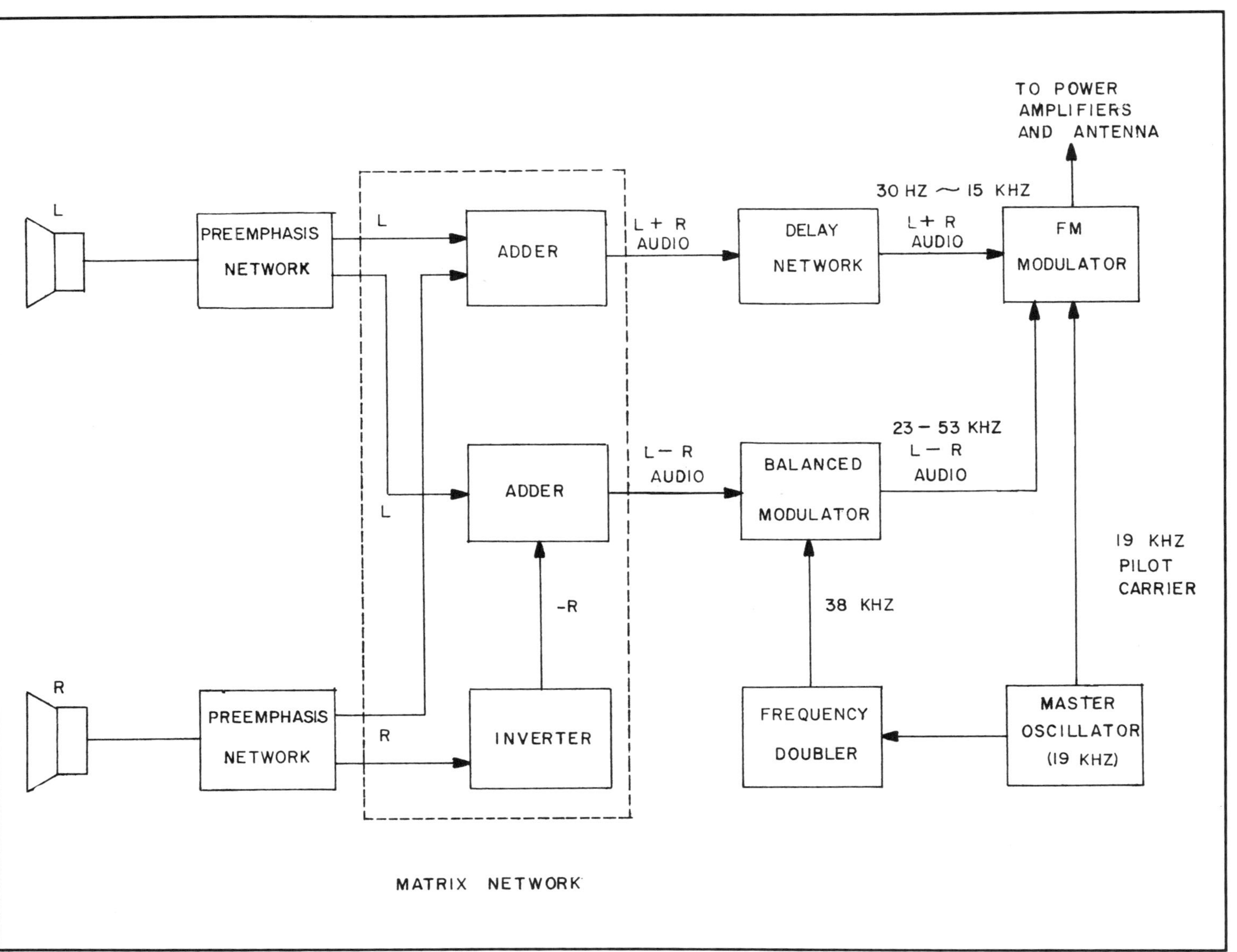

Fig. 7-9. FM stereo transmitter block diagram.

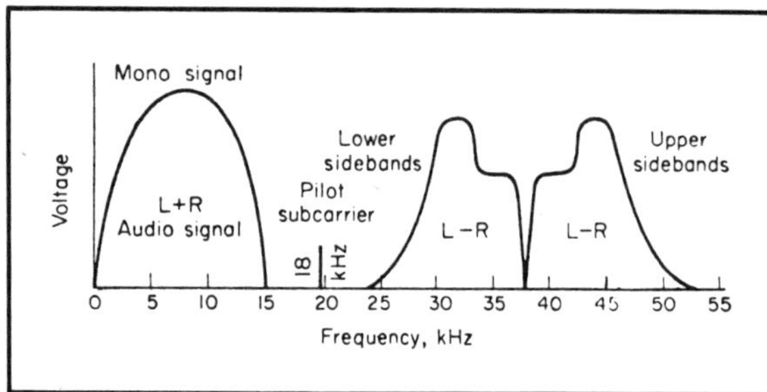

Fig. 7-10. FM stereo frequency spectrum.

audio signal to reconstruct the composite stereo wave. A switching bridge circuit consisting of CR3, CR4, CR5 and CR6 is turned on by the instantaneous polarity of the 38 kHz subcarrier voltage and the two sides of the bridge alternately conduct. Thus, the positive and negative envelopes of the composite waveform are sampled at a 38 kHz rate and the bridge output yields the demodulated L and R signals.

Tape players, stereos, and other sound equipment require one or more amplifiers to increase the signal so it can be heard in the speaker. Figure 7-15 illustrates a

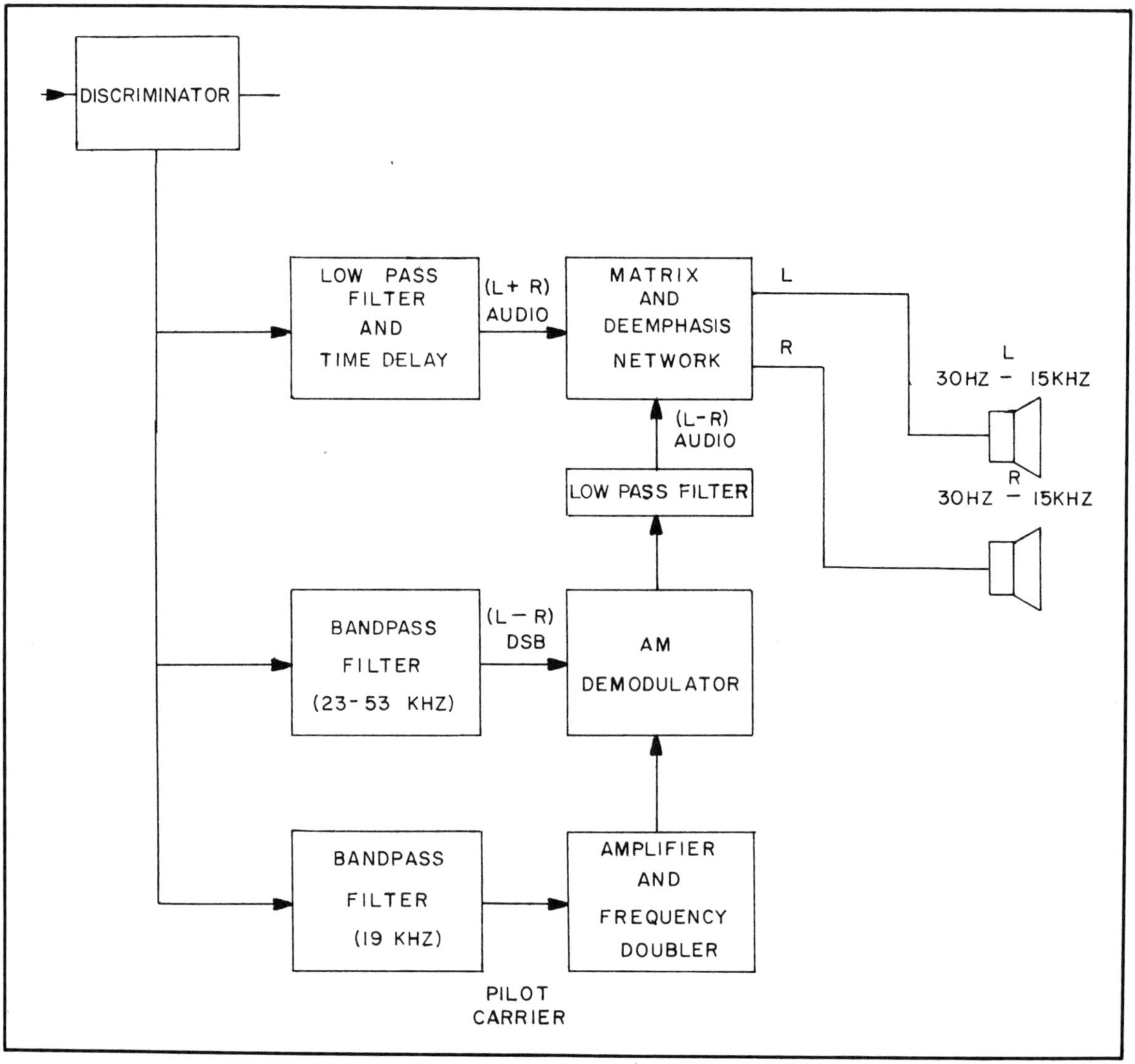

Fig. 7-11. Bandpass and matrix demodulation block diagram.

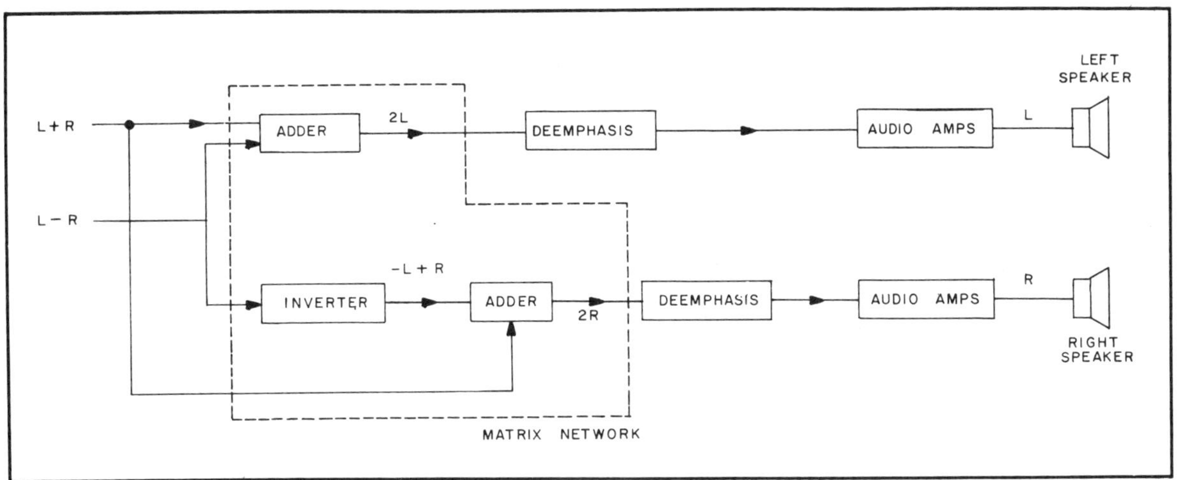

Fig. 7-12. Matrix network block diagram.

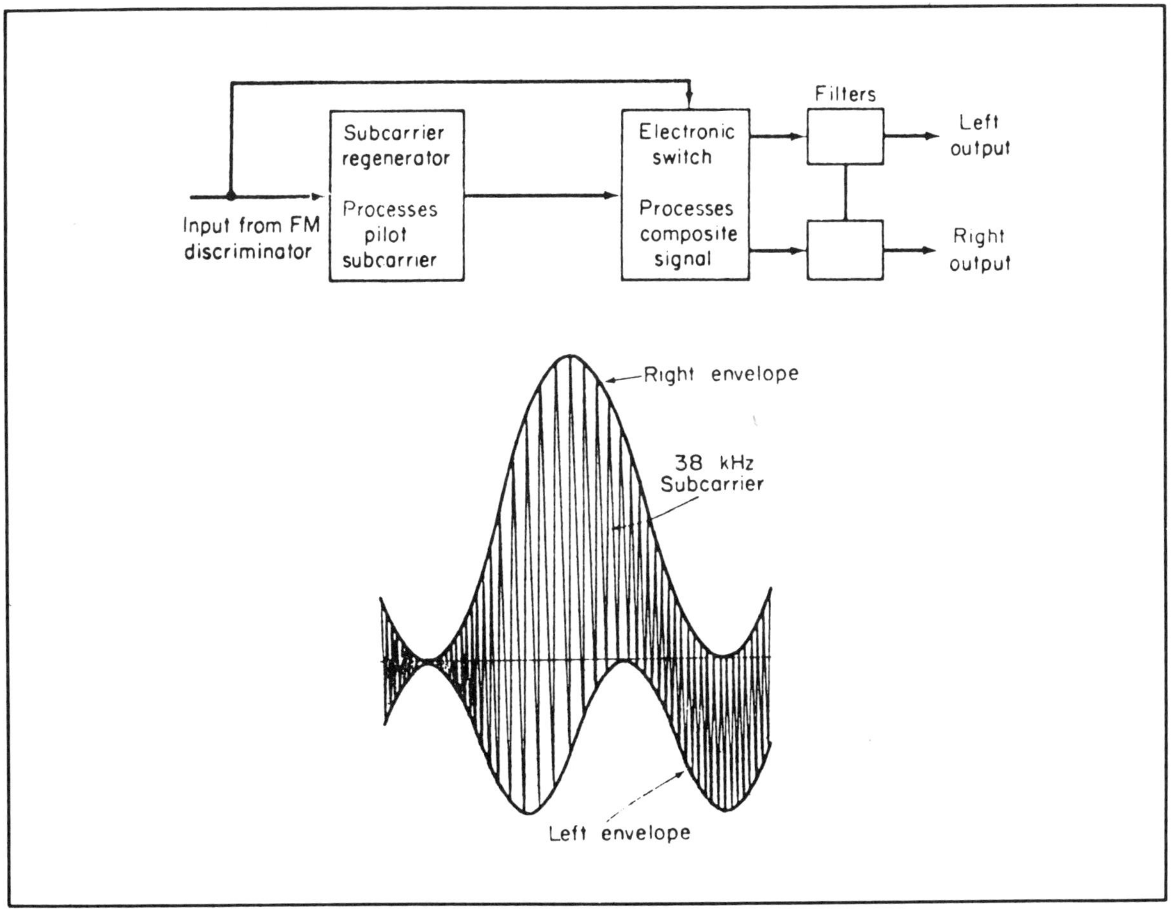

Fig. 7-13. Electronic switching demodulation block diagram and FM signal.

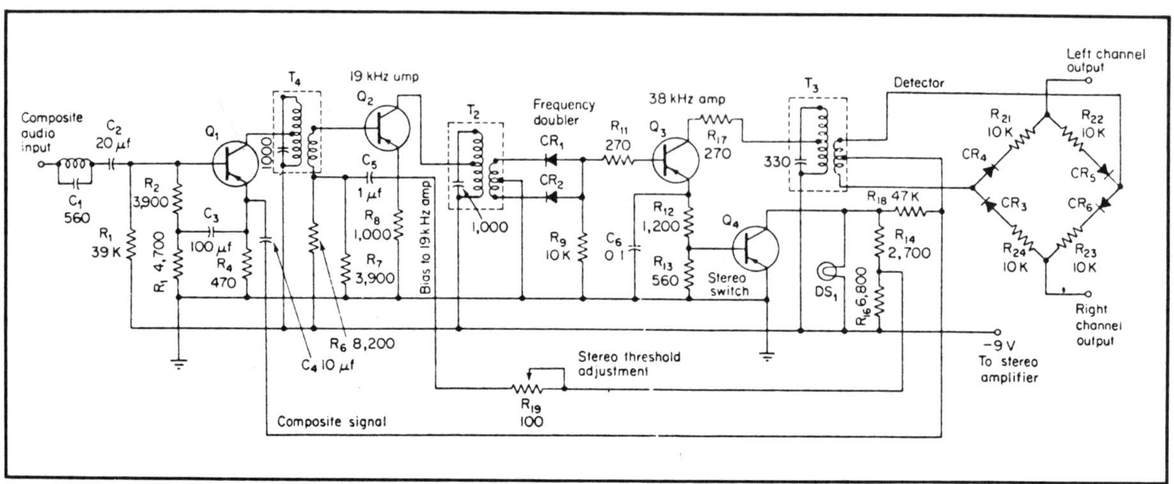

Fig. 7-14. Schematic of electronic-switching multiplex unit.

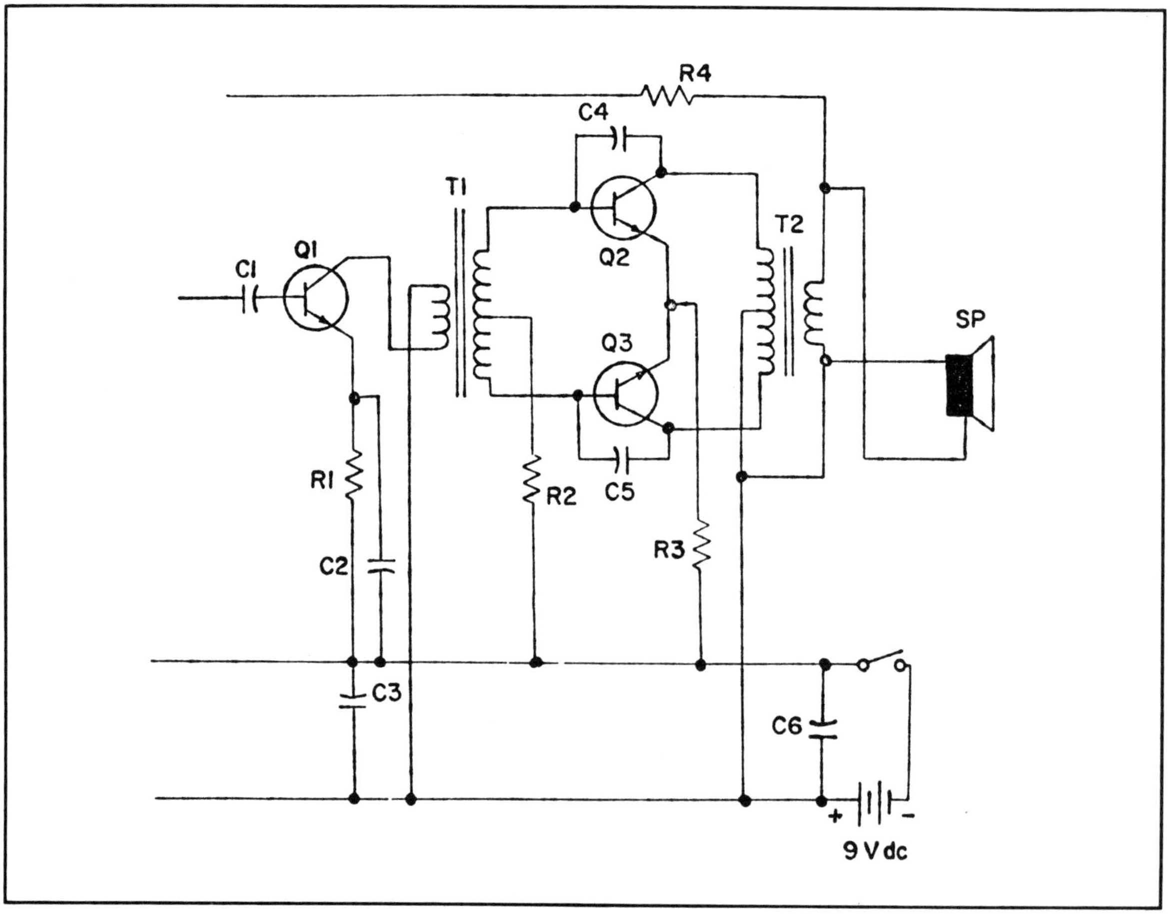

Fig. 7-15. Typical push-pull amplifier (courtesy Howard W. Sams & Co., Inc.).

typical push-pull amplifier common to many types of sound equipment. Transistor Q1 is the driver stage for the push-pull amplifier. Capacitor C1 is a coupling capacitor that passes the signal to the driver from the preamp stage. Capacitor C2 and resistor R1 are the RC emitter network that properly bias the transistor for amplification. Transformer T1 splits the signal phase by 180 degrees for Q2 and Q3. These two balanced transistors receive the signal alternately for each half cycle, adding the total power together at the output. Capacitors C4 and C5 couple part of the collector signal back into the base of the transistor. This negative feedback prevents the transistors from oscillating and reduces distortion. Transformer T2 matches the impedance of Q2 and Q3 with the speaker. Capacitors C3 and C6 are filter capacitors used to isolate each section from stray, unwanted signals in the line.

Another type of push-pull amplifier which eliminates the need for the transformer is the quasi-complementary type. It is the most popular one used. A schematic of it is shown in Fig. 7-16. The upper pair of N-P-N transistors, Q2 and Q4, conduct when the signal at the output of Q1 is positive. The P-N-P transistors, Q3 and Q5, conduct during the negative portion of the signal's cycle. Both portions of the cycles are reconstructed as a complete cycle across the load resistor RL, thereby implementing the push-pull.

TESTING AND REPAIR PROCEDURES

Several techniques are used in troubleshooting

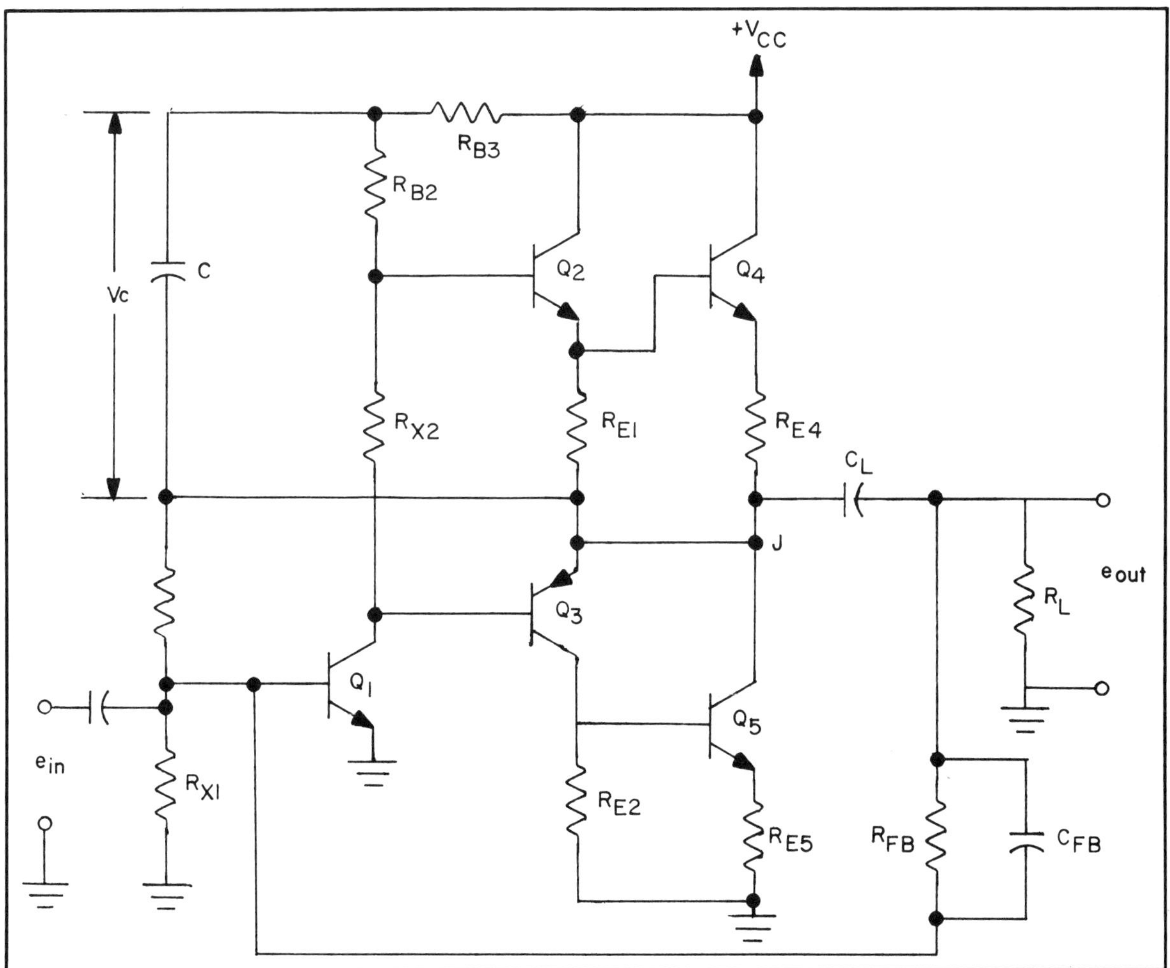

Fig. 7-16. Quasi-complementary amplifier circuit.

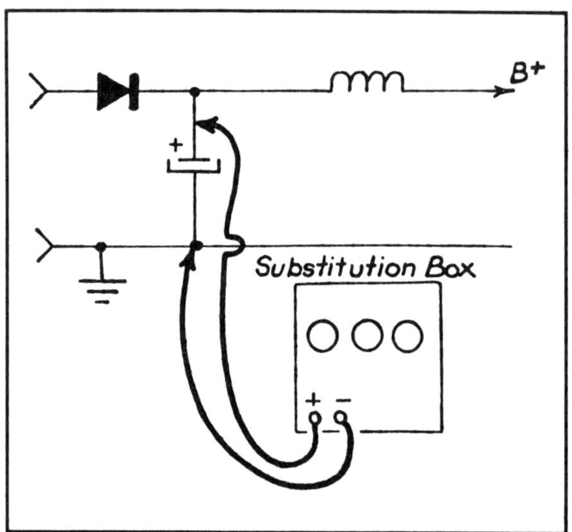

Fig. 7-17. Using a substitution box to bridge a capacitor (courtesy Howard W. Sams & Co., Inc.).

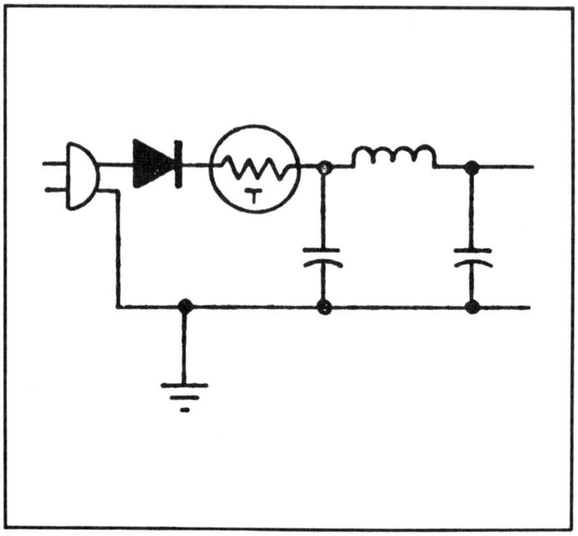

Fig. 7-19. Simple power supply (courtesy Howard W. Sams & Co., Inc.).

sound equipment. For example, when troubleshooting a superheterodyne receiver, start with a visual/hearing inspection. Look for obvious signs of damage. If the receiver has a hum, most likely it has a defective filter capacitor. As mentioned before, test the suspected filter capacitor by bridging it with a known good one of equal value or with a capacitor substitution box as shown in Fig. 7-17. If the hum disappears, replace the filter.

If the receiver is dead, check the fuse or circuit breaker. In a tube type receiver with series connected filaments, check each tube for an open filament since this will also open the circuit. Figure 7-18 illustrates the open filament.

In a dead set check the switch with an ohmmeter. The switch should show continuity one way and not the other. Also, use the ohmmeter to check the power supply

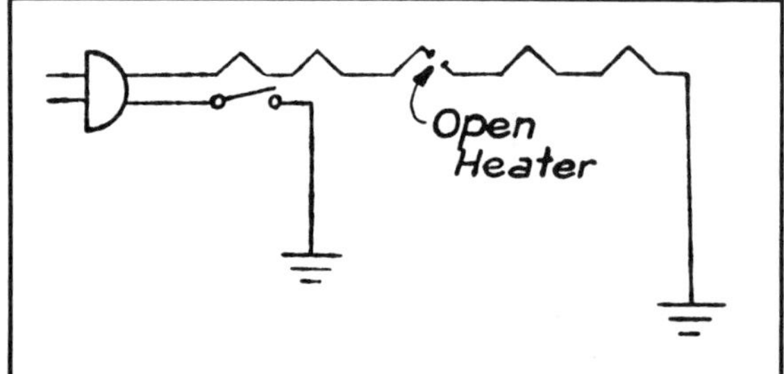

Fig. 7-18. Schematic diagram of a series string with an open filament (courtesy Howard Sams & Co., Inc.).

diode, thermistor, and filter coil. Any one of these components could open the circuit. A simple power supply showing the above components is shown in Fig. 7-19.

If a radio or stereo plays for a while, then stops, and comes back on later, check for thermal intermittent components. Using a combination hot/cool blower, heat the suspected transistors. When the defective transistor is heated, the receiver will stop playing or become badly distorted. Now cool off the transistor using the cool-air blower or a chemical freeze mist. The receiver should once again operate normally. When the thermal intermittent transistor has been found, replace it with a recommended replacement.

If a receiver has power but fails to produce sound, you should first try to isolate the defective stage. Signal injection can be used to locate the faulty stage as shown in Fig. 7-20. First, use an rf signal generator or a noise generator by connecting the ground lead to chassis ground and the other lead to each stage of the receiver. Start by injecting an af signal around 400 Hz into the speaker; this should produce a tone. If you hear a tone the speaker is operating. Proceed to the amplifier stage and inject a signal into the grid of this tube or the base of this transistor. If a tone is heard, proceed to the next stage toward the detector. When injecting a signal into the detector stage or back to the antenna, use an rf signal. If no tone is heard when the rf signal is injected into the detector stage, you can assume this is the faulty stage. At this time, proceed to take resistance checks of each

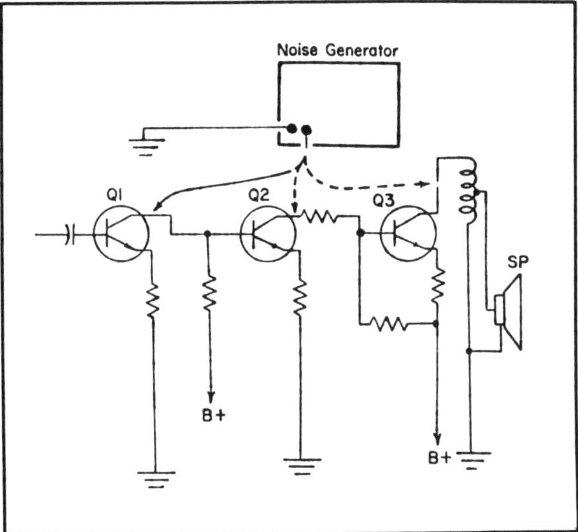

Fig. 7-20. Isolating a defective stage using a noise generator (courtesy Howard W. Sams & Co., Inc.).

component until the faulty component is located.

Should the faulty stage be the quasi-complementary amplifier circuit, the problem usually exists in the output and/or driver transistors, Q2 through Q5. To troubleshoot the circuit, check the resistance between ground and point J and between +Vcc and point J without power applied. Refer to Fig. 7-16 for these reference points. Connect the negative lead to ground and the positive probe lead of the voltmeter to point J. Note the measured value. Then reverse the voltmeter leads. The second resistance measure should be considerably less than the first. A dead short indicates that transistor Q3 and/or Q5 are shorted.

Repeat the above procedure, but this time check the resistance between +Vcc and point J. You will get a lower reading when the positive lead is at +Vcc. A short indicates that Q2 and/or Q4 is defective. Now short R_{X2}. Turn on the amplifier. A voltage measured at point J should be ½+Vcc. If the measurement is within 25 percent of this condition, then try different values of RB1 or RX1 to achieve a more acceptable balance. Should the voltage at point J be within 10 percent of ½+Vcc, then the amplifier is within reasonable tolerance. Sometimes RX2 is substituted with diodes to improve bias stability with varying temperatures. If your audio equipment has diodes, check them for shorts or opens.

Check the voltage across Q3. If it is +Vcc then Q4 is shorted. Similarly if the voltage measurement across Q2 is +Vcc then Q5 is shorted.

Voltage checks can be a very effective technique to determine a problem. For example, if a power supply is "loaded down" due to a shorted component, it will draw excessive current. This excessive current will bring down the voltage. If the zener diode shown in Fig. 7-21 were suspected of being shorted, you would lift one of its leads from the circuit. If the voltage goes back to normal, the diode is shorted and should be replaced.

If a receiver squeals, howls, or motorboats, the signal feedback is appearing in one of the circuit stages. Check filter capacitors and battery connections; most likely, a defective filter capacitor will be the problem.

Intermittent operation can be caused by other things besides transistors. For example, loose connections and soldered connections can cause intermittent operation. To locate the intermittent problem, wiggle the wires and tap on the printed circuit side of the board. Sometimes it may be necessary to resolder several soldered connections in hopes of locating the problem. Remember, always turn off the power before resoldering. The heat from the soldering iron will cause a transistor to draw more current and destory itself. Also, always use a heatsink when resoldering diodes and transistor connections.

If a receiver continues to blow fuses, turn the power off and start taking resistance measurements. Keep in mind that if a shorted capacitor is suspected, do not bridge it with another. Instead substitute another capacitor in its place. Check for shorted rectifiers by using the ohmmeter and the manufacturer's resistance specifications. Check each specification with the reading obtained with the ohmmeter until the short circuit has been identified. A short circuit has a near zero or zero reading as shown in Fig. 7-22.

An oscilloscope can also be effective in locating a

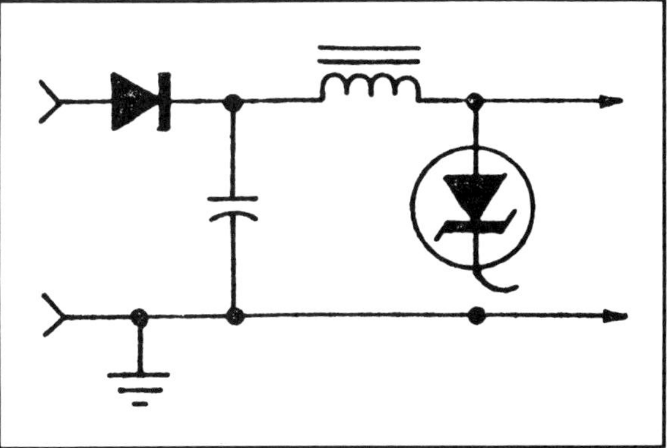

Fig. 7-21. Lifting one lead of a zener diode to determine its effect on the circuit (courtesy Howard W. Sams & Co., Inc.).

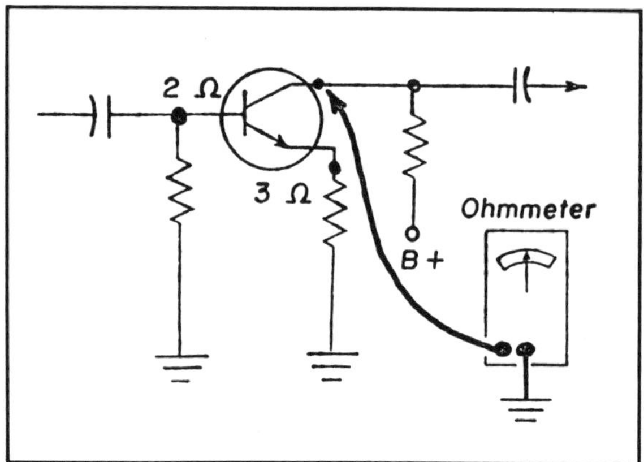

Fig. 7-22. Measuring the collector resistance of a transistor by using an ohmmeter (courtesy Howard W. Sams & Co., Inc.).

defective stage. The oscilloscope allows the service technician to actually see the signal waveform. Frequency response, gain, phase shift, noise, and hum can all be identified by the oscilloscope.

When repairing *tape recorders,* a common problem that causes low volume, lack of treble control, and distortion is dirty tape heads. Use isopropyl alcohol or tape head cleaner to clean the head of the tape recorder. Also, use a cotton swab and the cleaner to clean off the capstan shaft as shown in Fig. 7-23.

After a tape head has been cleaned, it is common

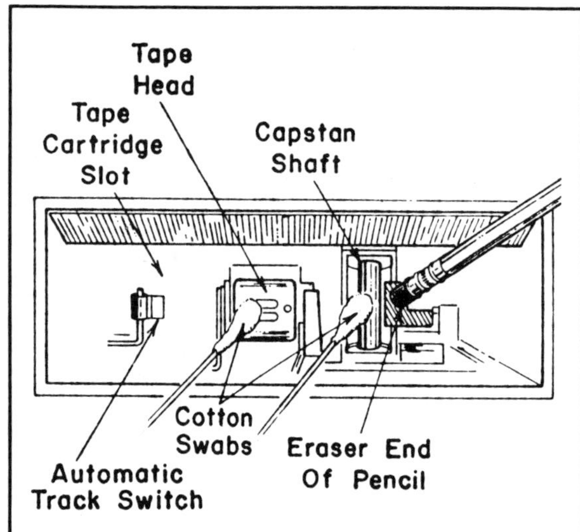

Fig. 7-23. Cleaning a tape player head using a cotton swab and isopropyl alcohol (courtesy Motorola, Inc.).

practice to demagnetize the head. Use a head demagnetizer that eliminates any possible unwanted magnetism in the head. Unwanted magnetism will cause distortion, especially at high frequencies. Using the head demagnetizer, stroke the demagnetizer near the head a few times, and then slowly remove the demagnetizer.

If a tape player fails to turn or is erratic, check the motor and drive mechanisms. Look for dirty or jammed idler wheels, shafts, belts, clutch, or spindle.

Often a belt will become stretched and need to be replaced. Use the exact replacement since the actual size, width, and diameter are very critical. Also, if the tape player sounds uneven (wavy) but a belt fails to correct the problem, the most likely cause may be a bad motor or unbalanced (warped) flywheel. When necessary, replace the defective part.

If the start/stop switch fails to operate correctly, check the lever arm and spring tension. Don't overlook the possibility of a defective tape.

If the tape player operates well, but the sound is weak or distorted, the head may be dirty.

Common problems in *record changers* include a misadjusted tone arm, incorrect needle pressure, and dirty idler wheel and turntable rim. Also, clean all mechanical sliding surfaces. Lubricate each surface using a light machine oil. Never lubricate the idler wheel or rim!

To check the speed of the turntable, most electronic parts distributors sell a stroboscope kit that can be used to check the correct rpm of the turntable. Carefully follow the directions that come with the kit. If cleaning does not correct the turntable speed, the idler wheel is probably worn and should be replaced.

The tone arm can be adjusted by a screw under the base of the arm. Adjust the screw until the needle rests in the center of the lead-in groove of a 12-inch record. Screw B2 can be used to adjust the tone arm height. Adjust the arm so that it clears each record. Also, the needle pressure can be adjusted by turning screws B3 and B4 until the needle rests with sufficient pressure without sliding on the record.

Often a record player cartridge becomes defective. When this happens, the amplifier and speaker will seem to have power but will not produce sound. If this is the case, tapping or injecting signal near the cartridge should produce noise out of the speaker. If the cartridge is found defective, replace it with an exact replacement.

When servicing stereo sound equipment, the basic test for a multiplex unit is to measure the decibel (dB) separation between the L and R signals that it produces. The setup for a stereo multiplex separation test is illustrated in Fig. 7-24. A stereo multiplex signal generator supplies the L and R signals which are the same but different in phase relationship. Therefore, when the R signal is applied to the multiplex unit, the R output channel should be maximum while the L channel should theoretically be zero.

On the other hand, when the L signal is applied to the multiplex unit, the L output channel should be at maximum and the R channel should theoretically be zero. In practice, about 20 to 30 dB difference should be observed between vtvm readings for proper operation of the multiplex unit. If a separation of 10 dB is obtained, the multiplex unit is borderline and is probably defective.

Inadequate separation is usually caused by a component failure in the multiplex unit such as unmatched diodes or diodes with poor forward and reverse bias ratios in the switching bridge, a collector leakage in a transistor or a shorted or open capacitor. Also misalignment can cause poor separation, especially in a matrix network unit. You should consult an appropriate service manual for multiplex unit alignment procedures.

An off value resistor in a matrix network circuit can reduce separation. If you find that the stereo multiplex unit has appropriate separation when tested with a composite audio signal from a generator but does not when it is operated normally, the problem would appear to be poor alignment of the high frequency signal channel in the receiver.

Distortion of the output signal can be caused by a defective diode or an off-value resistor in the switching bridge. A faulty electrolytic capacitor in the subcarrier regeneration circuit can cause a distorted output signal (i.e., humming).

If the left channel is distorted and the right channel is good, compare voltage, resistance, and waveform measurements and try to isolate the defective component. For example, in Fig. 7-25 if Q15 has a collector voltage of 7.38 volts, this transistor is not operating and could be defective. Also, if the voltage at this same point is down to 4 volts, a shorted transistor is suspected. If the output is low from the driver transistor Q15, a faulty emitter bypass capacitor may be suspected. Check this by bridg-

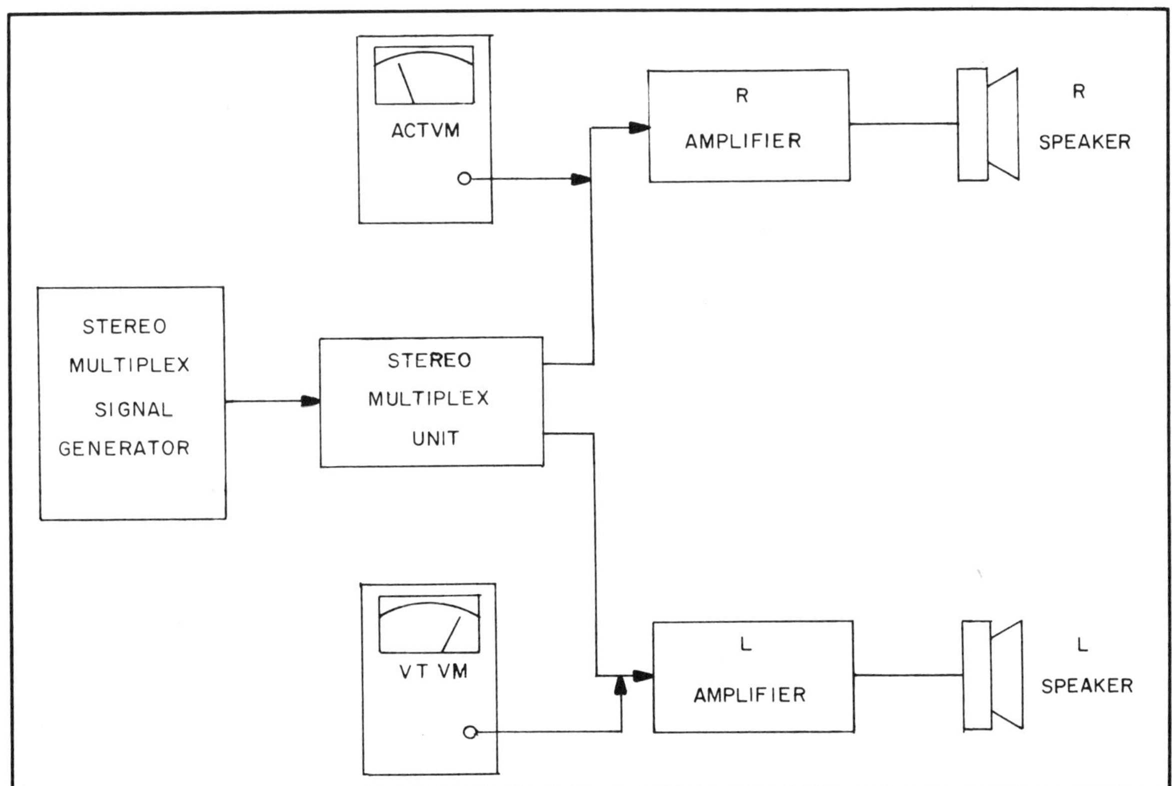

Fig. 7-24. Stereo multiplex separator test setup.

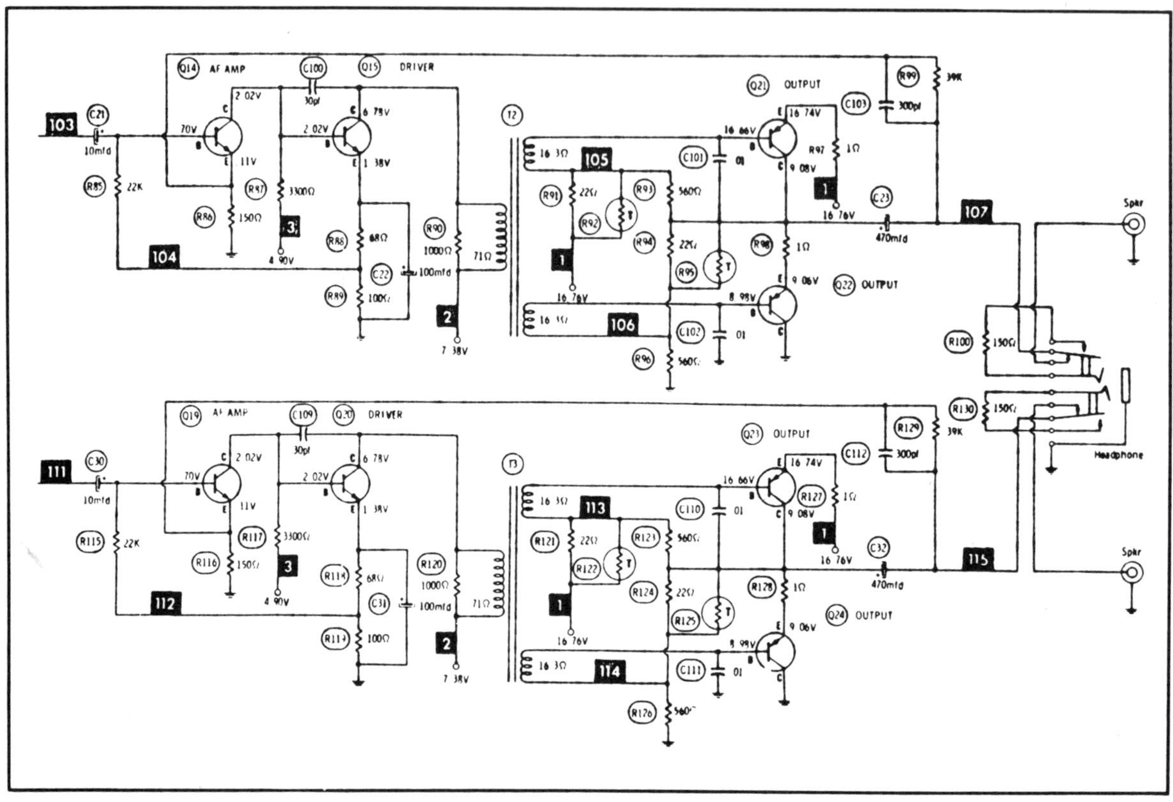

Fig. 7-25. Component values and operating voltages of a typical stereo system (courtesy Howard W. Sams & Co., Inc.).

ing it with another capacitor of the same value. Also, if any one of the resistors changed in value, the bias of the transistor will also change affecting the output. Resistors can be checked by resistance readings and, if faulty, be replaced.

If no signal enters the af amp stage, an open coupling capacitor may be suspected. Transformer T2, which acts as a splitter, will cause this channel to go dead if it is shorted. Often a shorted transformer will have a charred appearance and will easily be identified by its "burned" smell. By using an ohmmeter you should read 16.3 ohms across each half of the secondary coil. A shorted coil will read near zero.

If there is low volume and high distortion in this channel, one might suspect Q21 and/or Q22 to be defective. If only one of these transistor outputs is found defective, replace both of them with a matched pair. It is important that these transistors be properly balanced for best signal reproduction. Remember when replacing power transistors to always use a heat sink, and insulating mica. Also use heat sink silicon compound if necessary, as shown in Fig. 7-26.

Any time a radio receiver requires bench servicing, correctly connect the radio to a well filtered power supply. Use the correct voltage and correct polarities. If it is an automobile radio, remember to apply a voltage from 12 to 13 volts dc since the automobile electrical system typically operates at around 13 volts. Also, remember to arrange all wires neatly. Shorted speaker wires can blow out the output transistors and the like. Never operate the receiver without having speakers hooked up since this could have the same effect. Figure 7-27 illustrates a bench servicing setup for an automobile radio.

Figure 7-28 shows a typical three-stage amplifier commonly used in a radio receiver. If Q3 draws excessive current, this will cause Q1 to conduct more heavily. The collector voltage at Q1 will decrease because Q1 draws more current. Since the collector voltage of Q1 drops, this will in turn reduce the conduction of Q2 and Q3. Together this circuit will have low output and high distortion. Replacement of output transistor Q3 should solve the problem.

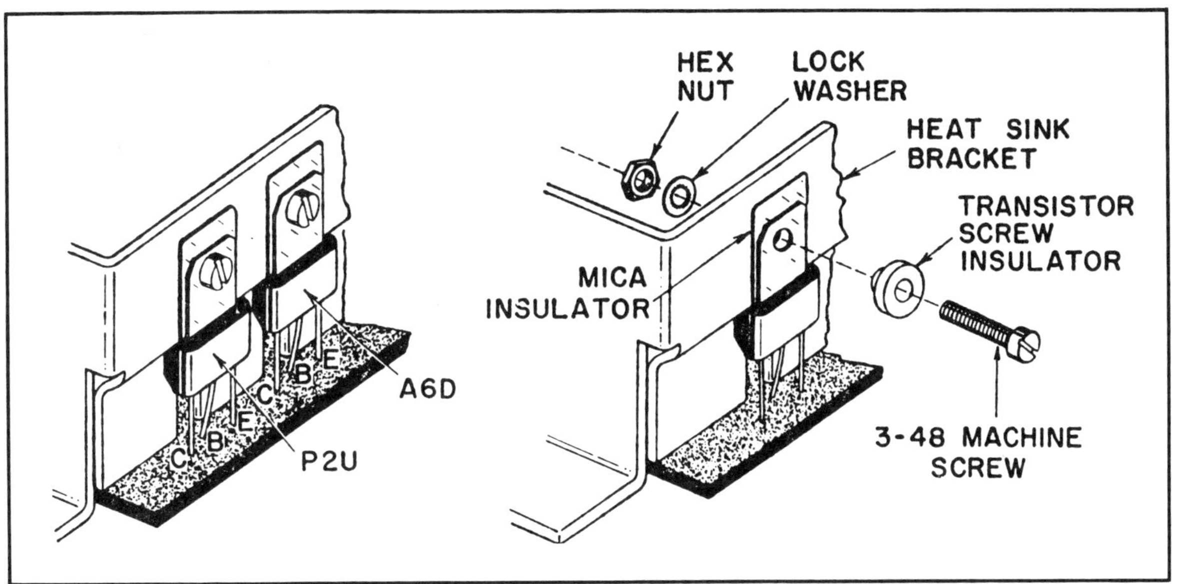

Fig. 7-26. Example of proper installation of an output transistor (courtesy Motorola, Inc.).

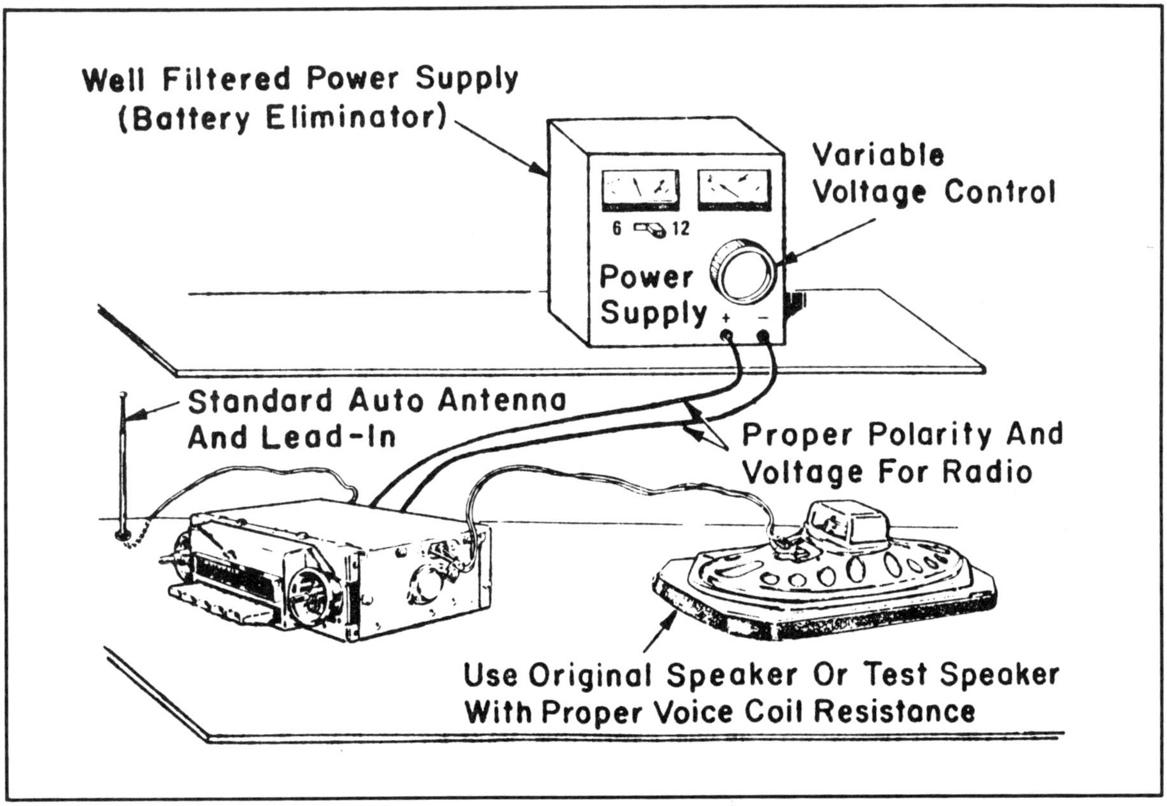

Fig. 7-27. Correct bench servicing setup for an automobile radio (courtesy Motorola, Inc.).

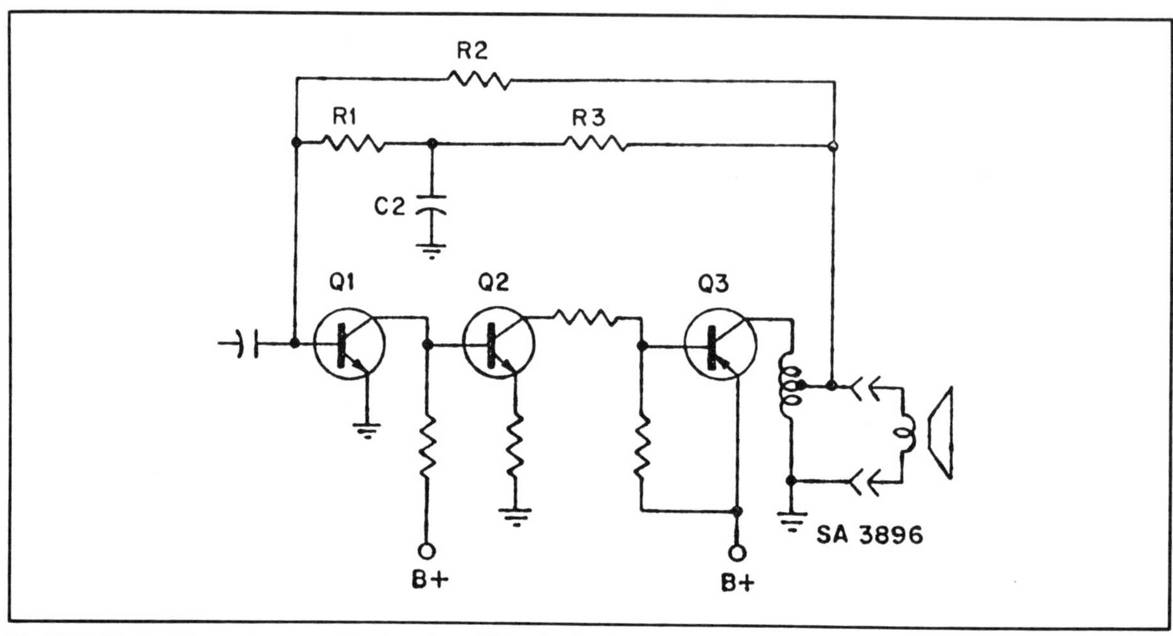

Fig. 7-28. Typical three-stage amplifier (courtesy Motorola, Inc.).

PREVENTIVE MAINTENANCE

Sound equipment breakdowns can often be directly attributed to owner abuse. Stereos should not be played at full volume or for an excessive length of time. Heat can cause premature breakdown of the speakers and amplifier outputs. Never play a stereo with the speakers removed or with the incorrect number of speakers. A mismatch impedance can break down the unit. Therefore, always match the impedance of the speakers with that of the output amplifiers.

Once a turntable has begun its cycle of changing, never force it to do something out of cycle. Likewise, always handle the unit with utmost care. Keep a dust cover over the unit when not in use to prevent dust and dirt from collecting. Also, use good records that are free from scratches. This will give the needle a long life.

Periodically clean and demagnetize tape heads. This will not only prevent breakdowns but will also increase your listening pleasure.

If a radio receiver fails to move the dial indicator when the dial is turned, the dial cord probably needs to be restrung as shown in Fig. 7-29.

When installing a radio in an automobile, be sure to ground one side of the antenna to the car frame. Tighten each nut and bolt to ensure good response, which is free from unnecessary noise and vibration due to sloppy installation.

Remember to tighten the speakers. Loose speakers can cause annoying noise and rattling. Also, make sure that the radio is installed with all wires and harnesses neatly secured. Sloppy wiring can cause interference and possible short circuits.

When installing the radio in an automobile, do not overlook the noise suppression equipment. This equipment usually consists of an alternator noise filter (that has a typical value of 0.5 mF to 0.8 mF), a distributor noise suppressor, and an antenna noise filter. Sometimes it may be necessary to add a noise suppressor to each spark plug wire, ignition coil, and/or heater and defroster fan motor.

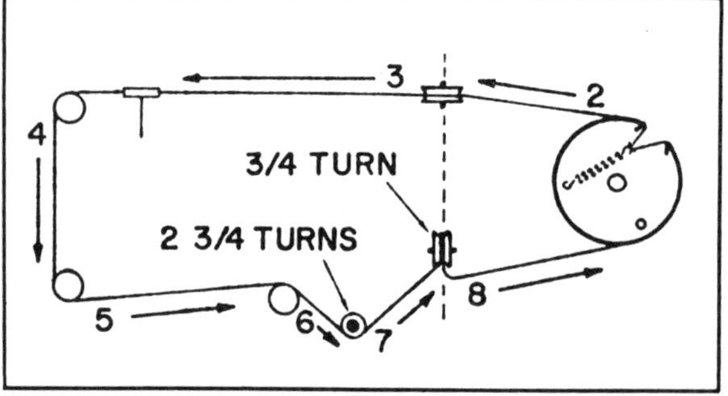

Fig. 7-29. An illustration of dial cord stringing (courtesy Howard W. Sams & Co., Inc.).

SELF-EXAMINATION

Select the best answer:

1. An audio frequency is said to be within:
   A. 400 Hz to 40,000 Hz.
   B. 20 Hz to 20,000 Hz.
   C. 100 Hz to 10,000 Hz.
   D. None of the above.
2. A modulated rf wave consists of:
   A. rf wave and cw.
   B. rf wave and radio wave.
   C. cw wave and rf wave.
   D. None of the above.
3. Which of the following demodulates an rf modulated wave?
   A. rf amp.
   B. i-f amp.
   C. Mixer.
   D. Detector.
4. In most mixers, the oscillator frequency is ______ than the carrier frequency of the input signal.
   A. Lower.
   B. Higher.
   C. The same.
   D. 10 kc above.
5. A push-pull amp requires a:
   A. Splitter at 180 degree phase shift.
   B. Inverter at 90 degree phase shift.
   C. Both A and B.
   D. None of the above.
6. A suspected shorted zener can be checked by:
   A. Bridging.
   B. Lifting one end of the diode.
   C. Shorting across the diode.
   D. None of the above.
7. What receiver stage is used to reduce signal fading and keep a constant volume?
   A. Detector.
   B. i-f amp.
   C. af amp.
   D. rf amp.
   E. AGC.
8. In FM multiplex reception, a monaural receiver uses only:
   A. The 19 kHz pilot signal.
   B. The (L−R) sideband.
   C. The (L+R) carrier signal.
   D. All of the above.
9. Hum is most likely caused by a defective:
   A. Diode.
   B. Transistor.
   C. Filter.
   D. None of the above.
10. Thermal intermittent components can often be checked by:
    A. Heat/freeze.
    B. Tapping.
    C. Bridging.
    D. None of the above.
11. When examining a dead set, which item(s) should be checked?
    A. On/off switch.
    B. Power supply diodes.
    C. Fuse.
    D. Open filament.
    E. All of the above.
12. The decibel reading for proper separation between channels of a multiplex unit is:
    A. 5 dB.
    B. 10 dB.
    C. 20 dB.
    D. 40 dB.
13. A dirty tape head should be cleaned with:
    A. Gasoline.
    B. Etching solution.
    C. Isopropyl alcohol.
    D. Any of the above.
14. If only one of the transistors in a push-pull amplifier is bad:
    A. Replace only the bad one.
    B. Replace both of them.
    C. A or B.
    D. None of the above.
15. A shorted capacitor can be checked by:
    A. Bridging.
    B. Substitution.
    C. Both A and B.
    D. All of the above.
16. A common problem that causes low volume, lack of treble control and distortion is:
    A. Defective AGC.
    B. Defective volume control potentiometer.

C. A dirty tape head.
D. Defective motor and drive mechanism.

17. An oscilloscope is an effective instrument for locating defective stages because it can display:
    A. The waveform of the signal.
    B. The frequency response of a stage.
    C. Noise on a signal.
    D. All of the above.

18. If a receiver squeals, howls or motorboats, the most likely cause is a:
    A. Transistor.
    B. Filter capacitor.
    C. Weak battery.
    D. Resistor.

19. When resistance checking a quasi-complementary amplifier,
    A. There should be a low resistance reading between common and V7CC.
    B. There should be a low resistance reading between common and the ground.
    C. There should be a difference between resistance measures between the common and ground and the common and Vcc.
    D. There should be no difference between resistance measures between the common and ground and the common and Vcc.

20. When using an ohmmeter to identify a short, the ohmmeter reading should indicate:
    A. Zero.
    B. Infinite.
    C. 100 kilohms.
    D. 1 megohm.

21. If an FM receiver signal drifts, a possible cause is a
    A. Defective AFC circuit.
    B. Defective oscillator circuit.
    C. Shorted or open component in the AFC.
    D. All of the above.

22. Which of the following stages are common to both AM and FM receivers?
    A. Tuner, local oscillator, detector, af amp.
    B. rf amp, mixer, i-f amp, af amp.
    C. Local oscillator, rf amp, frequency discriminator, detector.
    D. Tuner, i-f amp, detector, af amp.

23. The AM detector performs two basic functions in the receiver:
    A. Amplifies and filters.
    B. Buffer and amplifier.
    C. Buffer and detector.
    D. Rectifies and filters.

QUESTIONS AND PROBLEMS

1. Explain the development and characteristics of a modulated radio wave.
2. What is a crystal detector?
3. Draw a block diagram of a superheterodyne receiver.
4. Why is a splitter needed in a push-pull amplifier?
5. Explain the procedure in repairing a "dead" radio receiver.
6. Generally, what could cause a receiver to squeal, howl, or motorboat?
7. How can thermal intermittent components be located in a receiver?
8. What is a head demagnetizer and why is it used?
9. What is record stroboscope?
10. Explain the procedure in correcting a slow turning record turntable.
11. Explain one procedure used to check a suspected shorted zener diode.
12. Explain how to clean a tape recorder head.
13. What is a noise generator?
14. List some typical noise suppression equipment.
15. What is the difference between the method of bridging a capacitor and substituting a capacitor?

Chapter 8

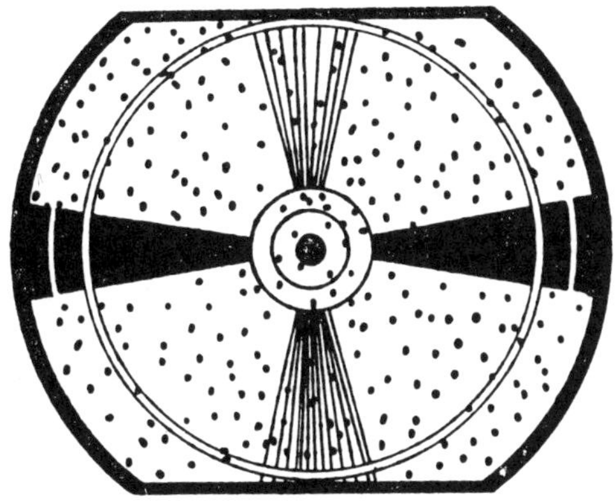

Troubleshooting Television

The field of television is a very special branch of electronics which employs many principles. The transmission and reception of pictures is a tremendous feat in black-and-white and marvelous when considering color. In this chapter we will examine some principles of transmitting a composite signal, then review the fundamentals of a black-and-white receiver, followed by troubleshooting of a black-and-white. We will conclude with a review of the fundamentals of color receivers, followed by troubleshooting of a color receiver.

TRANSMITTER PRINCIPLES

The TV transmitter is actually two separate transmitters. The video or picture signal is amplitude modulated onto a carrier, while the aural or sound transmitter is actually an FM system very similar to a broadcast FM station. Therefore, the composite transmitted signal is a combination of both AM and FM principles. A simplified block diagram of a TV transmission and reception system is shown in Fig. 8-1. The TV camera acts as a transducer which converts light energy into electrical energy, while the picture tube (CRT) is a transducer that converts the electrical energy back into light. The microphone and speaker are related transducers for the sound system.

The TV camera has a thin electron beam that is moved horizontally across a light sensitive surface producing a voltage proportional to the light. This creates a video line. The electron beam is made to retrace and scan another line. This is done 525 times per second. The TV receiver must have some means of synchronizing the traces made at the camera. Therefore the transmitted signal carries synchronization pulses.

BLACK-AND-WHITE FUNDAMENTALS

The television receiver actually reproduces a series of dots. These dots travel at such high speed that the viewer sees the total overall sequence as a picture on the screen. The degree of speed and intensity of the dots vary. The electron gun produces a stream of electrons which magnetically scan left to right and from top to bottom. Specific phosphors on the screen luminesce when struck by electrons.

The scanning system used in television is the interlace system which starts at the top left, scanning the odd lines from left to right completing 262½ lines. The interlaced scanning is illustrated in Fig. 8-2. The scanning returns from the bottom of the screen back to the top center and completes the scanning of the even lines. Each

SYNCHRONIZING
SIGNALS
TV
CAMERA
AM
TRANSMITTER
ELECTROMAGNETIC
WAVES
TRANSMITTING
ANTENNA
RECEIVING
ANTENNA
DIPLEXER
MICROPHONE
FM
TRANSMITTER
PICTURE
TUBE
AM
RECEIVER
FM
RECEIVER
LOUDSPEAKER

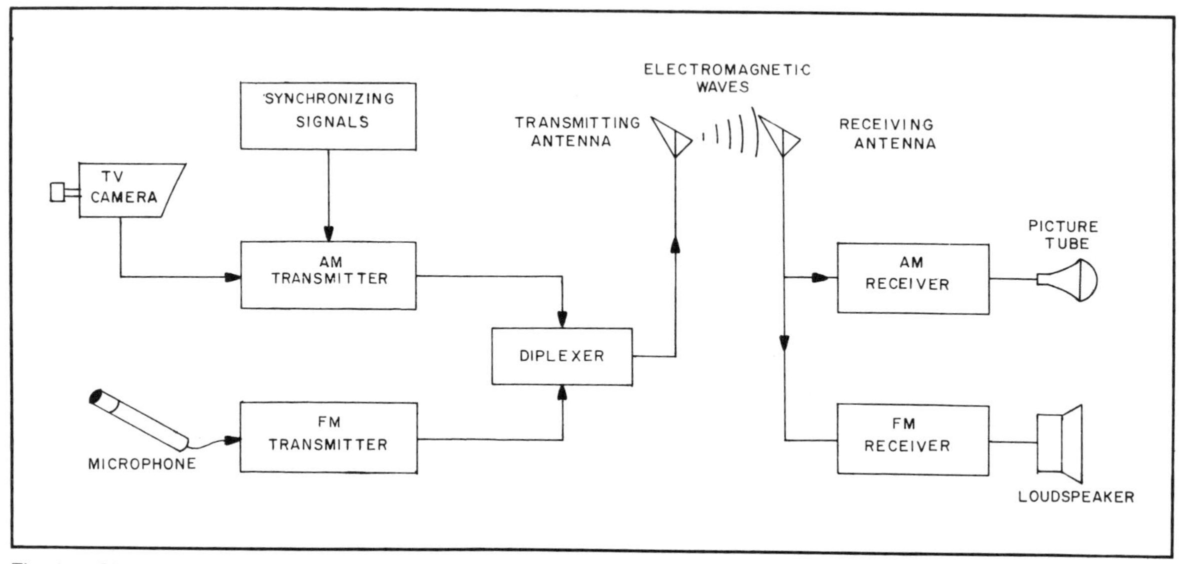

Fig. 8-1. Simplified TV transmission system.

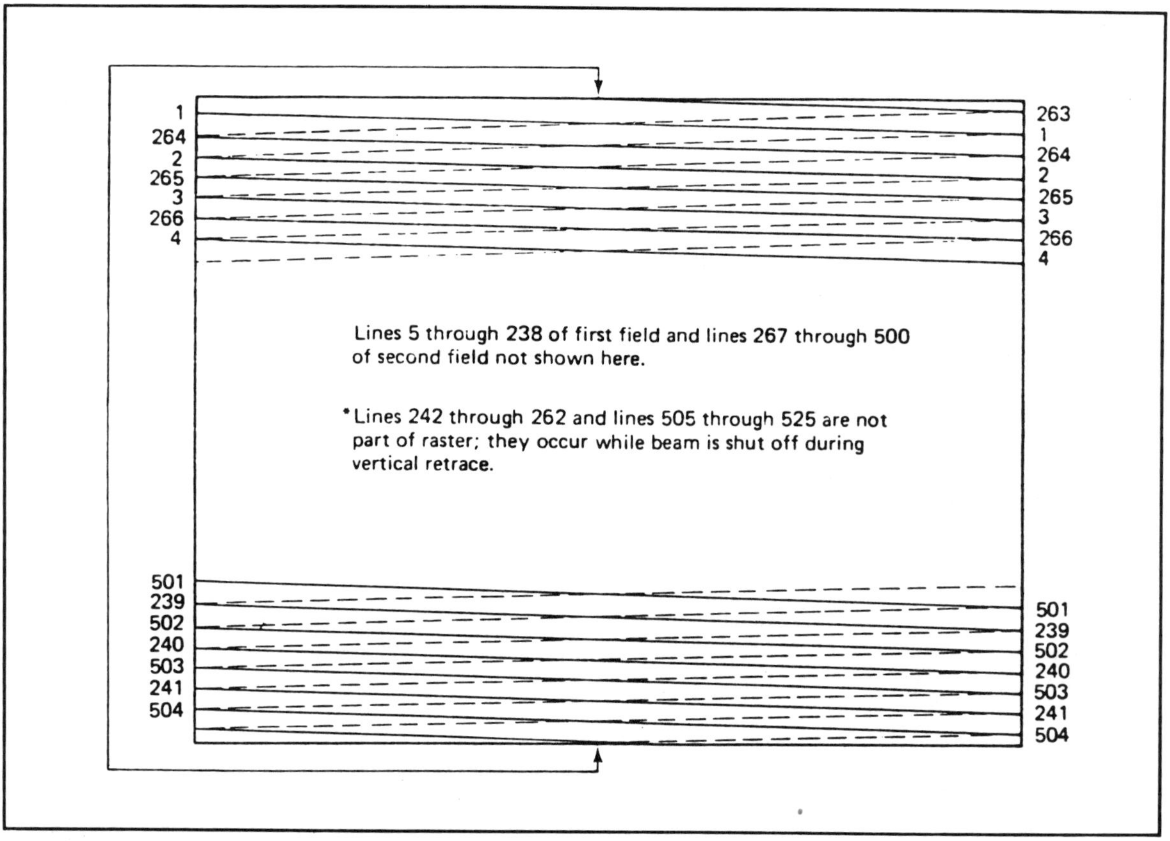

1
264
2
265
3
266
4
263
1
264
2
265
3
266
4
Lines 5 through 238 of first field and lines 267 through 500
of second field not shown here.
*Lines 242 through 262 and lines 505 through 525 are not
part of raster; they occur while beam is shut off during
vertical retrace.
501
239
502
240
503
241
504
501
239
502
240
503
241
504

Fig. 8-2. Interlace raster scanning pattern.

odd or even set of scanning lines represents a *field*, and both an odd and an even set represents a *frame*. Therefore, there is a total of 525 lines per frame, and the frequency is 30 frames per second.

Each time the scanning beam moves from the left side to the right, it must quickly return. This is called *horizontal retrace*. When the scanning beam gets to the bottom of the screen, it also must quickly return to the top of the screen. This is called *vertical retrace*. During retrace the picture is black. Only during the trace scanning period is a picture visible. The vertical oscillator running at 60 Hertz (Hz), deflects the scanning beam upward. The horizontal oscillator runs at 15,750 Hz and deflects the scanning beam from left to right across the screen.

Each time the scanning beam finishes a line, a high amplitude pulse is introduced. The pulse "syncs in" each transmitted line with the television receiver. Figure 8-3 shows a simplified video signal.

Figure 8-4 shows a block diagram of a black-and-white television receiver. The signal from the antenna is amplified in the radio frequency (rf) stage, mixed with a continuous wave of a predetermined frequency from the oscillator, and sent to the intermediate frequency i-f stages at 45.75 MHz (megahertz) where it is amplified. The video detector then demodulates the signal and sends the audio part of the signal to the audio stages and the video portion of the signal to the video stages. The sound detected from the video signal is FM. The audio signal is amplified in the audio i-f amp, demodulated in the FM detector, again amplified in the AM amp, and reproduced as sound by the speaker. Meanwhile, the video signal is amplified by the video amp and sent to the grid of the picture tube (cathode ray tube).

The automatic gain control (AGC) maintains the signal at a constant level. The sync separator removes the vertical and horizontal pulses and applies them to integrating and differentiating circuits. The integrating circuit shapes the vertical sync pulses into a series of triangular-shaped pulses and applies them to the vertical deflection oscillator. The vertical deflection amplifier drives the vertical deflection yoke and produces vertical scanning.

The differentiating circuit shapes the horizontal

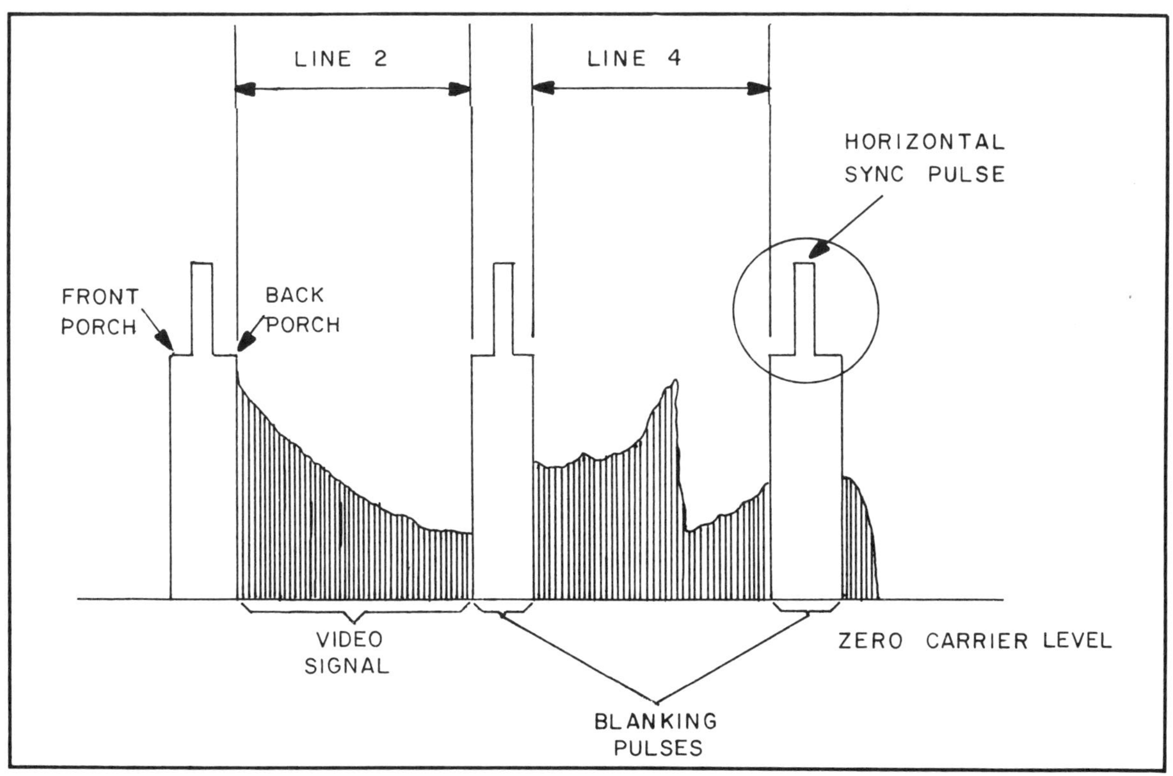

Fig. 8-3. Simplified video signal showing lines of video, blanking pulses and horizontal sync pulses.

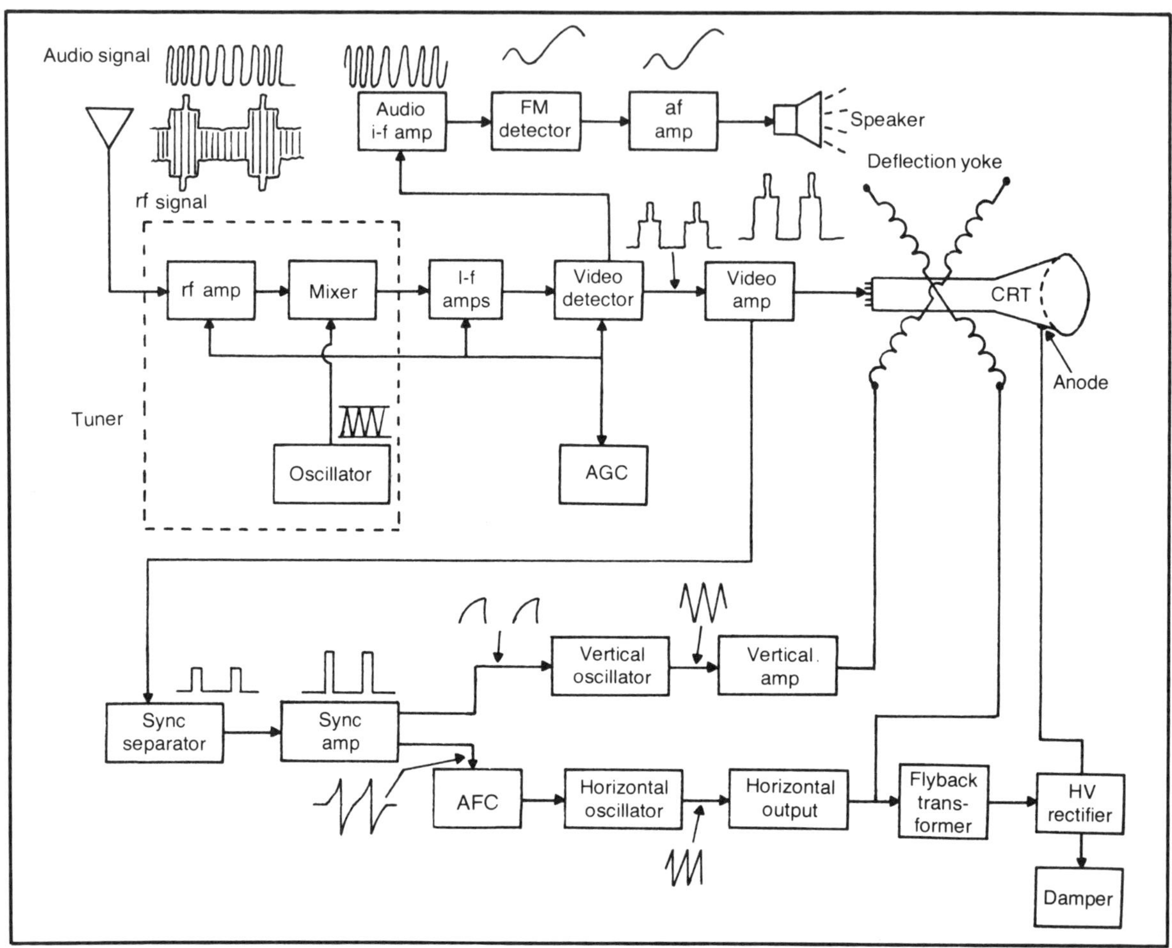

Fig. 8-4. Block diagram of a black-and-white receiver.

sync pulses and applies them to the horizontal deflection oscillator and automatic frequency control (AFC) circuit. The horizontal pulses are "locked in" by the AFC. They are amplified and drive the horizontal deflection yoke which produces horizontal scanning.

The high voltage needed to light the picture tube receives its power from the horizontal amplifier. This voltage from the horizontal amplifier is stepped up to approximately 30,000 volts or more by an auto-transformer (often called a flyback transformer). The voltage is rectified by the high voltage rectifier and finally sent to the anode of the picture tube. The damper is a diode placed in the path of the *kickback* pulse (ringing) from the yoke. The function of the damper is to prevent the kickback action from happening more than once for each applied "kick."

BLACK-AND-WHITE TV TROUBLESHOOTING

Since each stage of a TV receiver has a specific function to perform, it is likely that certain symptoms can easily be diagnosed as a problem in a specific stage or stages. The video and audio qualities can be used to track down the defective stage or stages. You will have to use a voltmeter, an oscilloscope, or both and, at times, other special TV testing equipment. The voltages and signal at critical stages are found on the manufacturer's schematic.

If many stages are involved, use the *half-splitting* method. That is, examine the signal at a stage that is halfway between a known good stage and the output. Should the results be proper, then move forward to another stage that is halfway between the previous test point and the output. However, if the first test indicated a defective signal, you should use the half-split method in the opposite direction until the defective stage is isolated. This technique will minimize the number of test measurements.

Weak Picture and Sound

If both the sound and picture are weak and distorted, a possible cause is a defective antenna system—such as a bad antenna, loose connection, bad cabling, or improperly oriented antenna. Check for the proper signal using a signal level meter or substituting another TV. Should the antenna system prove good, the problem is likely to be in the tuner section. Since selection of channels is accomplished generally by changing frequencies of the oscillator, you should check it for proper adjustment. If the tuner is operating properly, the problem may be in the i-f, video detector or AGC. All of these stages are common to both the sound and picture; therefore, a bad transistor, a change in resistive value, or low voltage supply can attribute to the problem.

Figure 8-5 illustrates a typical i-f amplifier circuit. The input consists of traps to eliminate carrier signals of adjacent channels, three i-f stages, sound trap and detector and a video detector. Examine your set's schematic and then check the dc voltages and signals as indicated. For example, if an abnormal collector to emitter voltage reading is found, it indicates either a defective component in the collector circuit or a change in transistor current conductance. If we assume that the transistor and collector circuit is normal, then there has been a shift in the base to emitter voltage bias. That bias shift can be the result of resistor value change or leaky capacitors in the base or emitter circuit.

Notice that the AGC controls the bias of the first i-f stage. If the AGC circuit is defective, it could reduce the gain of the i-f stages and create a weak picture.

Good Picture, Weak Sound

Should the picture be good, but the sound is weak and distorted, chances are the audio i-f amp, FM detector, af amp or speaker is causing the problem. The FM detector is the most likely problem. You should check the FM detector for proper voltage and signal first. If the test voltage and signal match the manufacturer's specification, then the problem is the af amplifier or speaker. An easy way to check the audio stage is to increase the volume and note the noise. If there is noise as the volume increases, then the af amplifier output and the speaker are working; therefore, the problem must be in the signal or the previous stage. If not, then the problem is in the FM detector or audio i-f amplifier.

Weak Picture with Normal Sound

Should the picture be weak but accompanied by normal sound and a bright screen (raster), the probable

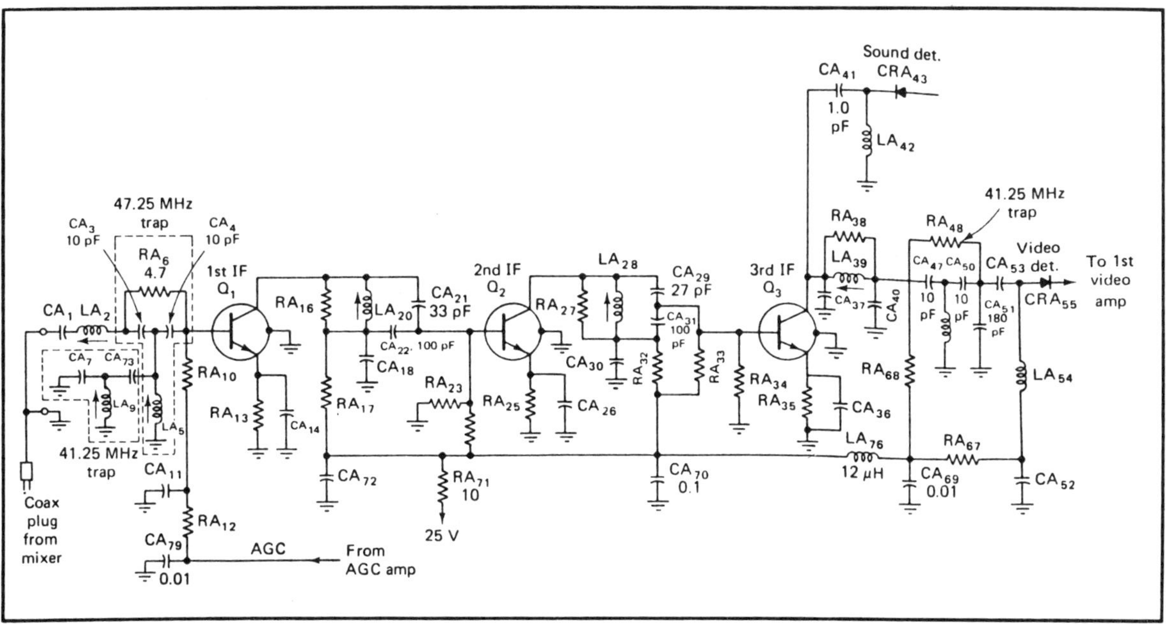

Fig. 8-5. Typical i-f amplifier circuit.

stages that can affect the picture are the antenna system, rf amplifier, converter, local oscillator, i-f amplifier, video detector, AGC system or, most likely, the video amplifier. The sound could be perceived as normal after leaving the video detector if the audio section has a number of amplification stages which would amplify a weak signal. Once again, a decrease in voltage from the low voltage power supply can be a source of the problem. Another possible cause could be the video detector. The video detector is the diode CRA55 at the output of the i-f amplifier shown in Fig. 8-5.

No Picture with Normal Sound

Should the TV receiver have no picture but a raster, or bright picture, is present, then the defect exists in the stages before the sound pickoff. However, it's possible that the video amplifier circuit is at fault. You should refer to Fig. 8-4 and note the stages. If there is no snow on the screen, the trouble is most likely in the video detector or the i-f amplifier stage. However, if the picture is snowy, the rf amplifier in the tuner or antenna/cable system is probably defective. See Fig. 8-6 for an illustration of a snowy screen.

To determine whether the tuner or the antenna/cable system is at fault, simply substitute a good TV in its place. If the snow disappears, the tuner is the problem. If the snow remains, the problem is in the antenna or transmission line. Often a defective rf amplifier causes the picture to be snowy. Also, many tuners using silver contacts have a tendency to tarnish and become dirty. If rocking the tuner channel switch causes the picture to shake, the tuner contacts should be cleaned using a commercial tuner wash. Take off the tuner cover and spray the tuner wash on the contacts of the tuner as you rotate the tuner selector five times in one direction, and then five or more times in the other direction.

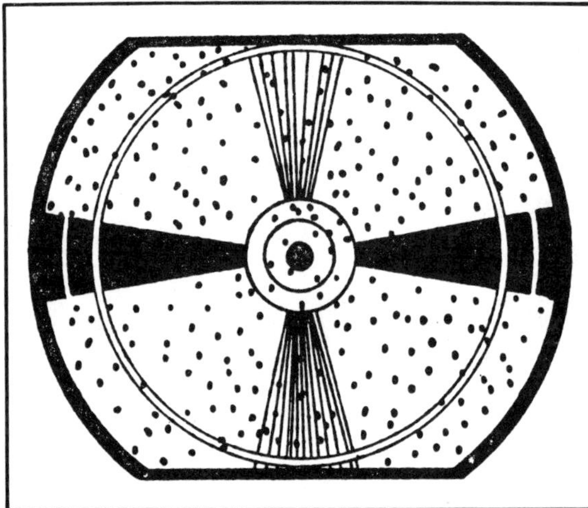

Fig. 8-6. Snowy, weak television reception (courtesy Howard W. Sams & Co., Inc.).

Sound Normal but No Raster

If your receiver lacks a raster, then a fault could exist in the high voltage power supply. There could also be a problem in the horizontal deflection stage and specifically, the flyback transformer or damper. Check the high voltage with a high voltage dc probe to determine if there is high voltage at the anode of the CRT. Be careful when checking this voltage because dangerous arcs can be drawn. If the voltages match the manufacturer's specification, then the picture tube is probably at fault. However, if there is a lack of dc voltage, check to see if there is ac voltage from the flyback transformer. This can be done on some TV sets by drawing an arc with a nut driver or screwdriver. Keep in mind that a blue arc indicates ac voltage; dc voltages produce a white arc. If an ac arc can be drawn at the flyback transformer, the high voltage rectifier is defective. Lack of an ac arc could indicate a faulty flyback transformer or horizontal circuit.

Another possibility for the lack of a raster is a faulty tuner or i-f stage which is sending only black signals; therefore, the screen is dark.

The picture tube, like any other kind of tube, works on the principle of *thermionic emission.* Weak emission causes the picture tube to become out of form and develop a silvery tint. An easy way to identify a weak picture tube is to turn the brightness control up. If the picture becomes silvery and out of focus, as opposed to when it is turned down, then the picture tube can be considered bad or going bad. If the high voltage is absent, the trouble is in the high voltage rectifier, damper, horizontal output, or the horizontal oscillator. Figure 8-7 shows a typical schematic of a horizontal system. Notice the horizontal oscillator, horizontal output, flyback transformer and the damper diode. Remember! Before checking the high-voltage rectifier or flyback transformer, *the picture tube must be discharged.* Using an alligator clip lead, clip one end to the chassis and the other end to the shank of a screwdriver. Touch the anode of the picture tube with the screwdriver. The anode is the conductor under the rubber-like grommet. Isolate the rectifier or flyback transformer and check for open circuits. Use the ac/dc arc test to identify if the rectifier is bad, or simply use a substitute diode. Check the flyback transformer by taking

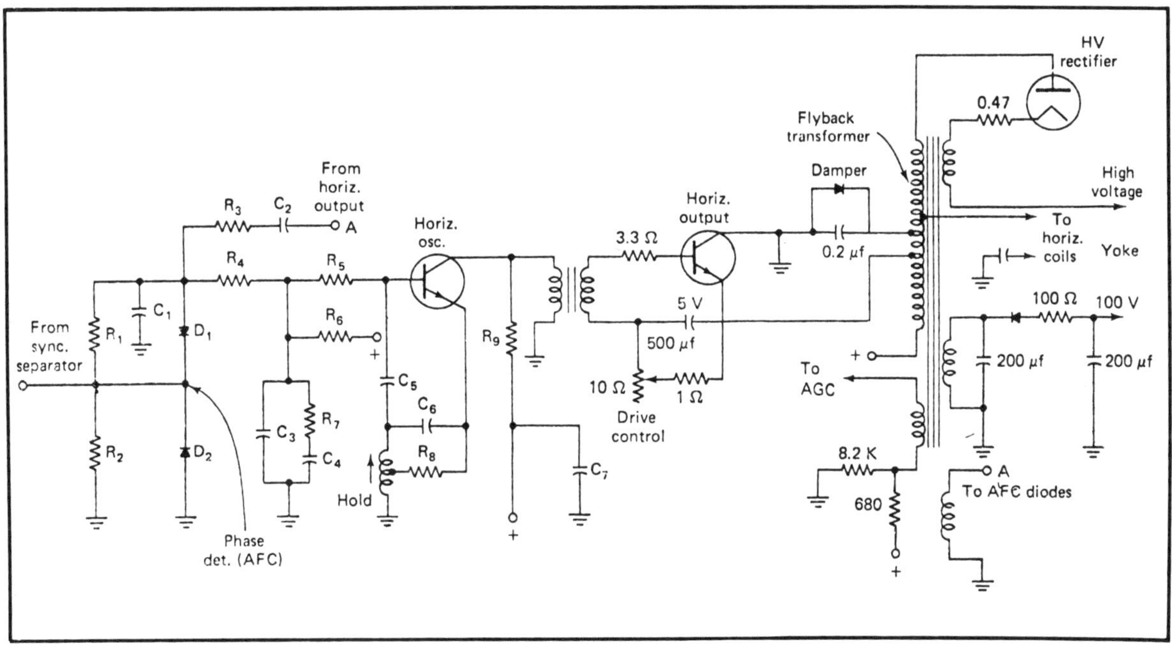

Fig. 8-7. Typical horizontal system.

a resistance test with an ohmmeter. Keep in mind that the horizontal oscillator must be running to get high voltage; therefore, check the horizontal oscillator output.

Sound Normal, Picture Out of Sync

Should the sound be normal but there is a tearing or heavy slanting streaks across the screen, then the problem is that the horizontal deflection is out of synchronization. Figure 8-8 illustrates the horizontal synchronization problem. You should check the horizontal deflection control to verify that it is set properly. If it is, then the fault exists in the horizontal oscillator or output stage. Should the horizontal oscillator stop oscillating, there would be no high voltage sent to drive the deflection coil in the yoke by means of the flyback transformer.

Sound Normal but Picture Tearing, Reduced Width

If your picture has tearing or heavy slanting streaks across the screen as well as vertical roll, first check for proper setting of the horizontal and vertical controls. Should these be properly set, then probable causes are defects in the sync separator or sync amplifier stages. There is a possibility that defects can exist in both horizontal and vertical deflection systems. Figure 8-9 illustrates a typical vertical deflection system. It consists of a vertical oscillator, a vertical driver and a vertical output stage which is tied to the yoke.

Sound Normal but Picture Rolls, Folds, Reduced Height

Other vertical system problems are shown in Fig. 8-10. If the picture rolls vertically a probable cause is a

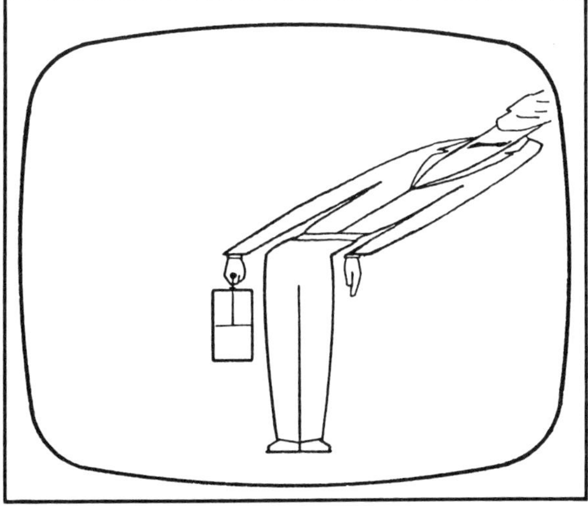

Fig. 8-8. Horizontal tearing.

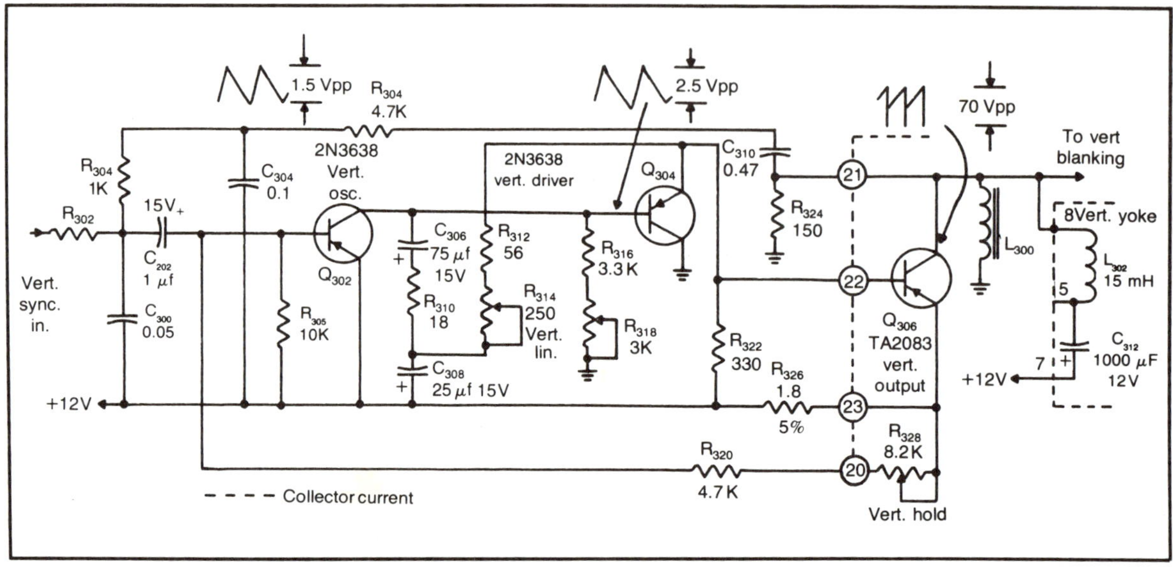

Fig. 8-9. Typical vertical deflection system.

defect in the vertical oscillator. Should the oscillator stop oscillating, there would be no vertical deflection and all that would be seen on the screen is a bright horizontal line. If you experience reduced height in the picture, the problem is a weak vertical output. Referring to Fig. 8-9, several likely causes could be a shift in the oscillator or output bias, low dc supply voltage, or a shorted or open

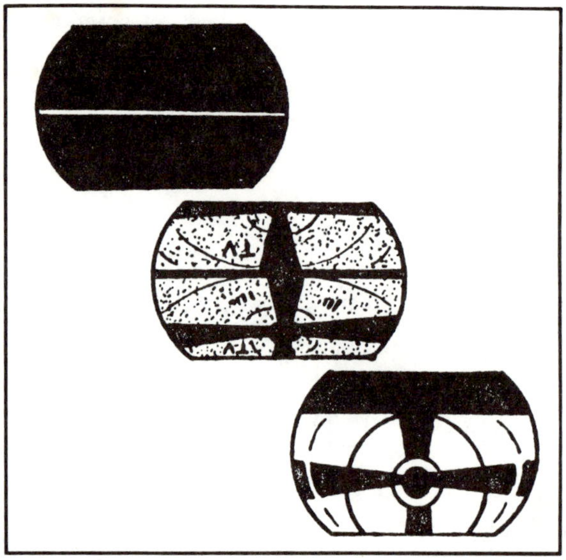

Fig. 8-10. Other vertical system problems (courtesy Howard W. Sams & Co., Inc.).

component. It probably isn't a shorted oscillator or output transistor if there is a partial picture.

However, if there was only a single horizontal line, then a shorted oscillator or output transistor would be suspected. In this case, the likely cause is a defective component such as C_{306}, C_{308} or R_{322}. For example, if C_{306} is shorted, then the sawtooth-waveforming process is interrupted and the bias of Q1 is shifted, which reduces the amplification and oscillation.

If C_{306} is open, the picture is likely to "fold over" with a white bar and compressed lines at the bottom of the screen. One way to check this component is to bridge capacitor C_{306} with a good one, or use a capacitor substitution box with the set on. If the picture height returns to normal, your problem has been identified. Incidentally, by checking the collector voltage of Q1 in this circuit, it would have been low because capacitor C4 was open and not placing appropriate charge on the collector.

Should the screen have a reduced height as well as trapezoidal distortion, the likely cause is a defective yoke, or trouble in the pincushion connection circuit in a color receiver. The problem is not in the vertical oscillator or output.

Often, when the service technician needs to determine whether vertical roll or horizontal tear is caused by the oscillator or sync stage, he/she can perform this simple test. If the picture can be made to hold when the oscillator control is turned, but fails to stay held, the problem is in the sync circuit. A defective diode in the horizontal AFC often causes the problem of horizontal tearing. See Fig. 8-7 for AFC diodes in the horizontal system. Looking at Fig. 8-7, if C5 is open, then the horizontal voltage is decreased which reduces the picture width. If the horizontal oscillator transistor is shorted or the 3.3 K resistor is shorted, then there will be no horizontal voltage.

Lack of vertical deflection could be caused by a bad oscillator or output transistor. An open emitter bypass capacitor or resistor could also cause insufficient gain that would decrease the height in the picture. Keep in mind, trapezoidal distortion is usually caused by a defective deflection yoke and not the vertical circuit. A "slanted picture" can be corrected by simply loosening the deflection yoke and turning it in the correct direction as shown in Fig. 8-11. Remember, never overtighten the deflection yoke, since the picture tube neck can be easily broken.

Picture Normal and Poor Sound

Should the picture be normal but the sound is missing, check the i-f amplifier, FM detector or af amplifier stages. There could also be a defective speaker coil. A weak sound suggests improperly adjusted fine tuning control or a shift in the local oscillator due to component values changing. Since multiple stages are involved, use the split-half troubleshooting technique to quickly identify the defective stage. Check for a defective IC, transistor, tube or component in the sound section. A change in a component value in an audio stage can affect the gain of an amplifier. Sound distortion can be caused by a defective interstage coupling capacitor.

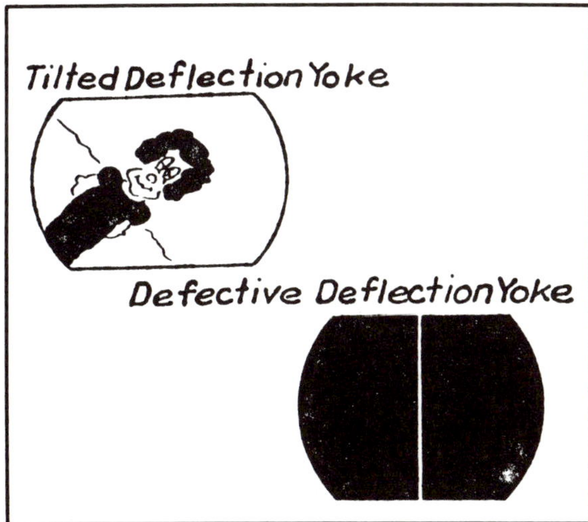

Fig. 8-11. Examples of defective and tilted deflection yoke (courtesy Howard W. Sams & Co., Inc.).

Dead Set

Like the radio, if a TV set is dead you should check the power supply. A list of possible causes of a dead TV set are as follows:

1. Fuse/circuit breaker.
2. Open heater tube, transistor or IC.
3. Open line cord.
4. Defective on/off switch.
5. Defective power supply transformer.
6. Defective diode or rectifier.
7. Defective thermistor.
8. Defective filament choke.

COLOR FUNDAMENTALS

At the TV station the scene to be televised is actually scanned by three separate cameras with each camera being sensitive to only one of the three primary colors of red, blue and green. In addition to these primary colors, complementary colors like yellow, orange, cyan, and magenta can be produced. Various combinations of these primary colors will produce any color to which the human eye is sensitive. (The hue or tint is the distinctive color itself. Saturated colors are vivid, strong colors. Lack of saturation in a color would result in a pale weak color. *Chrominance* refers to the combination of the hue and saturation. *Luminance* refers to the amount of brightness perceived.)

The three cameras scan the screen in unison. The primary colors of red, blue and green are fed into a matrix at the transmitter which creates a Y, or luminance, signal and chrominance, or color signals I and Q. The Y signal contains the proper proportions of red, blue, and green content so it can recreate a normal black and white signal. This signal is used to modulate the carrier. The I and Q chrominance signals are used to modulate a 3.58 MHz color subcarrier which is suppressed by the modulation process. The composite signal has a carrier, the Y, or luminance, and the I and Q, or chrominance signals along with the FM audio.

At the black-and-white, or monochrome, receiver only the Y signal is detected and processed. The chrominance signals, I and Q, cannot be detected and processed because the receiver lacks the 3.58 MHz oscillator needed to recover the I and Q signals. Thus a color receiver requires a 3.58 MHz oscillator to enable the detection of the I and Q signals.

A block diagram of the color section of a television receiver is illustrated in Fig. 8-12. The color signal comes from the video amplifier into the chroma amplifier where the signal is amplified. Notice that after the video amplification the Y signal is immediately available with the exception of a delay of 1 microsecond so that the Y signal and the I and Q signals arrive at the CRT at the same time. It requires about 1 microsecond for the I and Q signals to undergo additional signal processing.

After the chroma signal is amplified, it is sent into a bandpass amplifier of 2 to 4.2 MHz which separates the I and Q content from the Y content and is then sent to the I detector and the Q detector. These detectors have input from the 3.58 MHz crystal oscillator which aids the detection. Notice that the 3.58 MHz oscillator is phase-shifted 90° and then sent to the Q detector. The phase shift is necessary because it was phase-shifted 90° at the transmitter to separate the various signals.

Once the I and Q signals are detected, they are sent to their respective low pass filters and then processed by a phase inversion for both positive (+) and negative (−) chroma signals. The positive and negative chroma signals are necessary because:

$$\begin{aligned} \text{green} &= -I-Q+Y \\ \text{blue} &= -I+Q+Y \\ \text{red} &= +I+Q+Y \end{aligned}$$

The I, Q and Y signals are summed in the three-color adder circuits with resistor values providing the proper proportion of each signal. Each color signal is then sent to the appropriate CRT grid to control the beam's intensity. There is a rheostat at each adder circuit to allow for the intensity of each color to be varied proportionately to the other colors.

The subcarrier crystal oscillator is not precise

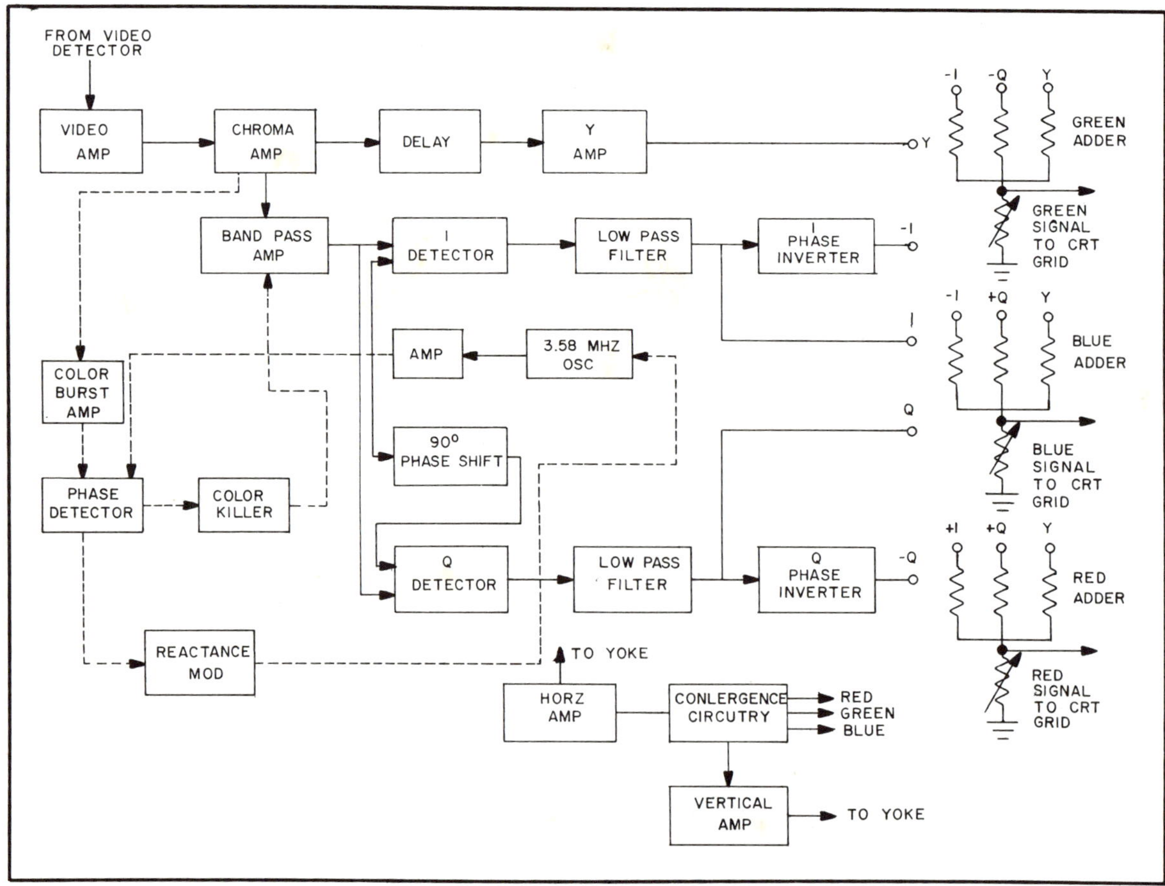

Fig. 8-12. Block diagram of a TV color section.

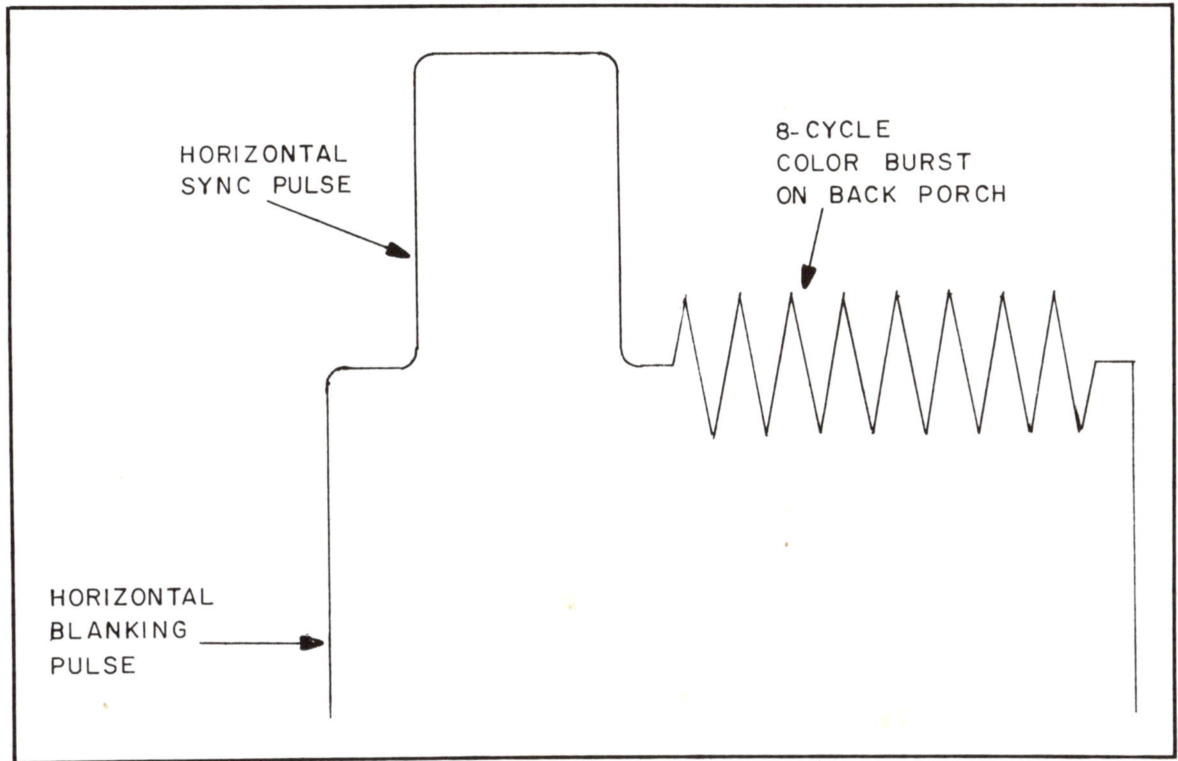

Fig. 8-13. Horizontal blanking pulse with back porch.

enough to permit proper detection of the chroma signals. Therefore at the transmission of the color signal, a sample of it is placed on the so-called back porch of the horizontal blanking pulse, as shown in Fig. 8-13. The color burst amplifier receives a portion and sends it to the phase detector. The phase detector compares the 3.58 MHz signal to the color burst and if the two are not equal, a dc signal is sent to the reactance modulator which "pulls" the two signals into precise synchronization.

The phase detector also sends a signal to the color killer. The purpose of the color killer is to prevent any signals from the chroma circuits during a monochrome broadcast. As long as there is a color burst present, then the color killer is off. However, when the color burst is absent, as it is during monochrome broadcasting, the phase detector sends out a dc signal which "kills" the bandpass amplifier.

COLOR TV TROUBLESHOOTING

Correct color reproduction requires undistorted chrominance and luminance signals. Any defects that alter a balck-and-white picture will have an effect on the color picture. A general localization of color troubles can be done in terms of the raster and monochrome picture quality. A good raster indicates normal dc voltages, and a good monochrome picture indicates normal Y signal and correct dc voltages for the picture tube.

Absence of Color

If there is an absence of color, look for a defective IC, transistor, tube or component in the I and Q signal processing stages. The color killer control may be improperly set or defective. Also the bandpass amplifier stage may be inoperative. Check for the presence of the 3.58 MHz at the oscillator and subcarrier, as well as the color burst signal.

Weak or Faded Color

Weak or faded color can result from an improperly adjusted bias control on the picture tube or improperly adjusted screen and drive controls. A weak IC, transistor or tube in the I and Q signal processing stages or defects in the bandpass amplifier may also cause the problem. The tuner and i-f stages may be slightly out of alignment and could contribute to the problem. Check for the presence of the 3.58 signal at the oscillator and subcarrier amplifier, as well as the color burst signal.

Screen Dominant Color

Should the receiver have a dominant blue color, then the probable cause is an improperly adjusted green and red drive or misadjusted screen controls. However, if the receiver has a dominant red color, then check for improperly adjusted blue and green drives. There is a possibility that the picture tube could be defective.

If some colors are more vivid than others, generally the problem is a misadjusted screen or color drive control.

Color Killer

The color killer cuts off the chroma amplifier during a black and white transmission. A defective color killer results in color noise, called confetti, which looks like snow but with larger color spots. If confetti is present during a monochrome reception, then check the adjustment or circuit of the color killer.

Color Bars

Another typical problem in color sets is the presence of color bars in the picture. Usually, the reactance transistor, the automatic frequency phase control or a defective color burst is the cause of the problem. Many of these circuits are contained in an integrated circuit called a chroma processor; therefore, the IC must be replaced.

Other Problems

A picture which is tinted black and white suggests that the purity is out of adjustment or the picture tube needs degaussing. Many late model TVs have automatic degaussing. Should the flesh tones vary with image position on the screen check for incorrect purity adjustments.

If the colors are not properly registered, check the convergence for proper adjustment.

Poor focus suggests a defective focus rectifier or focus circuit. Check the rectifier first, then the other circuit components.

If the colors are smeared, check for loss of the Y signal or a fault in the video amplifier system.

Total Convergence Setup

The need for making convergence setups is declining, since most modern sets have a pre-set convergence circuit. Although, when a new picture tube is installed in a television set, especially old sets, a total convergence setup is required. The standard procedure can be found in the television service manual. Generally, a total convergence setup consists of first adjusting the proper picture size, focus, and linearity. The brightness should also be adjusted to the typical viewing level. A dot/bar/crosshatch generator is generally used in the setup procedure.

Purity Adjustment

Bias the blue and green guns to cutoff and slide the deflection yoke forward. Adjust the purity magnets until the red focuses exactly in the middle of the picture tube. Now, push the deflection yoke backward until the raster is completely red as shown in Fig. 8-14.

Static Convergence

Turn on the green gun. Adjust the red and green static convergence magnets to merge the two colors in

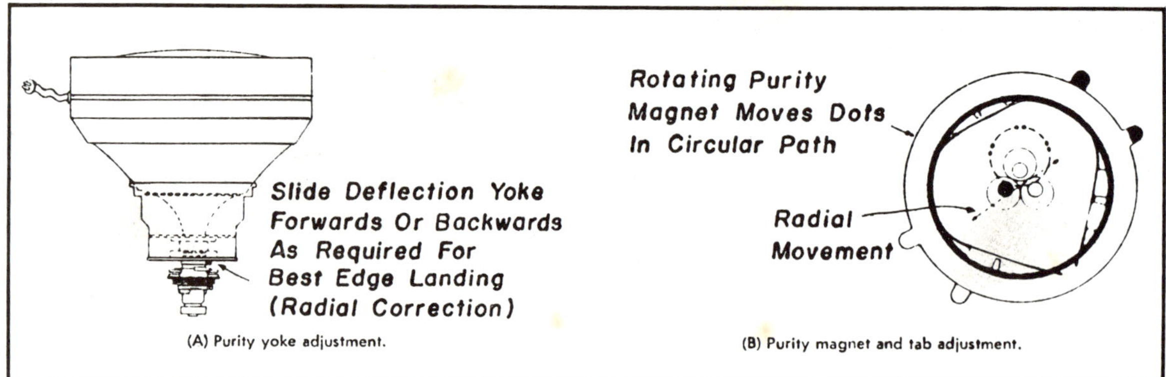

Fig. 8-14. Deflection yoke and purity rings of a color receiver (courtesy RCA Company).

the center of the picture tube until you get one yellow dot. Now, turn on the blue gun and merge all three colors in the center of the picture tube until the color white is formed. The static convergence assembly is shown in Fig. 8-15.

Dynamic Convergence

Adjust each dynamic control to converge the top, bottom, and sides of the picture tube as shown in Fig. 8-16.

Before completing the total convergence, it is important to adjust the picture for the best gray tracking. Gray tracking ensures the best possible black and white balance. While the TV is on a monochrome station, adjust the red, blue, and green drive controls until the best overall gray raster is obtained.

The last step is adjusting the screen controls. Set the service switch in the service position. This will display a horizontal line across the center of the screen. Now, turn all three color screen controls counterclockwise and then slowly adjust each one until the color is just visible.

If the convergence does not seem to hold during the operation of the TV, a defective convergence rectifier is probably the reason.

In troubleshooting color circuits, an oscilloscope should be used to trace the color signal through the progressive stages. Once a stage has been localized, voltage and resistance checks can be taken to determine the faulty component or components.

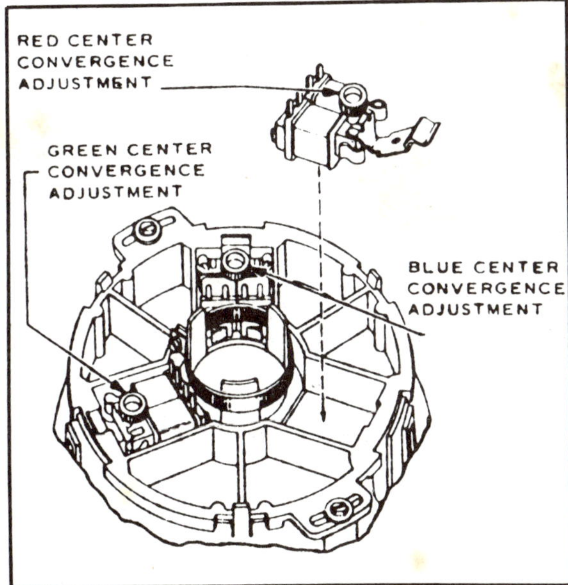

Fig. 8-15. Static convergence assembly of a color receiver (courtesy RCA Company).

Figure 8-17 illustrates the controls of a typical color television. The adjustment of these controls are as follows:

1. The color drives are adjusted to obtain the best black-and-white gray raster on a monochrome station.
2. The color screens are adjusted with the service switch in the service position until each color is barely visible.
3. The focus is adjusted for sharpest picture.
4. The pin phase and pin amplifier controls do not often need adjustment. Adjust the controls to obtain straight, centered horizontal lines.
5. The vertical linearity control is adjusted when a person's face appears to be too long or too short. Figure 8-18 illustrates excessive linearity. Adjust the control to obtain correct vertical shapes and figures.
6. The vertical height control should be adjusted if the picture appears shrunken as shown in Fig. 8-19.
7. Adjust the AGC when the picture appears to be overloaded or weak. Adjust the control to obtain the best overall picture on all channels.
8. The color killer should be adjusted whenever there is a lack of color or color interference on a monochrome station. Adjust the color killer control to the position where the color disappears from the picture during a monochrome broadcast.
9. Adjust the brightness limiter whenever too much brightness or lack of brightness is a problem. Adjust the brightness limiter control until the raster just begins to bloom, then adjust for the desired brightness using the brightness control.
10. The hi-low switch should not be adjusted unless a voltage higher than 120 volts ac is being used. Usually if the line voltage is between 125 and 130, the hi position should be used.
11. The rf bias control is used to obtain the best overall snow-free picture.
12. The tint control is used to adjust the desired level of flesh tones.

PREVENTIVE MAINTENANCE

Television sets, like any other type of receiver, should be operated with care. Never abuse the channel

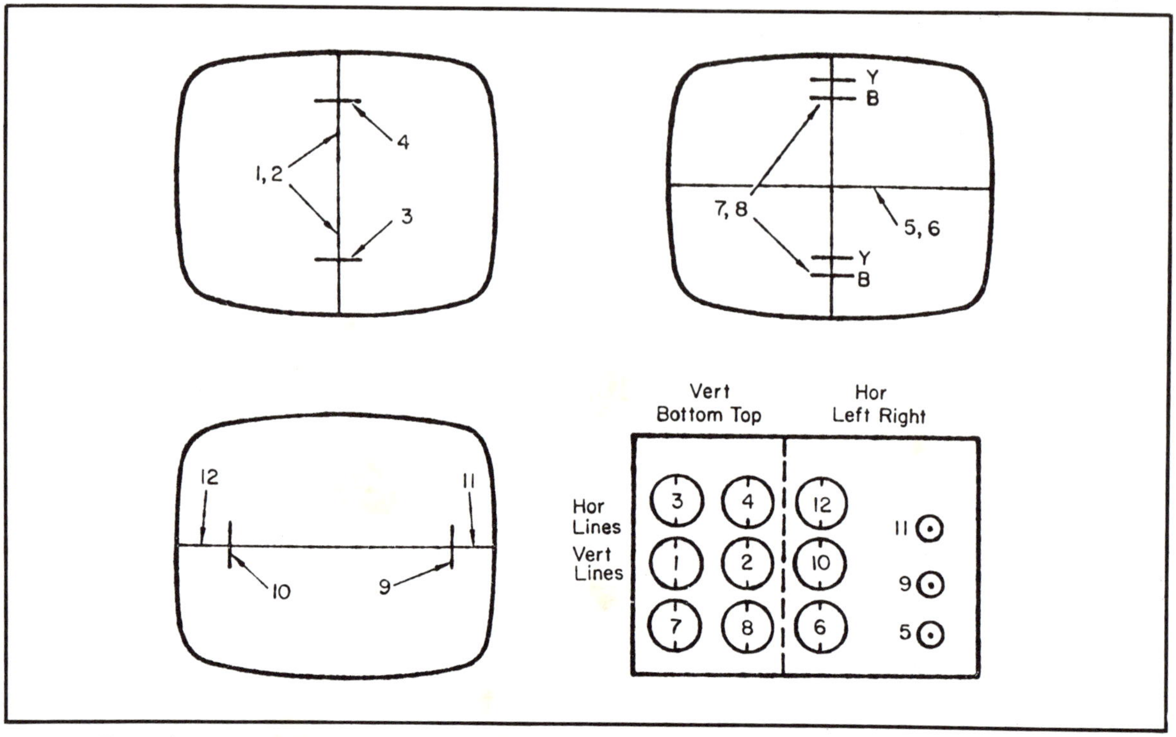

Fig. 8-16. Dynamic convergence controls and positions they control on a color receiver (courtesy RCA Company).

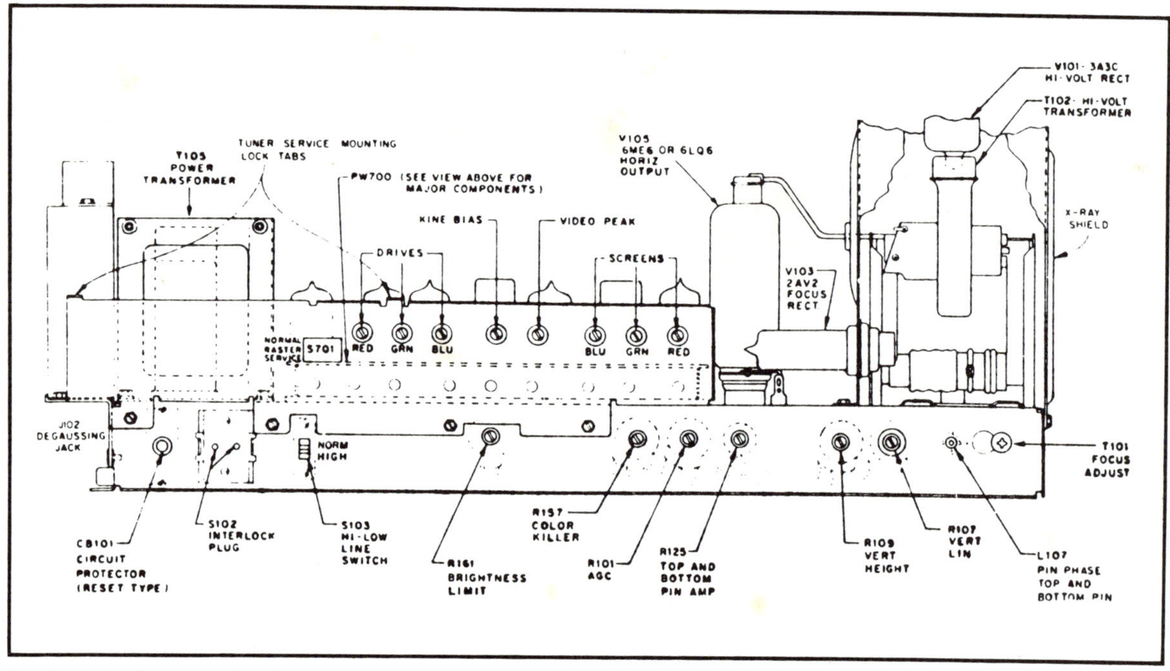

Fig. 8-17. Typical controls of a color receiver (courtesy RCA Company).

Fig. 8-18. Example of excessive linearity (courtesy Howard W. Sams & Co., Inc.).

Fig. 8-19. Example of insufficient height (courtesy Howard W. Sams & Co., Inc.).

selector or other controls. Maintain proper adjustment of controls and replace any broken parts.

Periodic cleaning of the tuner contacts and receiver can often prevent a breakdown. A dirty circuit can increase heat in the circuit, eventually wearing down the components. Also, a dirty, dusty CRT can cause high voltage arcing and cause a lack of brightness. Every chassis should be cleaned using an air gun or vacuum cleaner.

Never allow the TV to become overheated or be in contact with moisture. Both heat and moisture can destroy circuit components. Also, heavy moisture in the air can cause high voltage arc.

Never replace a polarized plug with a non-polarized plug. Severe shocks could be experienced from doing this.

Any time a circuit is being serviced with test equipment, always use an isolation transformer to prevent sparks and shocks.

SELF-EXAMINATION

Select the best answer:

1. One set of 262½ scanning lines represents:
   A. Field.
   B. Frame.
   C. Cycle.
   D. Interlace set.
2. A total of 525 scanning lines represents a:
   A. Field.
   B. Frame.
   C. Cycle.
   D. Interlace set.
3. The number of frames per second is:
   A. 20.
   B. 30.
   C. 60.
   D. 120.
4. The vertical oscillator runs at a frequency of:
   A. 20 Hz.
   B. 30 Hz.
   C. 60 Hz.
   D. 120 Hz.
5. The horizontal oscillator runs at a frequency of:
   A. 60 Hz.
   B. 15,750 Hz.
   C. 3.58 MHz.
   D. 45.75 MHz.
6. The sound and video signals are separated at the:
   A. i-f stage.
   B. Video detector.
   C. Video amp.
   D. Burst separator.
7. The vertical and horizontal pulses are separated at the:
   A. AFC.
   B. High voltage.
   C. Sync separator.
   D. AGC.

8. If both sound and picture are weak and distorted the problem is most likely in the:
   A. AFC.
   B. Tuner.
   C. Sound.
   D. Video.
9. Lack of a raster often indicates:
   A. No television signal.
   B. No video signal.
   C. No AGC.
   D. No high voltage.
10. A signal horizontal line across the middle of the screen indicates trouble in the:
    A. Tuner section.
    B. Vertical section.
    C. Horizontal section.
    D. Video section.
11. Slow rising white "hum bars" in the television are caused by a bad:
    A. Rectifier.
    B. Picture tube.
    C. High voltage transformer.
    D. Filter.
12. A silvery out of focus picture usually indicates a bad:
    A. Rectifier.
    B. Picture tube.
    C. High voltage transformer.
    D. Filter.
13. An overloaded picture can often be corrected by adusting the:
    A. Vertical oscillator.
    B. Horizontal oscillator.
    C. AGC.
    D. Color killer.
14. Vivid, strong colors are often referred to as:
    A. Hue.
    B. Luminance.
    C. Saturation.
    D. Chrominance.
15. The amount of brightness perceived is referred to as:
    A. Hue.
    B. Luminance.
    C. Saturation.
    D. Chrominance.
16. The alignment of all three color guns to a common point is referred to as:
    A. Demodulation.
    B. Confetti.
    C. Blooming.
    D. Convergence.
17. Color confetti can often be eliminated by adjusting the:
    A. Color chroma amplifier.
    B. Color detector.
    C. Color killer.
    D. Color oscillator.
18. The presence of color bars often indicates trouble in the ________ circuit.
    A. Reactor.
    B. Horizontal.
    C. Burst separator.
    D. Chroma amplifier.
19. Before a convergence setup is performed one should first perform:
    A. Gray tracking.
    B. Degaussing.
    C. Screen setting.
    D. Alignment.
20. When servicing TV circuits with test equipment use a/an:
    A. Nonpolarized plug.
    B. Noise generator.
    C. Isolation transformer.
    D. Degaussing coil.

QUESTIONS AND PROBLEMS

1. Draw a block diagram of a black and white television from memory.
2. Explain the basic function of each circuit in the black and white television.
3. Explain common problems associated with a power supply in a television receiver.
4. Explain what happens to the overall picture of a CRT when its emission becomes weak.
5. How can the picture be improved when the tube has weak emission?
6. How can a service technician tell if vertical roll is being caused by the vertical oscillator or the vertical sync?
7. What stage of the television receiver is probably at fault when the picture has trapezoidal distortion?
8. What is confetti?

9. What are the different colors between ac and dc high voltage?
10. Explain how the service technician can tell whether the snow is being caused by the tuner or the antenna?
11. What do the terms hue, saturation, chrominance, and luminance mean?
12. From memory, draw a block diagram of the color section of a television.
13. Explain the purpose of each section of a color television.
14. If color bars are present in the picture, what section should be investigated?
15. What section would most likely be causing a problem of weak color?
16. Explain the basic overall convergence of a television.
17. What is the general procedure in cleaning tuner contacts?
18. Why is an isolation transformer used in servicing television receivers?
19. Why is dirt harmful to a TV?
20. What might be the symptom if a convergence rectifier shorted?

Chapter 9

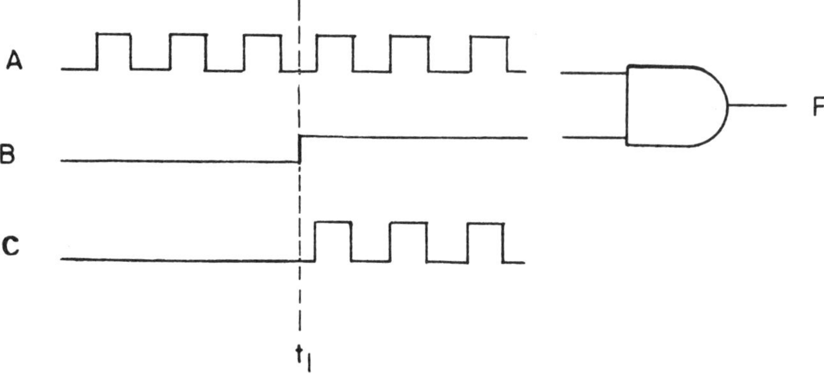

Troubleshooting Digital Circuits

The growth of digital equipment from basic measurement instruments to the computer has been explosive. An understanding of the binary concept, digital logic and the electronics of digital gates is necessary before you begin to troubleshoot digital equipment. In this chapter we will examine combination digital electronics

BINARY CONCEPT

The binary concept is easy to grasp. Just as a ceiling light has two states, either ON or OFF, the binary system has two states represented by a logical "1" or a logical "0". The logical level represents the two distinct dc voltage levels in a digital circuit. If a voltage is present on a wire, it is said to be high or H; and if a voltage is absent, it is said to be LOW or L. Figure 9-1 illustrates this concept. If a switch is open, the dc voltage at point A is zero or L. However, when a switch is closed, the dc voltage from the battery is present at point A and is said to be HIGH or H. If a light emitting diode (LED) with a limiting resistor is connected to the output of the circuits, the LED would be off when a LOW is present and on when a HIGH is present.

The above is true when dealing with a positive logic system. There is a logic system using negative logic and in this system the voltage level for HIGH and LOW are reversed. We will be dealing with the most common logic system—positive logic. In addition to the kind of logic, the type of digital family will dictate the dc level and polarity of a logical 1 and logical 0. We will examine various families and voltages later in this chapter.

LOGICAL DECISIONS

Decisions are often based on a number of YES or NO conditions or two-state conditions. For example, if the driver of a car sees a stop sign, red light, or obstacle in his/her path, he/she stops. The stop conditions and response can be represented by the symbology in Fig. 9-2. The condition to which the driver responds is either present or not present. These are two-state conditions in which there is or is not a stop sign; there is or is not a red light; or there is or is not an obstacle which forces the driver to make the response of stopping. If any one or all of the conditions are met, the response occurs. All of the possible combinations of conditions that can occur and responses to the conditions are shown in Table 9-1.

The table is called a truth table, and it is used to explain how a decision or a group of decisions work. The left-hand side of the table lists all the possible combina-

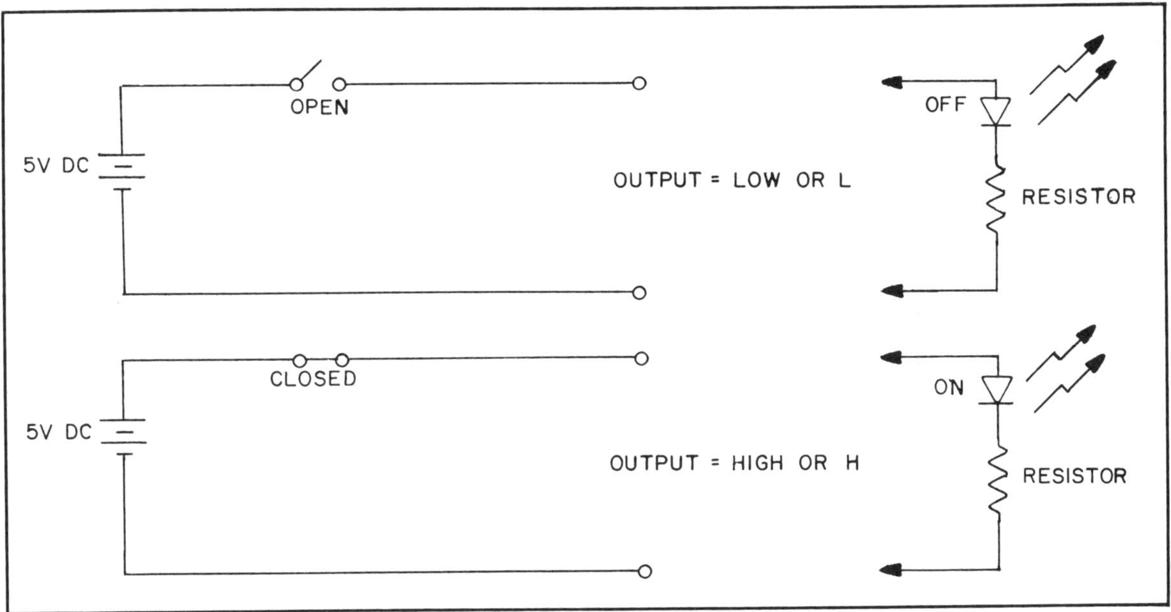

Fig. 9-1. An example of voltage and logic levels for digital electronics.

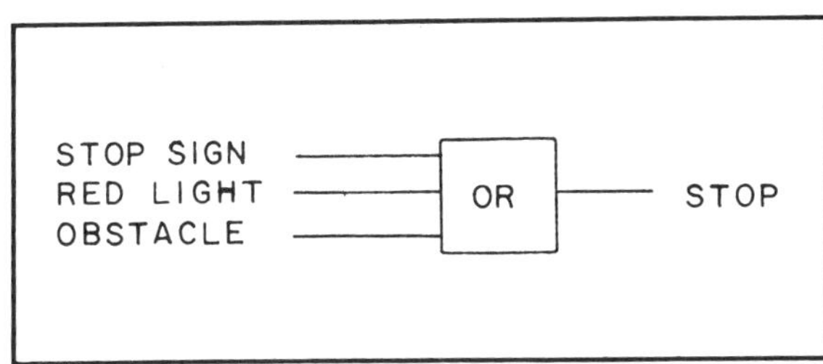

Fig. 9-2. Logical stop decision.

Table 9-1. A Truth Table Showing All the Input Conditions and Output Responses for a Logical Decision.

Conditions			
Stop Sign	**Red Light**	**Obstacle**	**Response (stop)**
No	No	No	No
No	No	Yes	Yes
No	Yes	No	Yes
No	Yes	Yes	Yes
Yes	No	No	Yes
Yes	No	Yes	Yes
Yes	Yes	No	Yes
Yes	Yes	Yes	Yes

tions of input decisions. The right-hand lists all the output responses for each of the input condition combinations.

Many situations can be represented in symbolic form. For instance, a mechanical situation, such as a door with two locks, is represented in Fig. 9-3. If either or both locks are bolted, the door is secured. All possibilities of input conditions that exist for the lock bolt and the output, a secured door, is represented in the truth table.

DIGITAL GATES AND LOGIC

Logic decisions can be made by people and many types of mechanical, hydraulic and electrical devices.

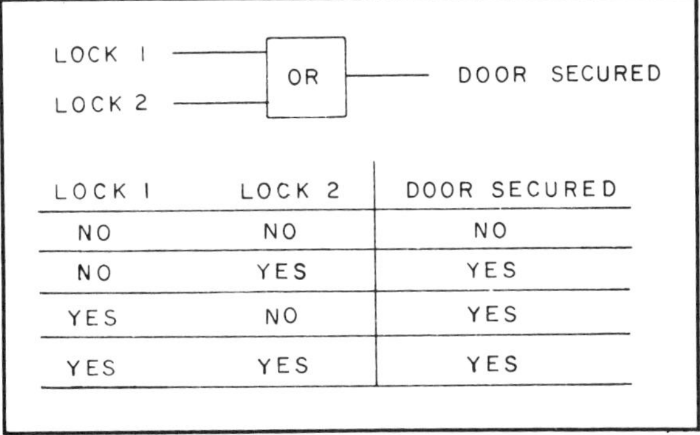

LOCK 1	LOCK 2	DOOR SECURED
NO	NO	NO
NO	YES	YES
YES	NO	YES
YES	YES	YES

Fig. 9-3. Secure door logic and truth table.

Regardless of the device, the decision can be represented in standard symbolic form. We will now examine the acceptable symbolic standards and truth tables for the basic digital gates.

The OR Gate

The OR logical decision has been illustrated in the previous examples. The standard digital symbol for an OR gate, truth table and switch analogy is shown in Fig. 9-4. The input conditions to the OR symbol are on the left and the results of the decision, or output, are on the right. The A and B input can represent any logical conditions, while the F represents the logic function output resulting from the gates decision.

The OR gate can have two or more inputs but only one output. When the input conditions are met, the gate is said to be enabled; that is, when any one of the inputs are present or all are present.

Electrically the OR gate can be represented by two parallel switches connected in series with a battery and the output indicator as shown. One switch is labeled A, the other labeled B with the output labeled F. Remember, in digital electronics the voltage level is either HIGH (H) or LOW (L). For this example, H and L are used to represent them in the truth tables. Many semiconductor manufacturers also present the logical function of complex digital circuits using a truth table and the letters H and L.

An X in the truth table simply means that the particular input can have either a HIGH or a LOW. Now examine the truth table in Fig. 9-4 and the switch analogy of the OR gate. If both switches, A and B, are open, the voltage across the light emitting diode (LED) is zero. Thus the first input conditions to the truth table is LOW (L), and the output is also LOW (L). However, when switch A or B is closed, or both switches are closed, the output is HIGH (H) and the LED will be lit.

The AND Gate

Another kind of decision involves an AND function. This gate requires both inputs to be HIGH for the gate to be enabled. The AND gate symbol, truth table and switch analogy is shown in Fig. 9-5.

Electrically the AND gate can be represented by two

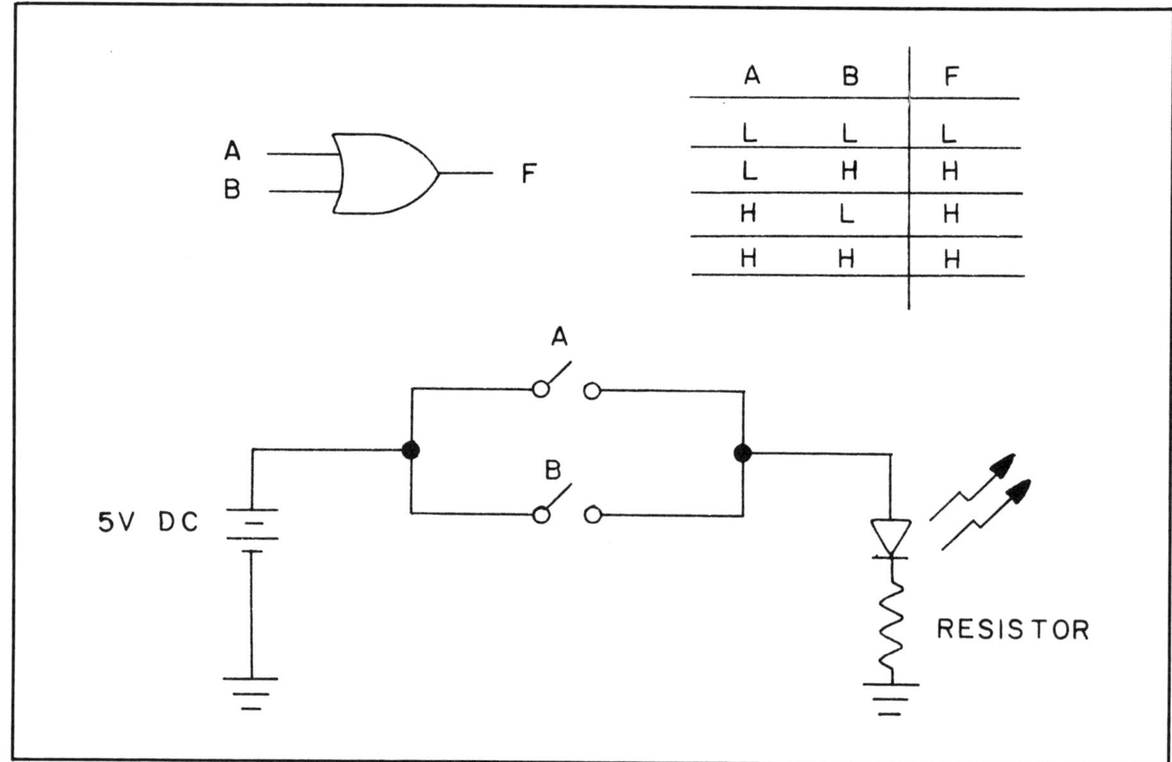

A	B	F
L	L	L
L	H	H
H	L	H
H	H	H

Fig. 9-4. OR symbol, truth table and switching analogy.

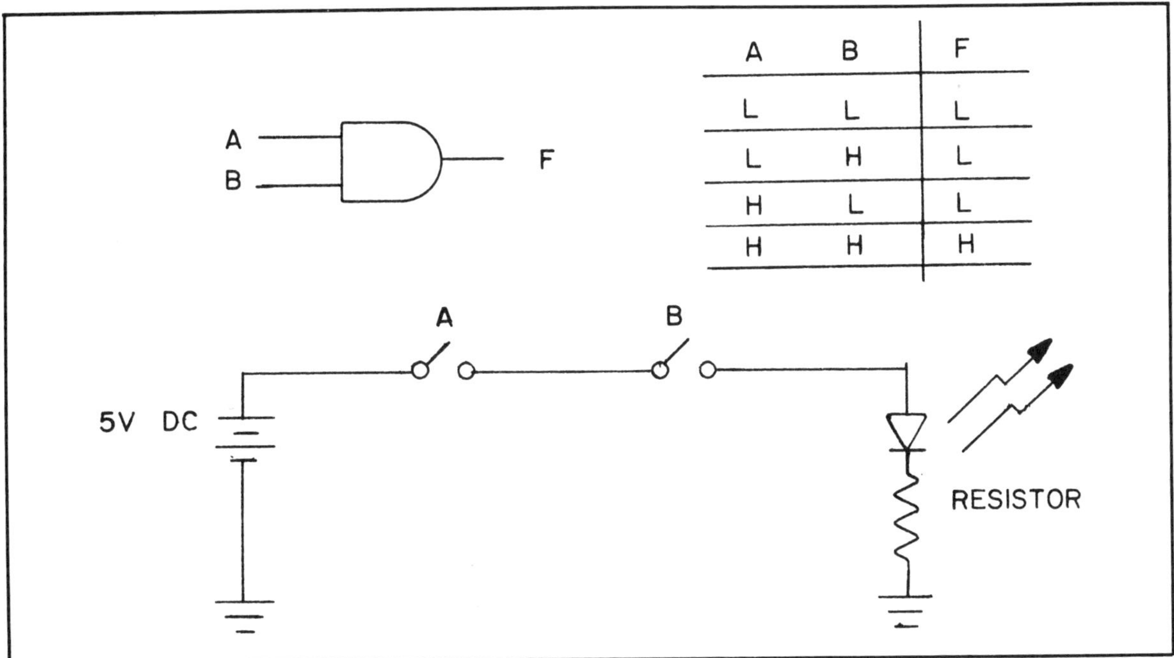

Fig. 9-5. AND symbol, truth table and switch analogy.

switches in series with the battery and the output indicator as shown. The switches are labeled A and B with the output labeled F. Examine the truth table and the switch analogy. In the first condition, if both switches are open, the voltage across the LED is zero: the LED is not lit. The same is true if either A or B switch is open while the other is closed. However, when both A and B switches are closed, the LED is lit. In other words, the only way to have a HIGH output is when both inputs have a HIGH voltage present.

The Inverter or NOT Gate

There are applications which require taking information and changing it into its opposite state. A gate

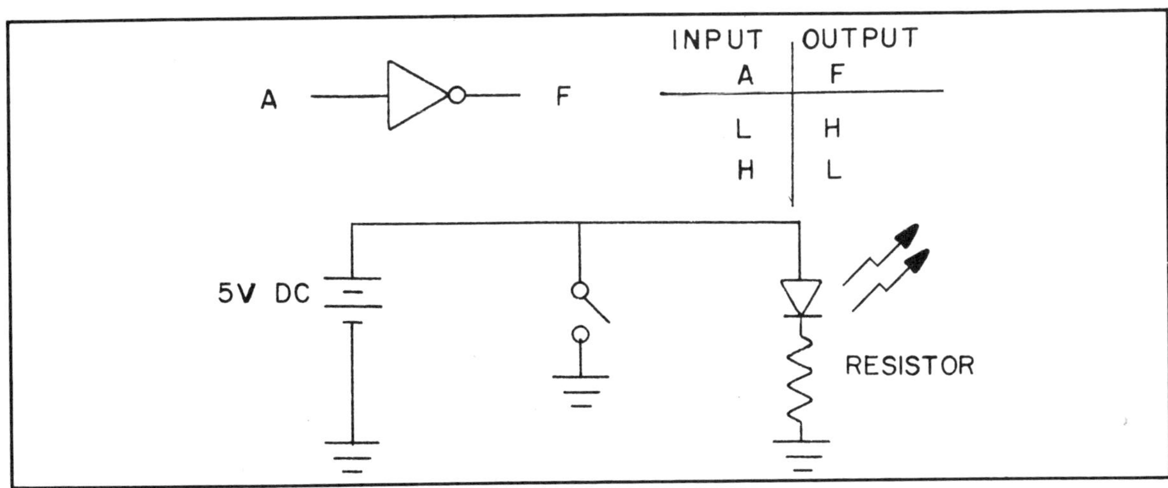

Fig. 9-6. Inverter symbol, truth table and switch analogy.

which is designed to do this is known as an inverter or a NOT gate. The symbol, truth table and a switch analogy is shown in Fig. 9-6.

Electrically the inverter can be represented by a switch connected in parallel between the battery and the output indicator. The switch is labeled A, and the output is F. Once again examine the truth table and switch analogy. In the first condition the switch is open representing a LOW signal and the output is HIGH. In the second condition the switch is closed representing a HIGH and the output is LOW.

Figure 9-7 illustrates a method for representing condition "A" and the inversion or complement of that condition "not A." Not A is represented by a bar above the A, like $\overline{A}$. Therefore if A is at the input of an inverter, its opposite, or $\overline{A}$, is at the output. And if $\overline{A}$ is at the input, then its opposite, A, is at the output.

Look at Fig. 9-7. There are two inverters, one with a small circle or bubble at the input while the other inverter has the bubble at the output. The bubble represents the appropriate input or output condition for enabling the gate. When the bubble is before the triangle, it says that the gate will be enabled HIGH with a LOW input condition. On the other hand, when the bubble is on the output, it indicates that the gate will be enabled LOW with a HIGH input condition.

The Gating Function

A gate is an electrical circuit which will pass or enable input signals to the output when certain input conditions are met. Examine Fig. 9-8. An AND gate will pass a signal if all input signals are logically HIGH when a control signal is applied. Notice that input A is actually a series of pulses commonly called a *pulse train*. In order to pass the pulse train to the output, we must allow input B of the gate to be at a logical HIGH. Notice that the first three pulses of the signal at input A do not pass through to the output. However, when input B goes high at time 1

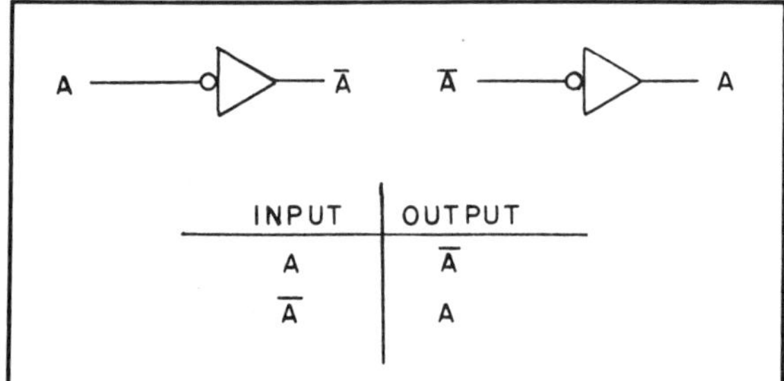

INPUT	OUTPUT
A	$\overline{A}$
$\overline{A}$	A

Fig. 9-7. Inverter symbol variations.

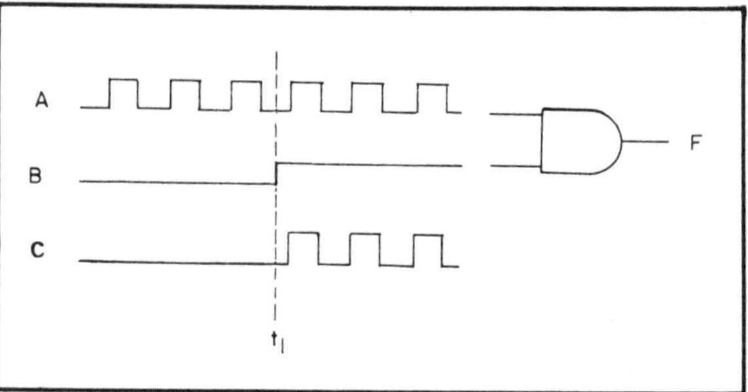

Fig. 9-8. AND gating action shows a pulse train on input A, the output F occurs only when input B is HIGH.

(t1), the input pulses at A pass through the gate. Thus, when the signal at input A is a pulse train and by controlling the "gating" input B, we can decide when and if to allow the signal to pass through the gate.

The NAND Gate

One of the most versatile gates is the NAND. It can be used to implement an OR, AND or NOT logical function. Figure 9-9 illustrates that a NAND gate consists of an AND function followed by an inverter or NOT function.

Basically the NAND gate operates by ANDing a number of input conditions together as previously described. The output of the AND is used as the input for the inverter, which complements the result of the AND logic. Figure 9-9 shows the standard symbol for the NAND gate as well as its truth table. The truth table states that the gate is enabled LOW only if both inputs are

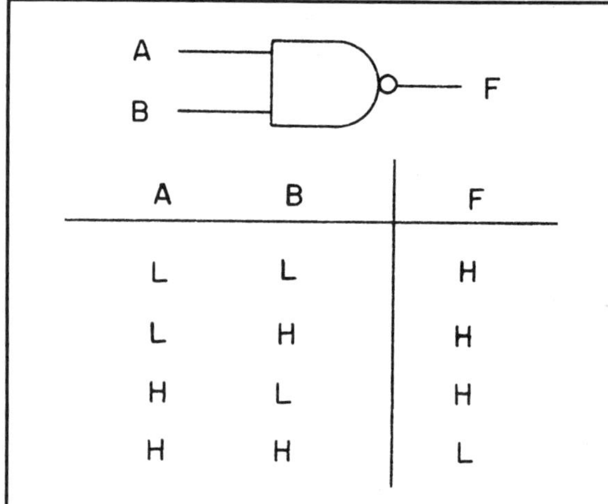

A	B	F
L	L	H
L	H	H
H	L	H
H	H	L

Fig. 9-9. NAND symbol and truth table.

HIGH. When any one of the inputs are LOW, the output is disabled and is indicated by a HIGH.

When analyzing any gate, observe the bubble. The bubble represents a LOW signal as the appropriate condition, whereas the absence of a bubble represents a high signal as the appropriate condition for enabling. Therefore, when examining the NAND gate and its truth table, notice there are no bubbles at the input. Thus, the appropriate inputs need to be HIGH. Look at the output. There is a bubble; thus the gate is enabled LOW and will pass the electrical signal when both input conditions are HIGH.

The NOR Gate

The logic of a NOR gate is essentially the opposite of an OR gate. The NOR gate is a combination of an OR gate followed by an inverter or NOT gate.

The gate operates by ORing a number of input conditions together. The output of the OR gate is used as the input to the inverter, which complements the result of the OR logic. Figure 9-10 shows the standard symbol for a NOR gate and its truth table.

Examining the truth table we see the NOR gate is enabled LOW whenever any or all inputs are HIGH. The NOR gate is disabled and the output is HIGH whenever all inputs are LOW.

The Exclusive-OR Gate

The exclusive-OR gate (EX-OR) is constructed from a combination of basic gates. This unique gate will produce a HIGH output whenever only one of the inputs is HIGH. If both inputs are HIGH, the gate is disabled.

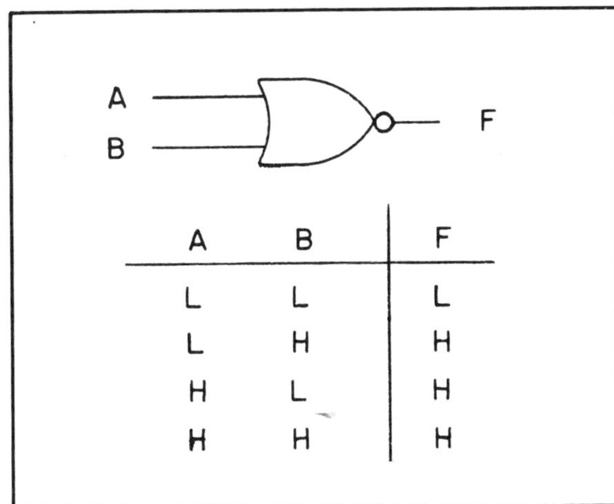

A	B	F
L	L	L
L	H	H
H	L	H
H	H	H

Fig. 9-10. NOR symbol and truth table.

Figure 9-11 shows the standard logic symbol, its truth table, logic circuit and switch analogy.

The exclusive-OR is used when examining two logic levels to determine whether the two input signals are not alike. The gating function is shown in Fig. 9-12. Notice that the output is enabled HIGH whenever only one of the input signals is HIGH.

The Exclusive-NOR gate

The exclusive-NOR gate (EX-NOR) is one whose output is the complement of the exclusive-OR gate. Figure 9-13 shows the standard symbol, truth table logic circuit and switch analogy for the EX-NOR gate. Notice the truth table of the exclusive-NOR shows the output is LOW whenever one of the inputs is HIGH. Its output is HIGH whenever the inputs are equal; that is, when both inputs are either HIGH or LOW. This gate is found in circuits that check for equality. A gating function of the EX-NOR is illustrated in Fig. 9-14. Notice that the output is HIGH whenever the inputs signals are equal.

More Input Lines

The previous gates described had only two input lines. There are basic gates which have more than two. Figure 9-15 shows the basic gate symbols for more than two input lines. The logic is the same for all the other gates. For example, if all the inputs to the NAND gate are HIGH, then the output will be HIGH. The reason for differences in symbol size rests with the companies' standards. Also the minimum spacing between the lines has to be controlled for readability as schematics are reduced for printing in a manual or storage on microfiche.

DIGITAL FAMILIES

There are three major families of integrated circuits. These are transistor-transistor logic (TTL), complementary metal-oxide semiconductor logic (CMOSL) and emitter-coupled logic (ECL). Each logic family is designed from a basic electronic circuit. This basic logic circuit is used to construct all integrated circuits within the families and therefore the electrical parameters are essentially equal between devices within a family. Within each family there exists a subfamily or logic series that have distinctive characteristics compared to other series contained in the family.

The TTL Family

The standard TTL has a 54/74 numbering series with a NAND gate as its basic circuit. The 5400 is used

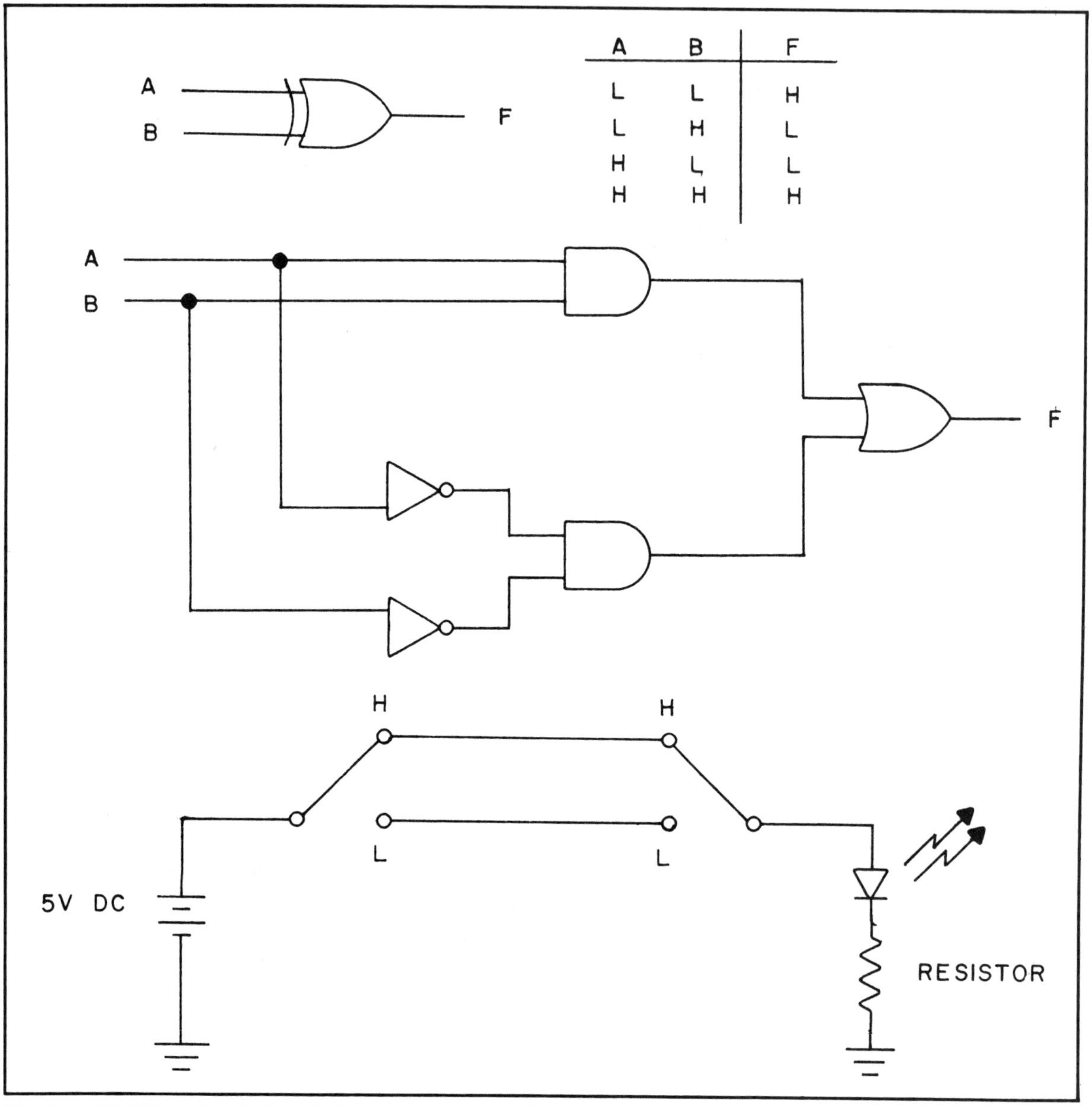

A	B	F
L	L	H
L	H	L
H	L	L
H	H	H

Fig. 9-11. EX-OR symbol, truth table, logic circuit and switch analogy.

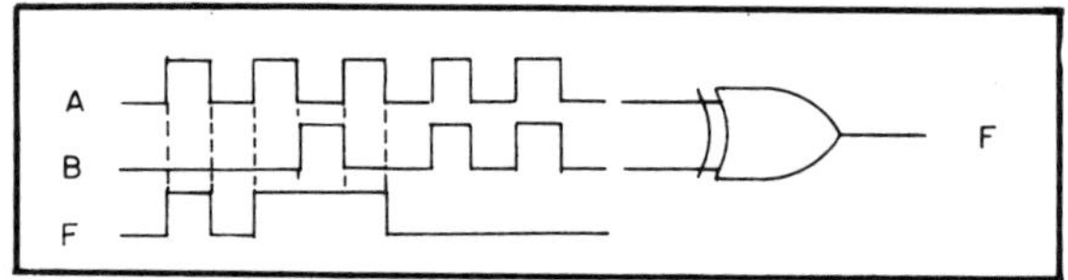

Fig. 9-12. EX-OR gating function shows a pulse train on input A and input B, the output F occurs only when either of the inputs are HIGH.

mainly in the military and has a higher operating temperature range. The basic TTL gate is shown in Fig. 9-16. Its name is derived from the fact that both the input and output are bipolar transistors. This is in contrast to older logic families, which are no longer in much use, called resistor-transistor logic (RTL) and diode-transistor logic (DTL) where the input circuitry was either a resistor or a diode.

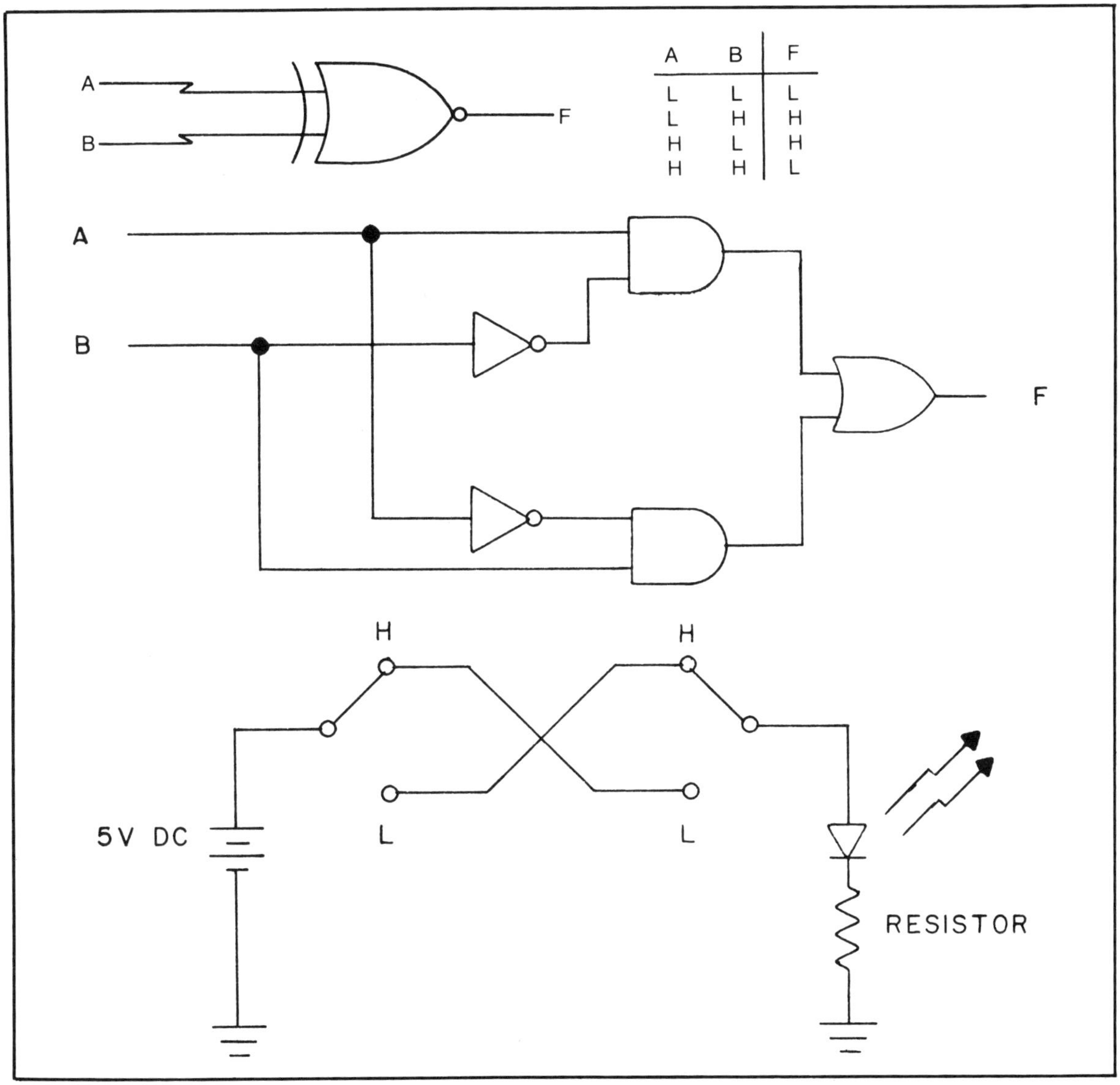

A	B	F
L	L	L
L	H	H
H	L	H
H	H	L

Fig. 9-13. EX-NOR symbol, truth table, logic circuit and switch analogy.

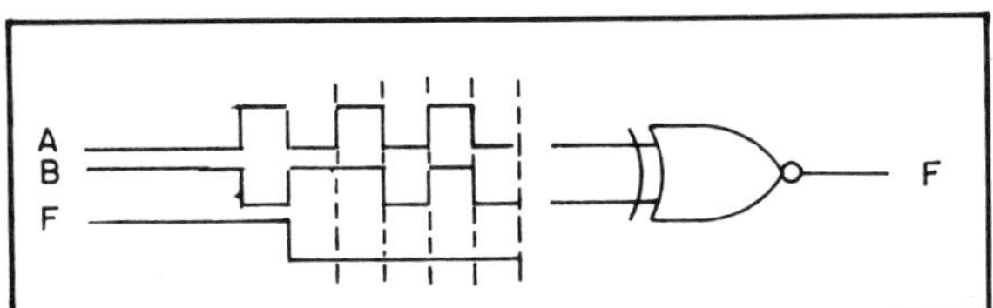

Fig. 9-14. EX-NOR gating function shows a pulse train on input A and input B, the output F occurs only when both inputs are HIGH.

Electrically, if both A and B inputs are LOW, a more negative signal is applied to Q1 which forward bias the base-emitter junction. Current flows from Vcc and out the input lines. This robs Q2 of base current. Since almost no current is flowing from the positive voltage to Vcc to the input of transistor Q2, it is turned off. No current flows from the collector base junction of Q2. Thus, when Q2 is turned off, there is no current flowing to the base of Q4; therefore it is turned off. When Q4 is

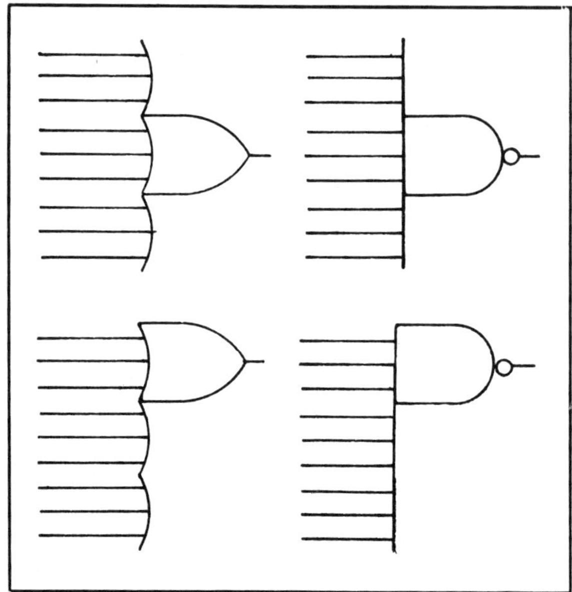

Fig. 9-15. Schematic symbols for greater than two input lines.

turned off, the output voltage, as measured with a voltmeter from output to ground, is HIGH. This situation is true if any or all inputs have a LOW level voltage applied. However, when both inputs are HIGH the base emitter junction is reverse biased, and current flows from the positive voltage Vcc to the base of Q2, turning it on.

When Q2 is turned on, current flows from +Vcc through its collector-emitter junction and supplies Q4 with enough base current to turn it on. When Q4 is turned on, the voltage across the output, as measured by a voltmeter, will be almost zero indicating a LOW output. When Q3 is turned off, the current flows from the output into the collector and to ground. This switching action is called "sinking." When Q4 is turned off, Q3 is turned on and current flows out the output. This switching action is referred to as "sourcing." Current sinking and current sourcing are illustrated in Fig. 9-17.

TTL Subfamilies

There are four subfamilies in the standard TTL 54/74 series which give wider range of power dissipation and speed characteristics. First is the 74L series which is a low-power version. This low power version is accomplished by increasing the resistor values of R1, R2, R3, and R4 of the standard TTL gate as shown in Fig. 9-16.

The second is the 74H series which has high speed switching capabilities. This high-speed switching is accomplished by decreasing the resistor values of R1, R2, R3 and R4 of the standard TTL gate shown in Fig. 9-16.

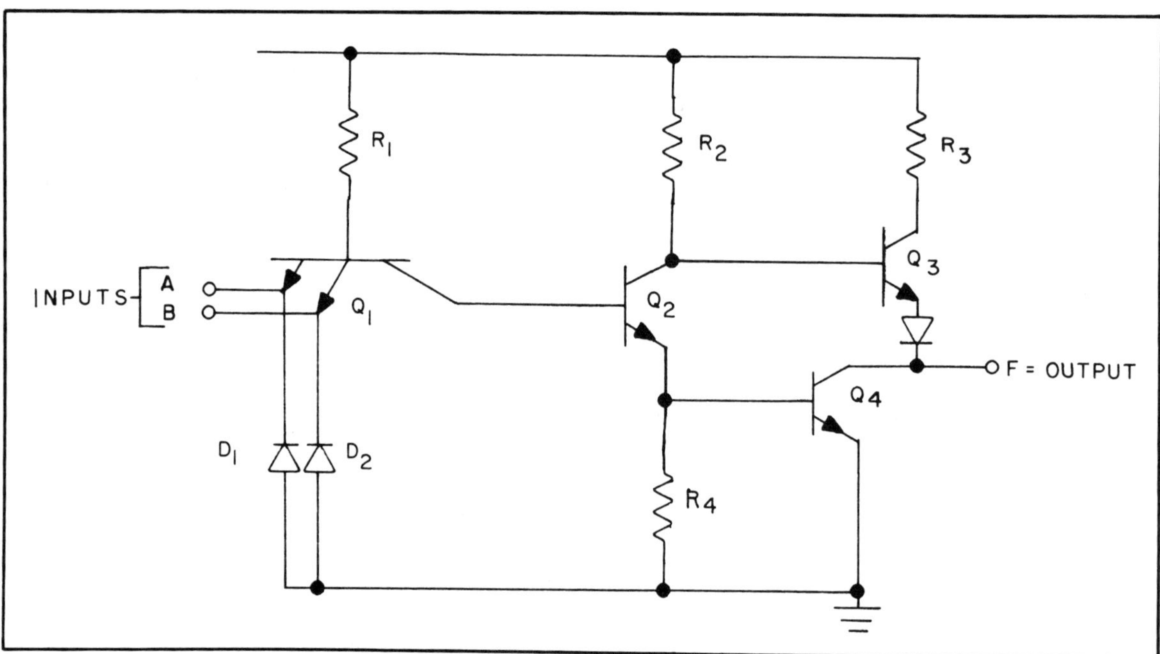

Fig. 9-16. Basic TTL NAND gate schematic.

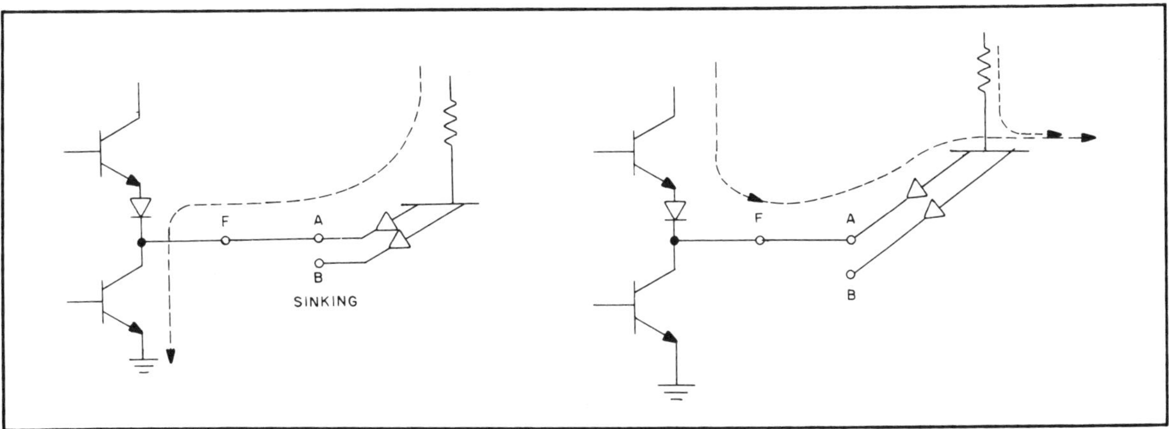

Fig. 9-17. Simplified TTL output showing current sinking into the gate when output is LOW and current sourcing when output is HIGH.

As switching speed increases, power dissipation also increases.

The third is the 74S series which uses Schottky barrier diodes across the two elements of the transistors as shown in Fig. 9-18. The Schottky diodes increase its switching speed. The fourth subfamily is the 74LS series which combines low-power dissipation by altering the resistors and adding Schottky diodes for speed.

Voltage Levels

Although we have assumed that HIGH and LOW voltage levels are 5 volts and 0 volts respectively, these

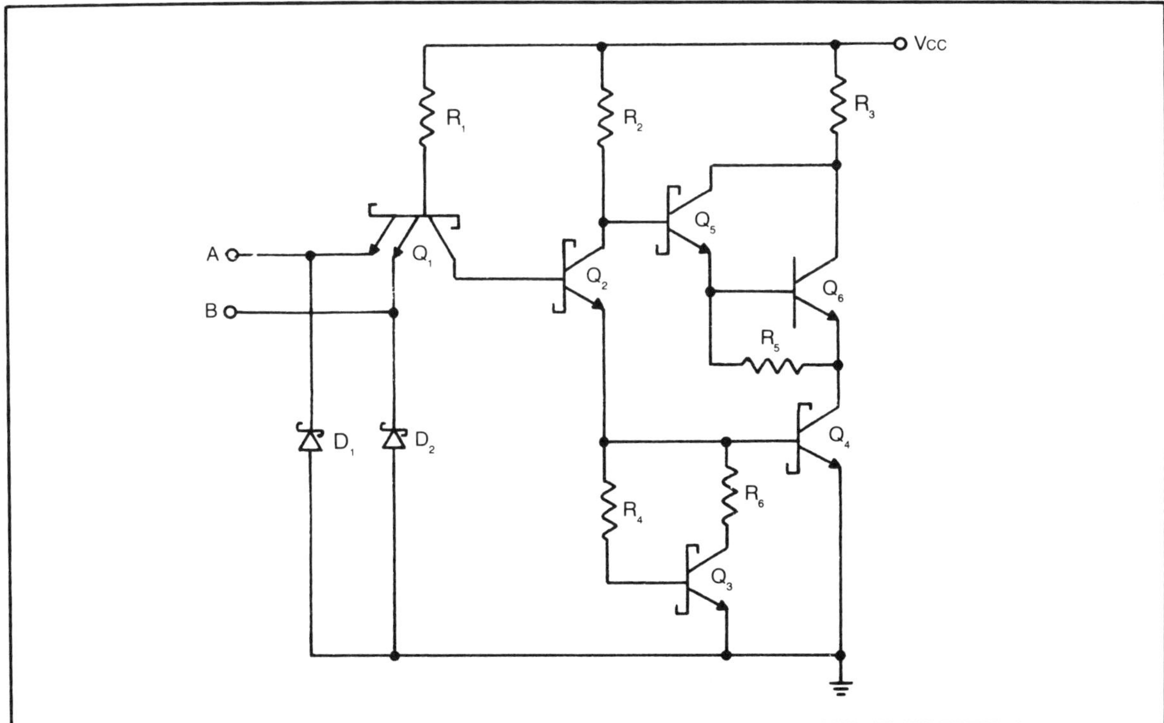

Fig. 9-18. Schottky clamped TTL.

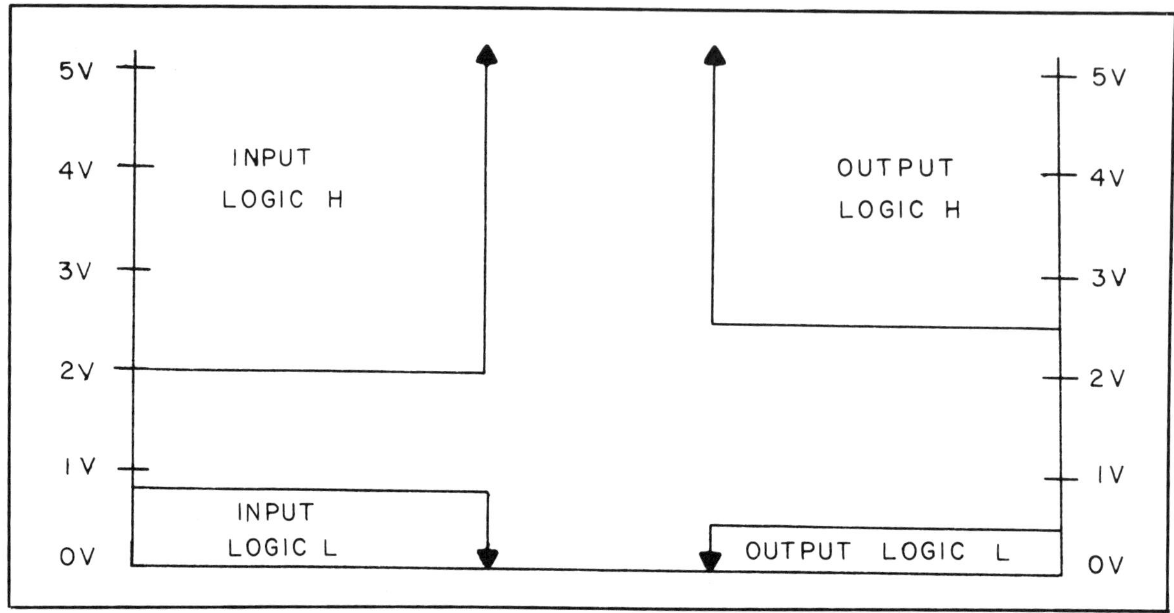

Fig. 9-19. TTL input and output voltage ranges.

two levels have a range of acceptability. The voltage range is designed to compensate for variation in operating temperature, capacitive loading and electrical noise such as poor power supply filtering or motor noise. For these reasons, binary voltage levels cannot be exact. The most positive voltage is always called a HIGH, while the least positive voltage is always called a LOW. Figure 9-19 presents the voltage level for input and output of a TTL integrated circuit. Thus, when using a voltmeter to measure statis signals of a gate, the input voltage for a HIGH can range from 2 to 5 volts, while the input voltage for a LOW can range from 0 to 0.8 voltage. The output of a gate for a HIGH can range from 2.4 to 5.5 volts, and a LOW can range from 0 to 0.4 volts.

Noise Immunity

Another characteristic of digital gates is noise immunity. It is a measure of the digital circuit's ability to prevent noise voltages from changing a given logic voltage level. For TTL the noise immunity is 200 millivolts, which means that a noise of less than 200 millivolts produced from an arcing motor or power supply ripple or current spiking caused by switching of the transistor, will not affect the transmission of digital information.

The Open Collector

The open collector is a special TTL gate designed to: (1) permit outputs to be connected together for increasing signal drive, (2) connect several devices to a common line (called a bus), (3) perform some logic function. A schematic of an open collector NAND gate is shown in Fig. 9-20. It is the same as a typical NAND except that transistor Q3, resistor R3 and diode D3 have been omitted.

Figure 9-21 shows a number of situations in which an open collector is used to drive devices. When the inputs to a NAND are HIGH the output goes LOW, sinking

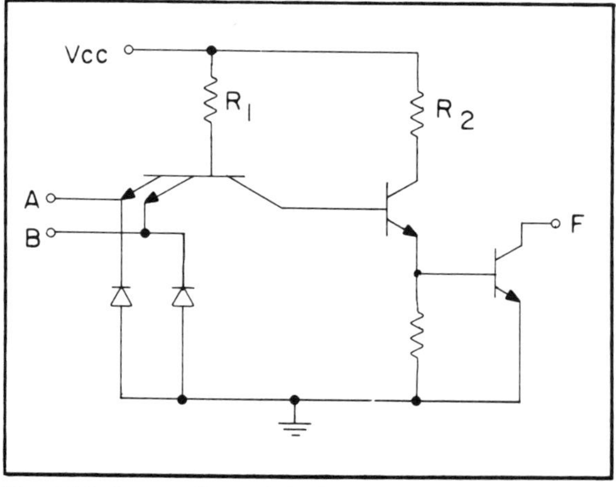

Fig. 9-20. Open collector NAND gate schematic.

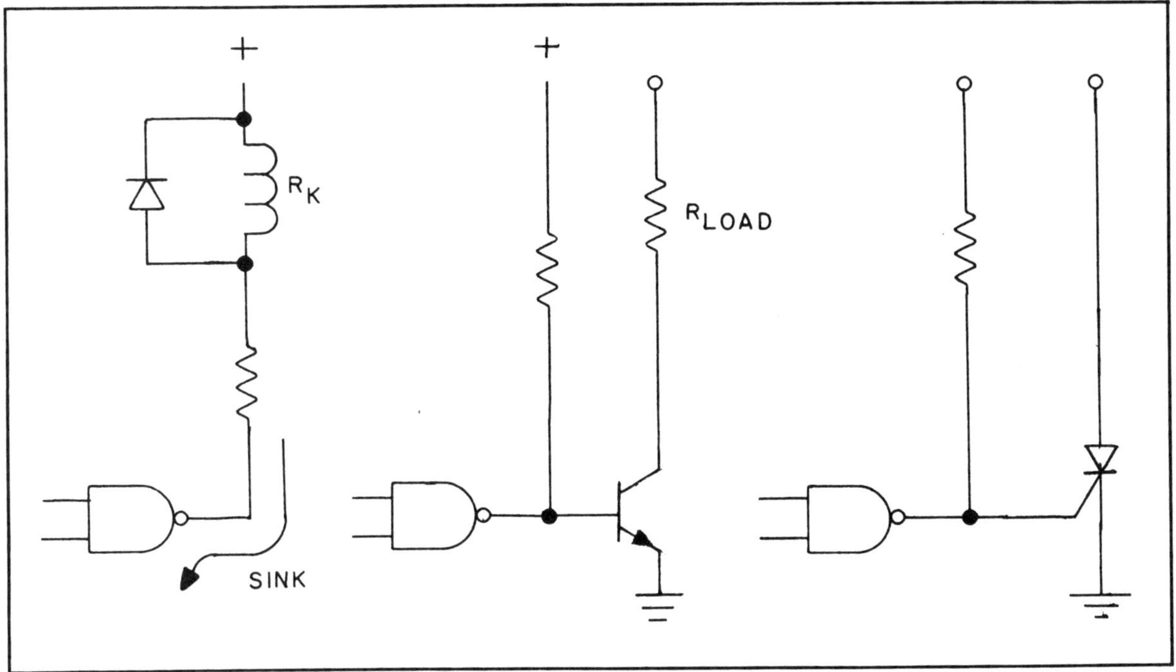

Fig. 9-21. Application of an open collector to drive a relay.

current from the higher external power source to activate a relay or turn on an indicating lamp. Lastly, Fig. 9-22 shows an open collector with its outputs connected together. The external resistor limits the current in the circuit when the open collectors are activated. The logic which is implemented at the common connection is a wired AND logic. The output at F ANDs the output of the logic of each open collector. The wired AND presents some very interesting troubleshooting challenges.

A wide variety of logic gates and other digital devices such as decoders and memories are available with open collector outputs.

A buffer/driver gate with higher than normal output current and/or voltage rating of the output transistor is available. The buffer/driver is available with open collector or totem-pole output.

The CMOSL Family

This family is based upon MOS technology (metal-oxide semiconductor) for manufacturing the transistors. The MOS transistor consists of four components called the source (S), the drain (D), the gate (G), and the substrate (B) as shown in Fig. 9-23. The MOS transistor controls the current between the source and drain by the application of a gate-to-source voltage. No current flows

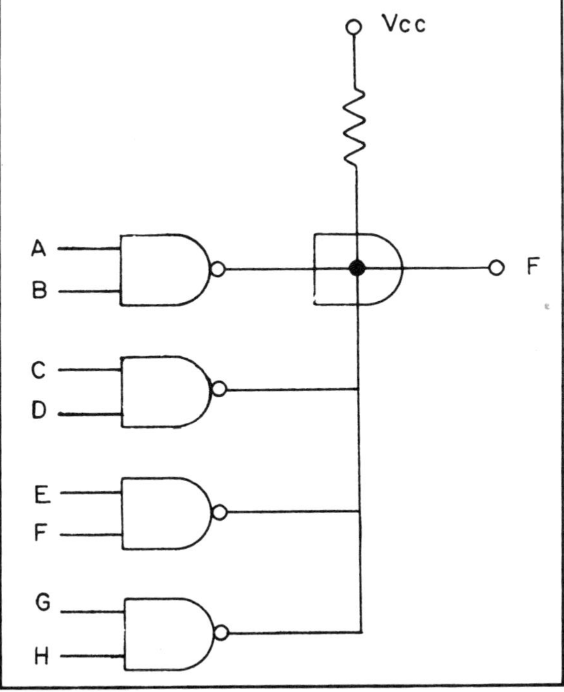

Fig. 9-22. Wire-AND logic diagram.

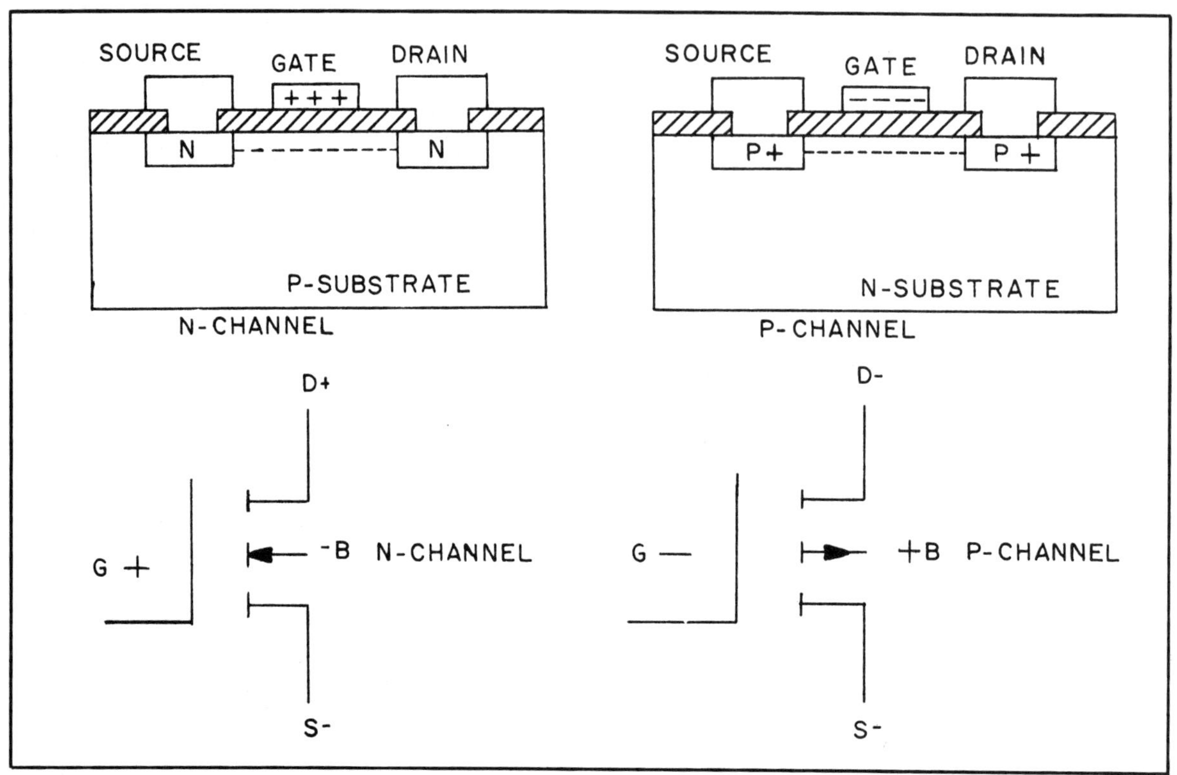

Fig. 9-23. Cross section and symbol of an enhancement mode N-channel and P-channel MOS transistor.

between the gate and source; thus the device is a voltage control switch as compared to the bipolar transistor which acts as a current control switch.

There are two modes of MOS transistor operation; the depletion mode and the enhancement mode. A depletion mode transistor is switched on when zero voltage is applied to the gate. The enhancement mode transistor does not conduct, or is off, until the gate voltage reaches a certain threshold and, therefore, is extensively used in digital devices.

The enhancement mode transistor has two types, the p-channel transistor (PMOS) and the n-channel transistor (NMOS) from which the term complementary is derived. In Fig. 9-23 notice the arrow pointing toward the dotted line indicating an n-channel, whereas the arrow pointing away from the line indicates a p-channel. The dotted line symbolizes an incomplete current path between source to drain. The dotted line is enhanced by the appropriate charge on the gate, i.e., positive gate charge for n-channel and negative gate charge for a p-channel.

The CMOS family has very low dc power dissipation and is used in applications requiring low power consumption. The standard series is 14000 with the basic gate being a NOR gate. Figure 9-24 shows a schematic of a CMOSL 14001 NOR gate. Notice that two MOS transistors are p-channel and two are n-channel. The positive supply voltage is connected to V_{DD}, and ground is at V_{SS}. If both A and B inputs have a LOW signal applied, then the output of the gate is HIGH. The LOW signal at A and B inputs will turn on Q1 and Q2 while Q3 and Q4 will be turned off. The voltage drop across Q3 will measure close to V_{DD} value. If either A or B or both have a high signal applied then Q1 and Q2 will be turned off and Q3 and Q4 will be turned on. The voltage drop across Q3 and Q4 will measure essentially zero. The diodes D1 and D2 at the inputs protect the MOS transistor gate against high electrical voltages of about 30 volts.

The CMOSL family can have supply voltages (V_{DD} of 5, 10 or 15 volts in contrast to the 5-volt level of the TTL family. The logic level HIGH voltage is approximately equal to 70 percent of the supply voltage, while the logic level LOW voltage is approximately equal to 30 percent of the supply voltage. A general rule for switch-

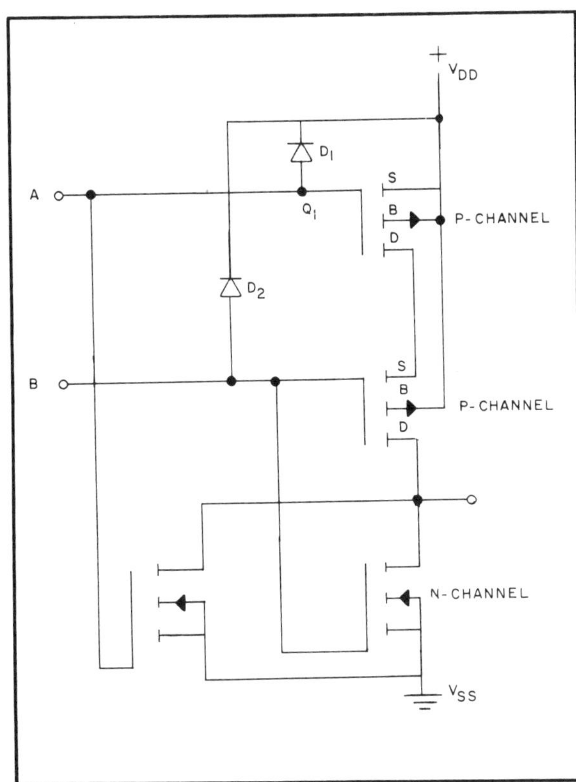

Fig. 9-24. CMOSL 14001 NOR gate schematic.

ing threshold is approximately 50 percent of the supply voltage. For example, if the supply voltage is 10 volts, then logic level HIGH is 7 volts or greater and logic level LOW is 3 volts or less while the switching threshold is approximately 5 volts.

Another characteristic of the CMOSL is noise immunity, which is typically 45 percent of the supply voltage, therefore making this family very good in noisy environments such as industrial plants and automobiles. The propagation delay time—the time required to transfer information from its input to its output—is much slower than TTL series. In terms of *fan-out*—the number of inputs a single output can drive—the CMOSL can drive 50 inputs compared to only 10 inputs of the standard TTL.

The ECL Family

Speed always seems to be a prime feature of gate operation. The emitter coupled family or ECL was developed to enhance this feature. A typical circuit is shown in Fig. 9-25. The operation of the ECL centers on the differential amplifier Q4-Q5. The diodes D1, D2 and Q6 develop a constant current through R7, thereby developing a constant voltage at the base of Q5 with about −1.3 volts. If input B has a signal which is more positive (−0.9V), then Q4 will turn on and its collector voltage will be pulled more negative toward the value of VEE (−5.0V). This action will cause Q8 to turn on, and the output C will be pulled high which is a −0.9V. Similarly, if any of the inputs go high, the same action will occur. Thus the logic level HIGH for ECL gate is −0.9 volts, while a logic level LOW is −1.7 volts. It should be noted that some ECL circuits have complementary outputs which will have an opposite output condition. Table 9-2 summarizes the major digital families and subfamilies, listing typical electrical performance characteristics. By studying the chart you can notice the difference that exists.

It should be noted that there is a 74C series which is a pin-for-pin compatible CMOS version of the standard TTL series. Although the device is pin-for-pin compatible, direct substitution should be done with caution, because the 74C series operates at a slower speed compared to the standard TTL series. The output signal from a 74C series may be delayed enough to cause an apparent malfunction of the driven gate or gates.

IDENTIFICATION OF INTEGRATED CIRCUITS

When troubleshooting a digital system, it may not be apparent which digital family is used in the product. The schematic diagram will contain a logic network composed of standard logic symbols regardless, whether the devices are TTL or CMOS. It is important to identify the appropriate digital family, because a static signal being monitored may be correct for one family but incorrect for another. Check the service manual to see if it specifies the digital family. If it doesn't, examine the power supply voltages. If the power supply has a +5 voltage terminal, it could indicate a TTL series is used whereas a −5 volts terminal could indicate a ECL series. A measurement of a few signal lines could confirm the exact one.

A digital family can be recognized by part number stamped on the integrated circuit. Generally a TTL series will have the manufacturer's prefix, followed by a series designator or temperature range, the family type, the particular part number or gate configuration, and the type of package. For example, let's examine a N74LS08N. The N indicates that the manufacturer is Signetics and it is a commercial product; an S would indicate military. The 74 tells us that the family is TTL, while the LS indicates low power Schottky. The 08 indicates the particular part number or gate configuration. In this case the integrated circuit contains four two-input

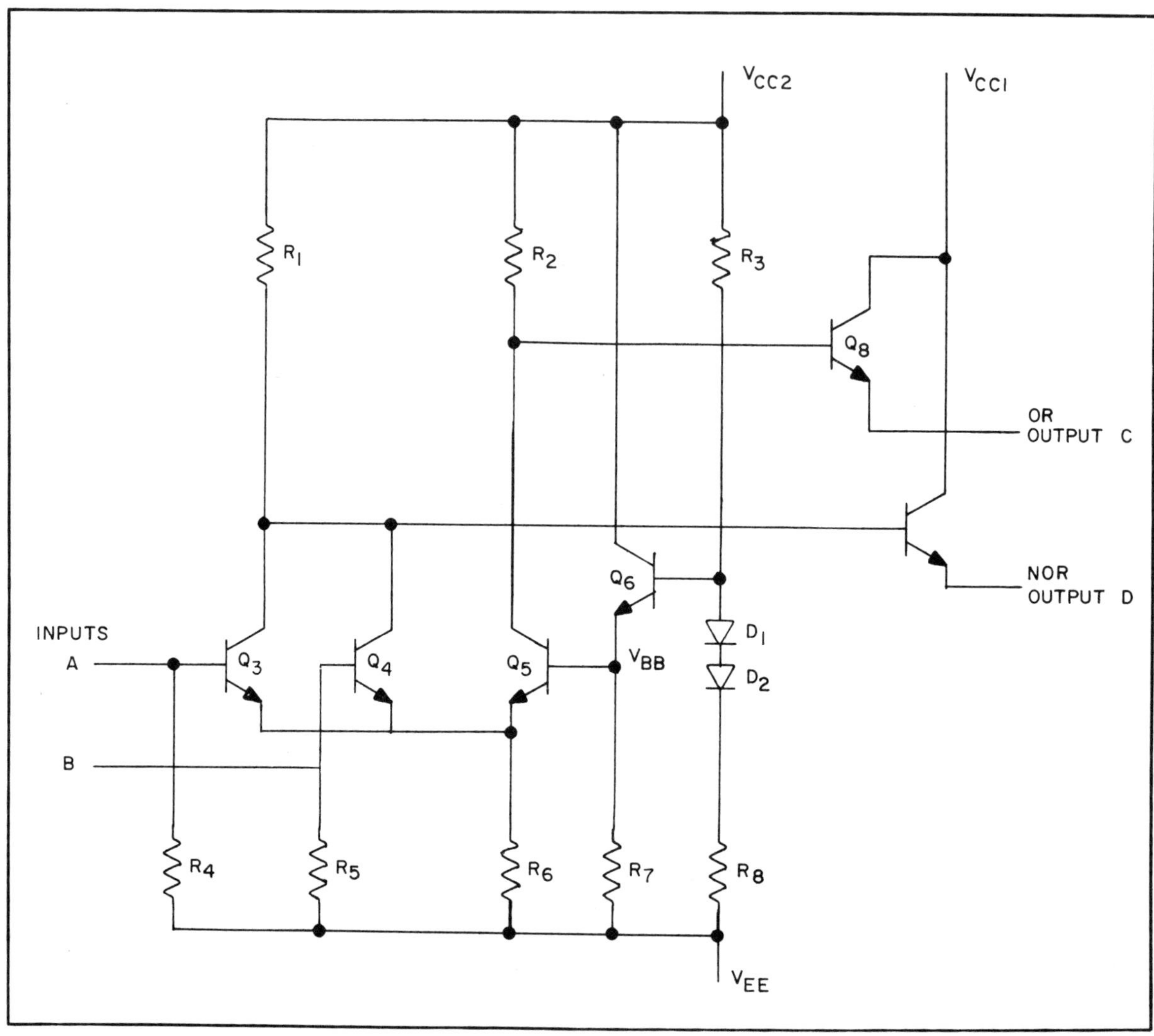

Fig. 9-25. Emitter coupled logic (ECL) integrated circuit.

AND gates. The suffix N indicates the package type which is a dual-in-line with a plastic case.

Many times the manufacturer's logo is also stamped on the integrated circuit which can direct you to their data book. That provides some very useful information. Table 9-3 is a list of common semiconductor manufacturers' logos. A list of prefixes which identifies the manufacturers is listed in Table 9-4. The series designator is made up of one to four digits and/or letters. Internal letters within the series designator indicate the following:

C = CMOS
H = high speed
L = low power
LS = low power Schottky
S = Schottky
T = interface chip (Signetics only)

The part number or gate configuration may be two to three digits. The suffix specifies the type of packaging. Acquiring various semiconductor manufacturers' data books will give specifics of their product's coding scheme.

PACKAGING OF INTEGRATED CIRCUITS

There are essentially four levels of integration,

Table 9-2. Typical Characteristics of Major Families and Subfamilies Electrical Characteristics.

	RTL	DTL	Standard TTL	High Speed TTL	Low Power TTL	Schottky TTL	Low Power Schottky	CMOS	MECL 10,000
Supply Voltage (volts)	3.6 & 0 or +5 & 0	+5 & 0	+5 & 0	+5 & 0	+5 & 0	+5 & 0	+5 & 0	+3 to +18 & 0	0 & −5.2
Logic High Level (volts)	2 to 5	2 to 5	2 to 5	2 to 5	2 to 5	2 to 5	2 to 5	>70% of supply	−0.9
Logic Low Level (volts)	0 to 0.4	0 to 0.8	0 to 0.8	0 to 0.8	0 to 0.8	0 to 0.8	0 to 0.8	<30% of supply	−1.75
Typical Fanout (within same family)	5	8	10	10	10	10	10	50	90
Typical Propagation Delay per gate	12 ns	30 ns	10 ns	6 ns	33 ns	3 ns	9.5 ns	10 to 35 ns	2 ns
Typical Power Consumption per gate	25 mW	11 mW	10 mW	22 mW	1 mW	19 mW	2 mW	0.01 μW to 10 mW	25 mW

small scale integration (SSI), medium scale integration (MSI), large scale integration (LSI) and very large scale integration (VLSI). The scale of integration relates to the number of gates. An integrated circuit (IC) containing 1 to 12 would be classified as SSI; 13-99, as MSI; greater than 99 as LSI; and greater than 150, as VLSI. The number of pins total 14, 16, 18, 20, 24, 36, 40, 48. The protective case is called a package. One type is called dual-in-line (DIP) with the connecting pins arranged in two (dual) lines.

The pin numbering system begins with 1 and goes clockwise from a top view and counterclockwise from a bottom view. To identify pin 1, the manufacturers have made a notch, a depression, or some other unique marking on the package. Notice the identification mark in Fig. 9-26. To know the function of each pin, you must identify the IC and consult the manufacturer's data book which will contain a pin connection, a diagram, and a description.

TROUBLESHOOTING

A digital system composed largely of integrated circuits of medium to large scale integration have very complex circuits inside the IC. Time does not permit you, as a troubleshooter, to study the details of the internal IC circuit. All you can do is check the function behavior of the IC. Therefore you may have from one to ten input signals and simultaneously monitor that many outputs. Digital instruments designed to aid the person in troubleshooting ICs consist of a logic pulser, current tracer, logic probe and logic monitor plus others for complex system fault detection.

There are essentially four general steps to follow in troubleshooting and repairing a digital system which has a failure or fault in it. First is failure detection, followed by failure isolation, then faulty device replacement, and finally operational testing.

Failure Detection

Faults or failures in a digital system can be detected in a variety of ways. Some systems have warning lights which signal a malfunction of a subsystem. Other clues are fixed outputs on a display or on the subsystem output

Table 9-3. Common IC Manufacturers' Logos.

Manufacturer	Logo
Advance Micro Devices	AM
American Micro Systems	Ai
Fairchild	FAIRCHILD
Harris	
HP	hp
Intel	intel i
Intersil	i
ITT	ITT
Monolithic Memories	
Mostek	MOSTEK
Motorola	
National Semiconductor	NS
Precision Monolythics	
Raytheon	RAY
RCA	RCA
Rockwell International	Rockwell
Signetics	S
Silicon General	SG
Solitron Device, Inc.	S
Teledyne Semiconductor	
Texas Instrument	
TRW Semiconductors	TRW
Zilog	Zilog

Table 9-4. Manufacturers' Prefixes.

Prefix	Manufacturer
Am=	Advanced Micro Devices (AMD) IC
CD=	RCA digital IC or National 40xx-series CMOS
DM=	National bipolar digital IC
DS=	National interface IC
HD=	Harris digital IC
HM=	Harris memory IC
HPROM=	Harris programmable read-only memory IC
MC=	Motorola IC
MCM=	Motorola memory IC
MM=	National MOS, CMOS, or Teledyne 74C CMOS
N=	Signetics commercial temperature range IC
RC=	Raytheon commercial temperature range IC
RM=	Raytheon military temperature range IC
S=	Signetics military temperature range IC
SN=	Texas Instruments bipolar IC
TF=	Texas Instruments CMOS military temperature range IC
TMS=	Texas Instruments MOS IC
TP=	Texas Instruments CMOS commercial temperature range IC

signals even when all possible combinations of system inputs are tried. Often you may experience the erratic behavior of a digital system, and therefore it needs repair.

Failure Localization

Once a failure has been detected, your next step is to localize or isolate the failure within the digital system. If you are unfamiliar with the digital system, obtain an operating and service manual. There is no substitute for familiarity with the system when attempting to troubleshoot. How far you are able to proceed in troubleshooting depends upon your knowledge of the system. Begin to localize the failure by asking yourself, "What are the malfunction symptoms, and what is most likely causing the failure?" By asking and answering these two questions, you will be able to follow a logical sequence.

The following general logical sequence may be used to localize the failure. First, the digital system should be adjusted or calibrated according to the manufacturer's specifications. Often this procedure will correct the failure. If not, a visual inspection should be made for obvious circuit or component failures such as blown fuses, open circuit breakers, burned or hot components, loose wires, loose or corroded printed circuit connectors or a loose IC in their sockets. However, if the digital system is a prototype undergoing its initial checkout, then additional items like wiring errors and correct interconnections between subsystems must be considered.

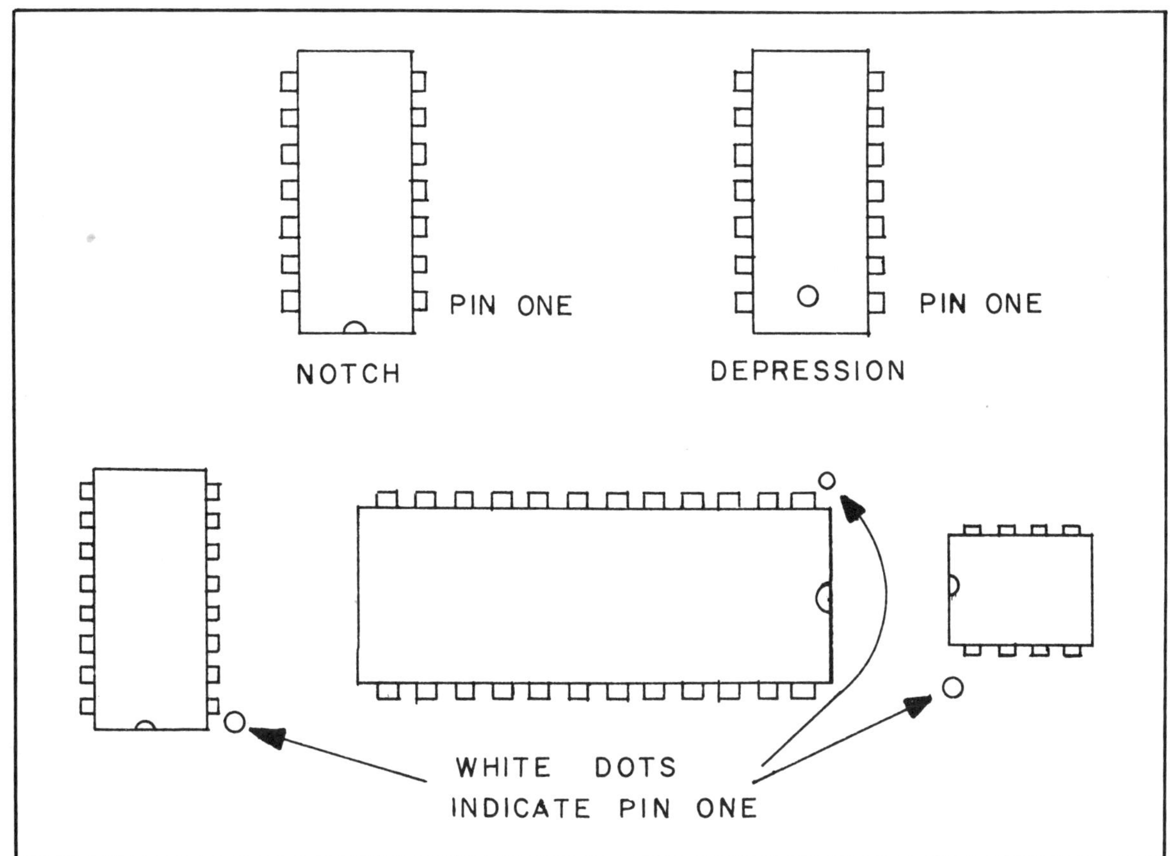

Fig. 9-26. Different methods used to identify pin one on a dual-in-line (DIP) integrated circuit.

Should the preceding procedure not yield a failure isolation, then check the power supplies for the presence of voltages and their correct values. A service manual will give you the correct power supply voltages and tolerances as well as test points.

Since most digital systems have subsystems which perform a specific function, you should attempt to locate the failure in one or more printed circuit boards that plug into a rack. A system block diagram supplied by the manufacturer in its service manual should be consulted.

Faulty Device Replacement

When the faulty printed circuit board(s) have been identified, you have two choices. First the faulty printed circuit board can be replaced with a good board. There are some manufacturers which have an exchange program, a defective printed circuit board for a good one. Your second choice is to isolate the specific IC device or component that has failed and replace it. This may be more challenging, but with the aid of the manufacturer's schematic and physical layout of the ICs on the printed circuit board, it can be done.

Operational Testing

After the defective IC device or discrete component has been replaced in the digital subsystem, you should operate and test it to verify proper operation. You may have found one fault and fixed it, but there could be others. Lastly, the digital systems should be calibrated and adjusted according to manufacturer's specifications.

IC AND CIRCUIT FAILURES

There are essentially four kinds of failure of an IC. One is the inputs or outputs could be open. Second, the inputs or outputs could be shorted to ground or to the supply level voltage. Third, a short can occur between

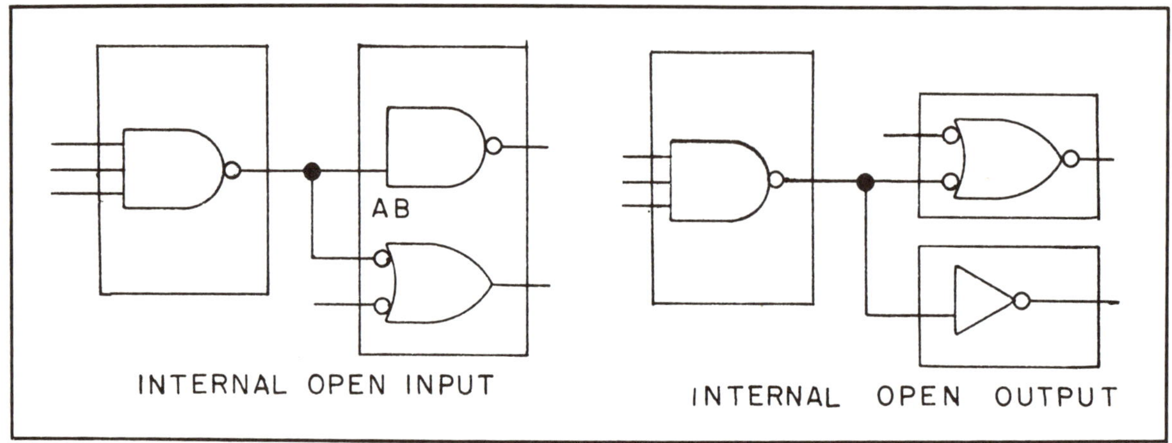

Fig. 9-27. Internal open input and open output.

two pins but not touching the ground or the supply voltage. The fourth is simply a failure of a discrete component.

Open Inputs or Outputs

If there is an internal IC open circuit in the output of a gate, the inputs which are driven by the output are also open. For TTL and CMOSL devices, the input will float either HIGH or LOW. There is no guarantee of either state. Also the open input can oscillate between a HIGH and a LOW due to noise voltages in the circuit. The behavior of ECL with an open input and which has input pull-down resistors results in LOW. Figure 9-27 illustrates an internal open input and an open output which results in a need to replace the defective IC.

There can also an external open. These open failures can result from pin(s) not firmly seated in a socket, a broken conductor on a printed circuit board, a broken wire-wrap wire, an IC inserted with one of its pins bent underneath the socket, or an improperly soldered pin of an IC to a printed circuit board. The signal behavior is the same as if it were an internal open. Continuity checks with a voltohmmeter can be made to locate this kind of external fault.

Shorted Inputs or Outputs

In digital ICs the defect may be in one or more of the gate components that have caused the trouble. The gate failure may be identified by the inputs or outputs being fixed at a HIGH or LOW logic level. For example, the input lead B as shown in Fig. 9-28 may be grounded internally or externally to the IC so that the input is always held LOW regardless of the input signal level applied. This is called stuck-LOW and produces a failure as shown in the truth tables in Fig. 9-28. Notice that the circled inputs cannot change states and the output for this particular gate remains HIGH regardless of input changes. All signal lines connected to the internal IC short are also stuck-LOW or stuck-HIGH depending upon the digital family. For an ECL device, a short to either

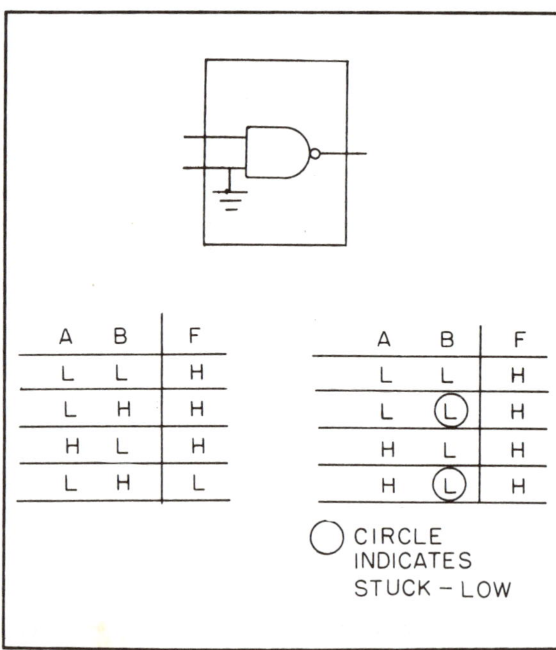

A	B	F
L	L	H
L	H	H
H	L	H
L	H	L

A	B	F
L	L	H
L	(L)	H
H	L	H
H	(L)	H

Fig. 9-28. An illustration of the effects of a shorted input (stuck-LOW) on a NAND gate and its inability to implement its truth table as noted by the circles.

ground or supply voltage represents a HIGH or LOW.

If a signal line is shorted to ground or power supply which is external to the IC, the failure cannot be differentiated from an internal IC short.

Short Between Pins

If a short between pins occurs internally in the IC, but not to ground or power supply, then the output of the previous gates are connected together. Figure 9-29 illustrates this situation. In the digital families of TTL and CMOSL, if the output of the gates are connected together as illustrated and if both outputs transition HIGH or LOW simultaneously, no fault will occur. However, if one output, say A is LOW, while the output of B is HIGH, the output will be pulled LOW. The only exceptions to the above circuit behavior are open collectors and tristate devices. In ECL devices the outputs can be connected together; therefore, no gate faults will be noticed, unless there is excessive current drawn which will damage the driving gates G1 and G2.

The behavior of shorts between signal lines external to the IC is identical to an internal short detween pins. This problem often occurs on printed circuit boards where solder brings two signal lines or wire wrap pins are bent causing a short.

Discrete Component Failure

Most digital systems have discrete external components connected to outputs of ICs such as resistors, capacitors, diodes or transistors which serve a variety of functions. Failures of discrete components can cause timing problems or oscillation of an IC as well as having a signal stuck LOW or HIGH. These situations should be checked using standard procedures.

Static Electricity

Before you begin work on any digital system, it is important to realize the danger of static electricity. The danger is very critical with MOS technology devices. The very act of touching a MOS IC with your finger can destroy it. Even though the IC manufacturers have a built-in diode for protection, they are often inadequate. Another point about static electricity is that it doesn't care who generates it. Suppose someone walks over to you and gives you a tap while you're working on a digital system. Zap! The static electricity your friend generated walking in goes from him to you and from you to the IC. So two rules should be observed. One is to discharge yourself to the same potential as the equipment you are repairing. Touch the chassis first before beginning your work. The second rule is not to touch anyone working on digital equipment.

TROUBLESHOOTING USING DIGITAL TEST EQUIPMENT

The oscilloscope and voltmeter can effectively be used to isolate many IC or circuit faults in simple digital systems. The voltmeter can check a static HIGH or LOW within a system. It also can be used to identify open circuits by making continuity measures. The oscilloscope can be used to trace static or dynamic signals. An oscilloscope with dual trace is very important for checking timing problems. A timing problem is one where input signals to a gate do not arrive at the precise time, thereby causing an inappropriate output signal.

Though these two types of test equipment are valuable, other new test equipment has been developed for troubleshooting digital equipment.

Logic Probe

A logic probe is used to detect the logic level of a HIGH or LOW in the digital system. An LED is used to denote the presence of a HIGH by being turned on while a LOW is denoted by being turned off. A pulse can be detected with the logic probe LED blinking ON and OFF or a visual display registering a "P." An open circuit is indicated by half brilliance of the LED, because the logic level is "floating" somewhere between a HIGH and a LOW.

Logic Pulser

A logic pulser is a device which is touched to a pin

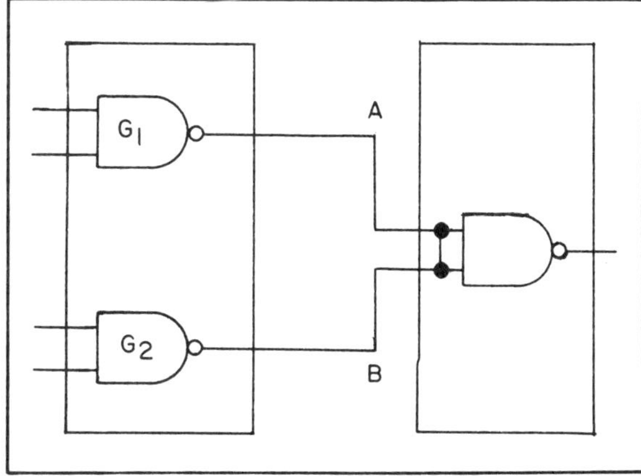

Fig. 9-29. Internal short between two pins which will pull the output LOW if the outputs are not transitioning at the same time.

and a button on the logic pulser can be pushed to cause an opposite signal to be transmitted to all lines connected to that particular output pin.

The logic probe and pulser are often used together. Figure 9-30 illustrates troubleshooting a digital network. The logic pulser is touched at one output of one gate which is connected to two or more gate inputs. The interconnecting signal lines represent a common electrical point called a *node*. Notice the pulser is transmitting a pulse train, and a logic probe is touched to each gate output of the network. Remember, a flashing LED which matches the pulse rate indicates a good gate, while no flashing indicates a defective gate in the network.

A logic probe, logic pulser and current tracer are used to identify a shorted gate. Power to the circuit must be off when using these instruments together. Figure 9-31 illustrates this procedure. Note that the probe LED is OFF indicating a LOW is present. Now touching a current tracer to the node, the LED action reveals that the node is stuck-LOW. By injection of a pulse to the node with the logic pulser and using a current tracer, the specific input that is shorted can be identified.

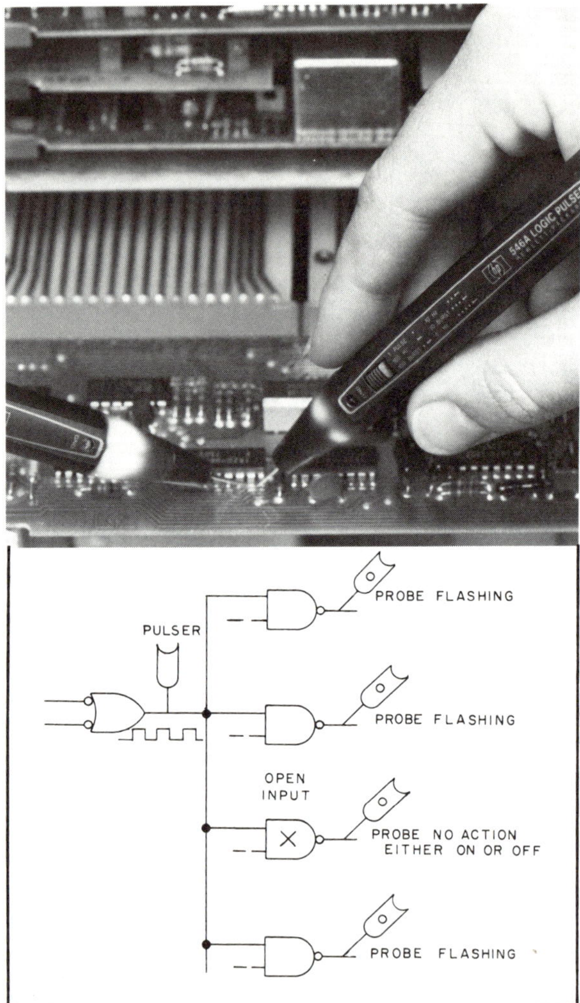

Fig. 9-30. Troubleshooting a gate network using a logic pulser and logic probe plus illustrated schematic application (courtesy Hewlett-Packard).

Logic Monitor

A logic monitor performs the same function as a logic probe, but it can indicate all the signals of a 16-pin DIP integrated circuit. Figure 9-32 shows a 16-channel logic monitor. Logic monitors can have as many as 40 channels and can be used to observe static or dynamic action of all signals of a large scale integrated circuit. Knowing the logical inputs and outputs of a specific IC, a mental comparison is made between what ought to be observed and what is observed on the LEDs of the logic monitor. Figure 9-33 shows how a logic pulse and a logic monitor are used together to determine if an input signal generates the appropriate outputs.

Logic Comparator

A logic comparator is a test instrument, shown in Fig. 9-34, which compares a questionable IC, called a device-under-test (DUT), with an identical reference IC known to be good. An LED display on the logic comparator indicates if the chip outputs and the reference IC outputs are equal. There is an LED for each pin. Any output pin paired comparison which is not equal, or does not have the same logic level, will cause the corresponding LED to be turned on indicating a fault. A simplified diagram of the logic comparator illustrating its basic principle is shown in Fig. 9-35. Notice that at pin 4 of the DUT, there is an incoming signal which also sent to pin 4 of the reference IC. The output of the DUT gate is pin 9, and it is sent to an EX-OR with the signal at pin 4 of the reference IC. If the two output signals are equal, then the LED is not lit. However, if there is a difference in output signal, then the LED is lit and identifies a defective DUT.

More sophisticated test instruments are available and are very useful in identifying faults on large complex digital systems. A *signature* is a test instrument that collects a stream of digital information from a test point and converts it into a code which is displayed. This code is then compared with the manufacturer's code or signature.

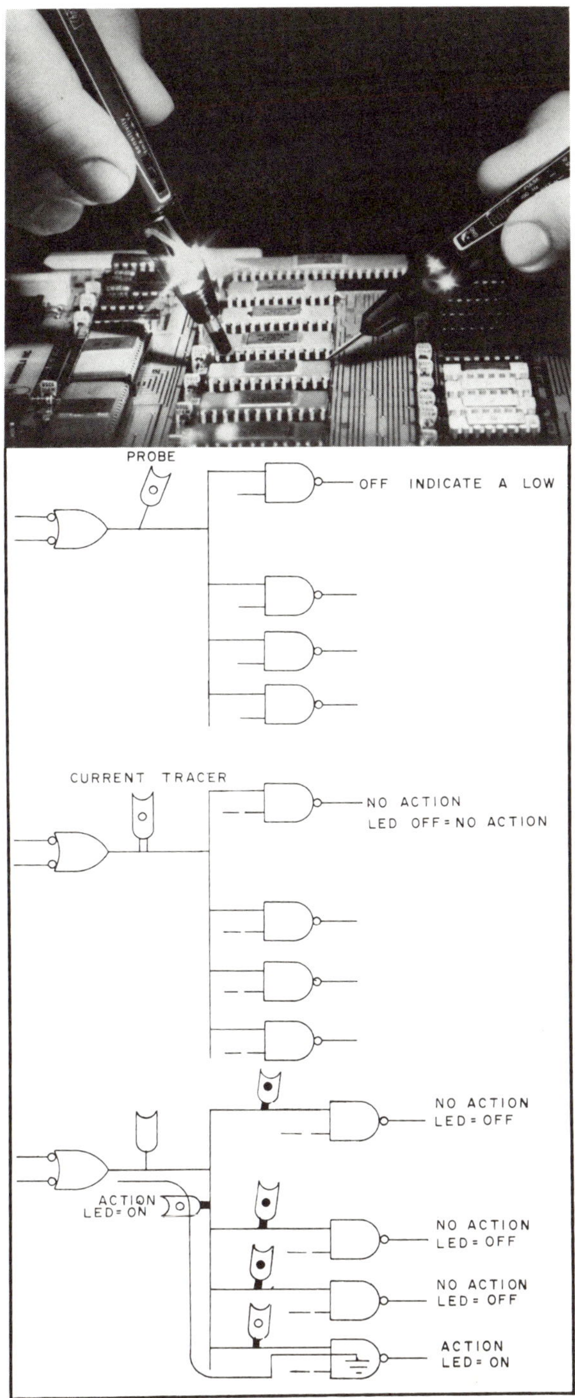

Fig. 9-31. Troubleshooting with logic probe and current tracer plus illustrated schematic application.

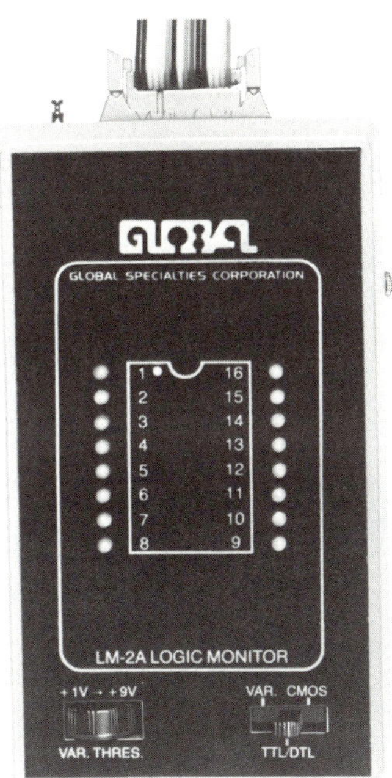

Fig. 9-32. A 16-channel logic monitor and a logic pulser to identify proper circuit operation (courtesy Global Specialties, Incorporated.).

Fig. 9-33. Using a logic monitor and a logic pulser to identify proper circuit operation (courtesy Hewlett-Packard Company).

Fig. 9-34. Logic comparator test instrument (courtesy Hewlett-Packard Company).

A logic analyzer is another sophisticated test instrument which can display many digital waveforms simultaneously, permitting a close look at selected digital waveforms, and it will be discussed later.

REMOVING AN IC

Once the defective device has been identified, remove the IC and replace it. Integrated circuits which are in sockets sometimes can be removed by hand. However, this is a very bad practice. Upon removal of an IC by hand, the pins can easily be bent and subsequently broken. Because the pins are rigid and slender, they can easily puncture a thumb or finger. The proper technique for removal is to use an IC remover as shown in Fig. 9-36. Hook the IC remover on the ends of the IC and gently rock the IC out of the socket.

If the defective IC is soldered in a printed circuit board, then remove it by cutting each pin with a diagonal cutter. This procedure is shown in Fig. 9-37. After the IC body has been clipped, unsolder the pins and remove them with needle-nose pliers.

Another way to remove a soldered IC is with desoldering wick. Desoldering wick is a braided wire specially treated so that the solder easily flows to it and away from the IC pin. It is simple to use. Place the wick between the soldering iron and the pin to be unsoldered. The solder flows onto the braid. For best results, dip the braid in non-corrosive liquid flux before using. However, its use on printed circuit boards with solder plated through holes is not as effective as the soldering iron and vacuuming

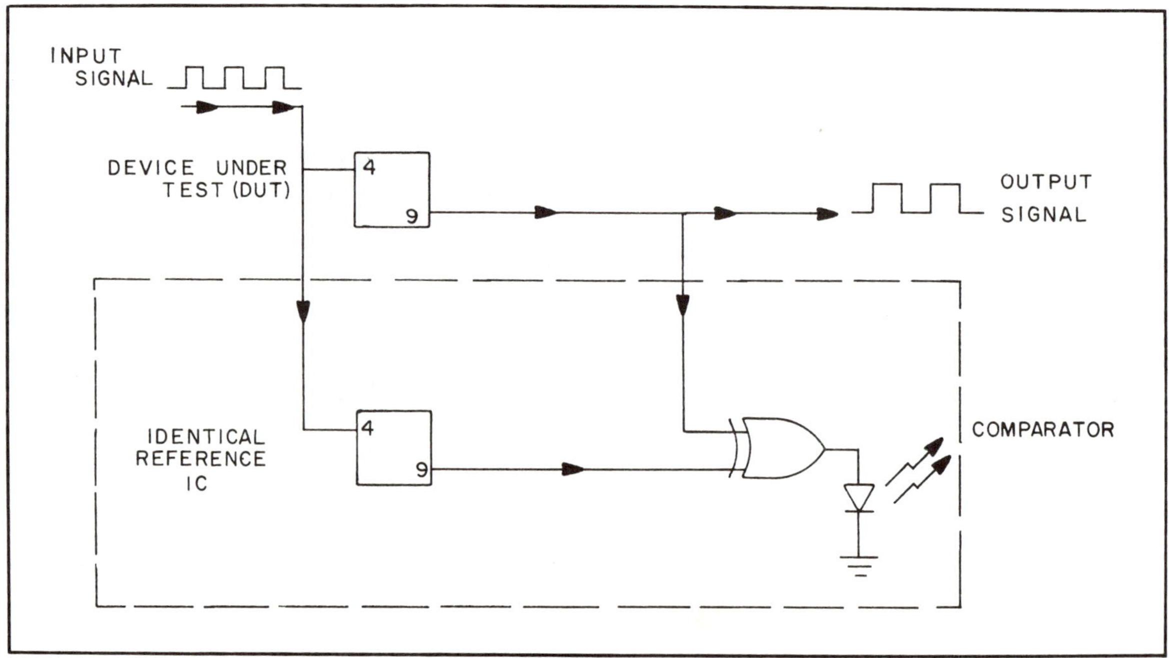

Fig. 9-35. Simplified diagram of a logic comparator test instrument which compares a device under test to an identical reference IC. A lit LED indicates a defective IC.

device. A 25 watt soldering iron is most often used. If the wattage of the iron is too high, excessive heat could cause damage to other parts of the circuit.

A word of caution should be made regarding repairing equipment. You should make sure that the power is disconnected when soldering. The use of a soldering gun is not recommended because the huge transients caused by turning the gun on and off may induce voltage to sensitive circuits. Also, many soldering guns are high wattage.

There are a few soldering manufacturers who offer attachments that heat all the IC pins simultaneously. An example is shown in Fig. 9-38. The clip on the opposite side is spring loaded and pulls out the IC as the solder melts. Technicians have mixed feelings about this attachment. Some are happy with the results, while others have difficulties and overheat the printed circuit board which removes the foil.

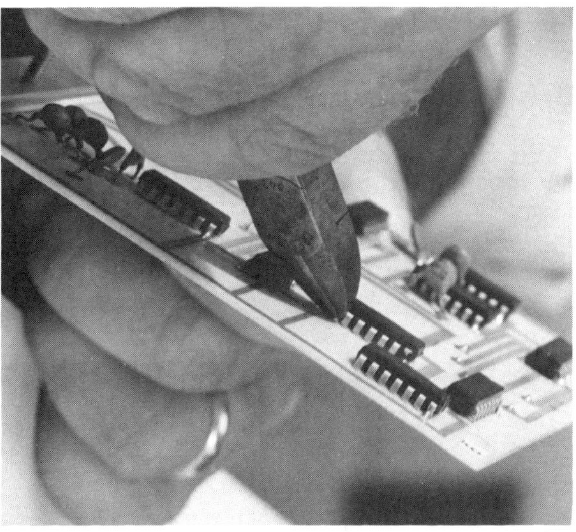

Fig. 9-37. Using a diagonal cutter to clip the pins to remove the body of an IC.

INSERTING AN IC

Inserting an IC in a socket requires a certain knack. If you insert the IC the wrong way, it can be disastrous. The chip will more than likely be destroyed, as well as other chips when power is turned on. Since the pins of the IC are not polarized, the chip will fit in backwards. Thus one common error is inserting the IC in the socket backwards. So the first step when plugging an IC in is to find pin 1. Recall that every IC has a notch or a depression which orients you to pin 1.

To find the socket for pin 1, many manufacturers will have screened (painted) a mark on the printed circuit board for identification. Another way of marking the socket for pin 1 is that some IC sockets have one corner cut

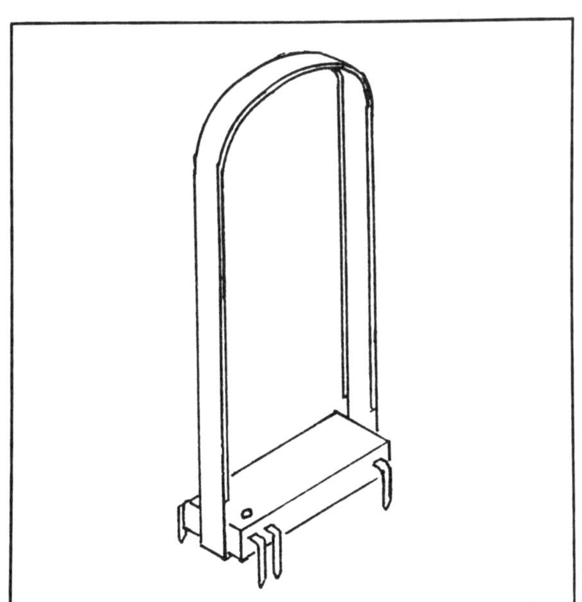

Fig. 9-36. An integrated circuit remover.

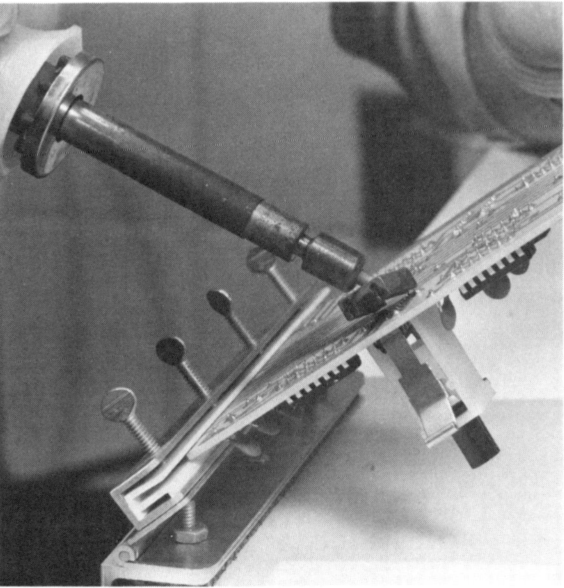

Fig. 9-38. Using an IC desoldering iron to remove an IC from a printed circuit board.

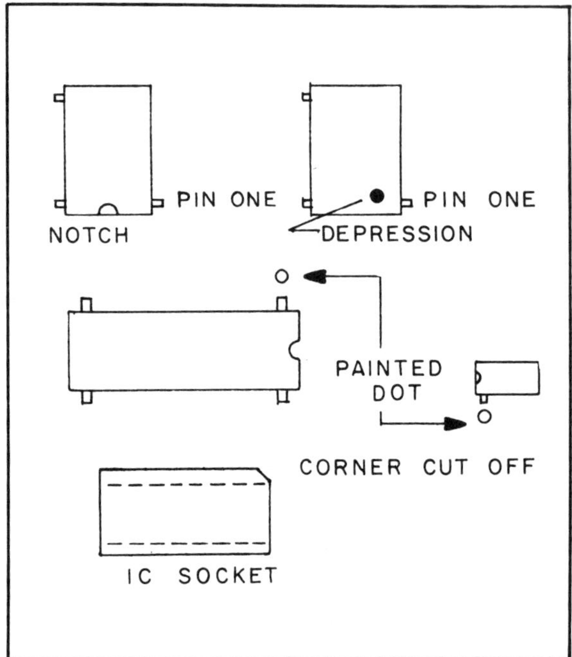

Fig. 9-39. Different techniques for identifying pin 1 on ICs and sockets.

off. Examine Fig. 9-39.

Another insertion problem is offsetting the IC by one pin as shown in Fig. 9-40. It is easier than you think to offset the IC when inserting. The last insertion problem is caused by forcing an IC into its socket. If you have to force it, something is wrong. The results can be a pin or two that will curl up underneath the IC as shown in Fig. 9-41.

The proper way to insert an IC is to first orient the chip's pin with the socket marking or printed circuit marking for pin 1. Then, with the chip held at an angle, line up the pins of one side with the sockets as shown in Fig. 9-42. You should press down gently but with sufficient force to bend the row of pins slightly. The opposite row of pins will then line up with its sockets and you will be sure no pins will be bent under. Now insert the opposite row of pins securely in place. Press down firmly on the IC to make certain it is seated securely.

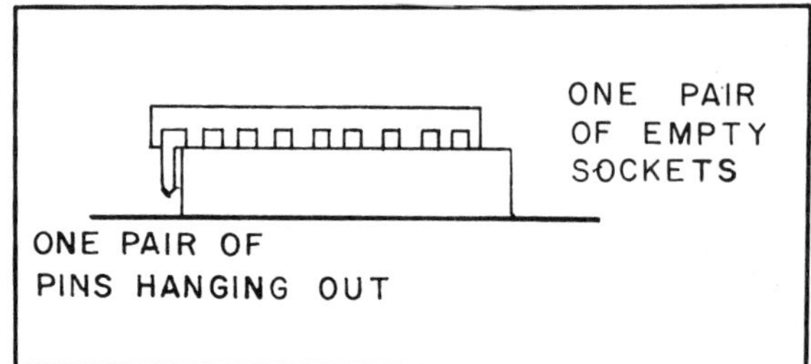

Fig. 9-40. An insertion error of offsetting the IC by one pin.

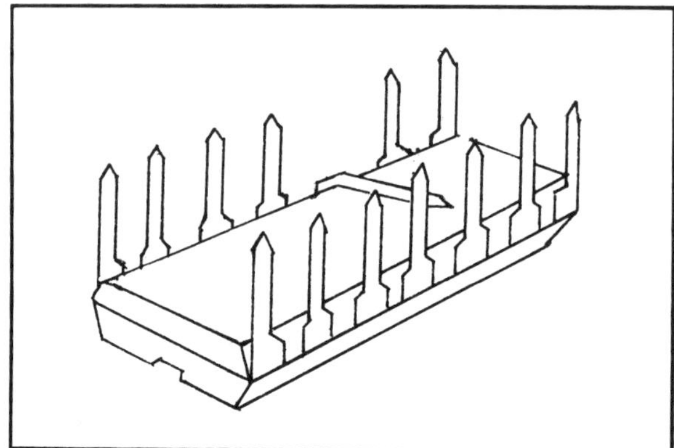

Fig. 9-41. Curling under of a pin as a result of forcing an IC into a socket.

There is a special tool designed for inserting ICs which is shown in Fig. 9-43. By clipping a lead or wire from ground to the terminal at the top, CMOS devices can be inserted.

READING SCHEMATICS

Schematics for digital systems contain the logic symbols for small scale integration, and rectangular blocks are used for larger scale integration. The schematic has numbers adjacent to each connection line that comes in contact with the logic symbols or function blocks. These numbers represent the pin numbers of the IC. Each IC is numbered and contains a letter. The IC number is referenced to a parts list, and the letter desig-

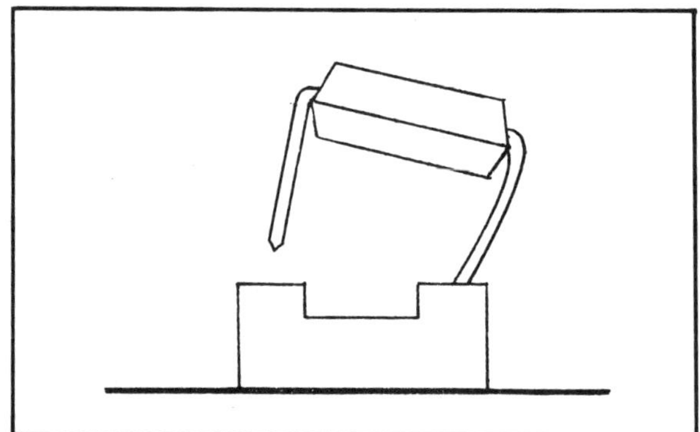

Fig. 9-42. Properly inserting an IC.

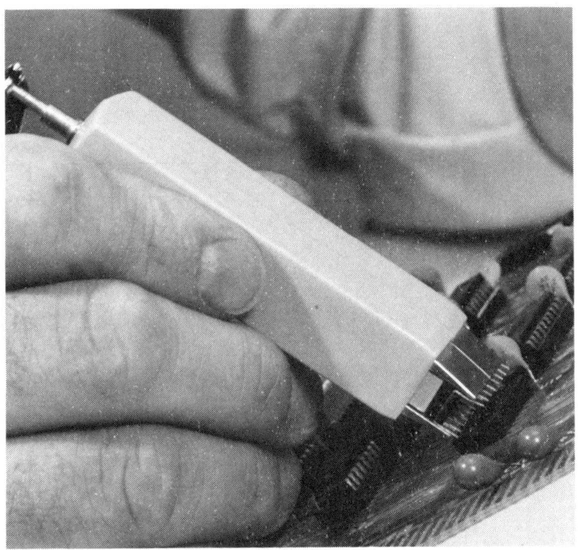

Fig. 9-43. Using a dual-in-line package (DIP) IC insertion tool.

nates which gate or flip-flop in an IC package is used for implementing the logic shown on the schematic. Stars or diamonds generally refer to test points. The solid arrows on one schematic sheet designate the schematic sheet that the signal is coming from or is going to. It also has a table for the signal such as ENA to represent the signal enable or FSET for the signal flag set. The symbol with a tubular shape is generally used to indicate external connectors and will bear the connector pin number and signal. Every company will have its own unique way of using symbols and generally will have their schematic convention described in the service manual. Figure 9-44 presents an example of a digital schematic and a logic diagram that may appear in a service manual.

SELF-EXAMINATION

Select the best answer:

1. The purpose of a truth table is:
   - A. To be sealed inside a gate or memory.
   - B. To explain the internal construction of a digital logic circuit.
   - C. To help analyze an analog circuit.
   - D. To prove which outputs are true and which are false.
   - E. To show all possible combinations of inputs to a digital circuit and the resulting outputs.
2. A gate that requires any or all of the inputs to be HIGH to turn off the output is known as:
   - A. An AND gate.
   - B. A NOR gate.
   - C. An OR gate.
   - D. A NAND gate.

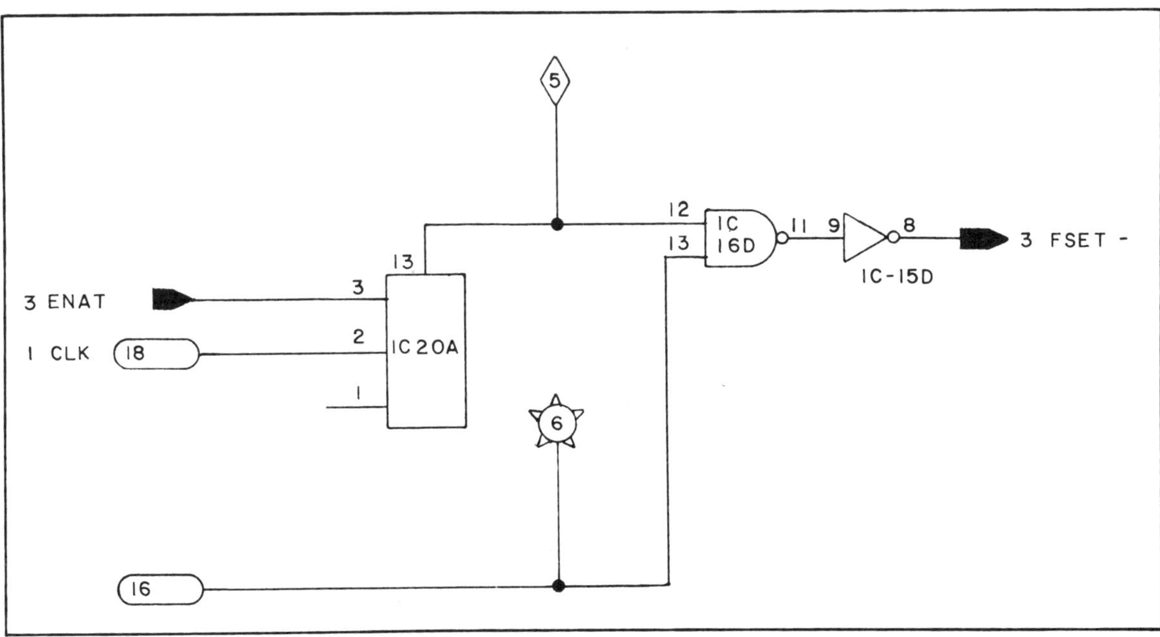

Fig. 9-44. An example of a digital schematic that may appear in a service manual.

3. Inverters are used in logic circuits to:

   A. Amplify signals.
   B. Produce a signal output which is opposite the input.
   C. Change ac to dc current.
   D. Change electronic signals to fluidic currents.

4. The following truth table represents what type of logic gate?

A	B	OUT
L	L	L
L	H	L
H	L	L
H	H	H

   A. AND.
   B. NAND.
   C. OR.
   D. NOR.

5. What type of logic gate corresponds to the following truth table?

A	B	OUT
L	L	H
L	H	L
H	L	L
H	H	L

   A. AND.
   B. OR.
   C. NOR.
   D. Exclusive-OR.
   E. Exclusive-NOR.

6. The following truth table represents what type of logic gate?

A	B	OUT
L	L	H
L	H	H
H	L	H
H	H	L

   A. AND.
   B. NAND.
   C. OR.
   D. NOR.

7. What logic gate does the following truth table correspond to?

A	B	OUT
L	L	L
L	H	H
H	L	H
H	H	L

   A. NOR.
   B. OR.
   C. Exclusive-OR.
   D. Exclusive-NOR.
   E. AND.

8. When measuring the voltage levels of a CMOS device, a rule of thumb for voltage level for:

   A. A LOW is equal to 0 to 0.8 volts dc.
   B. A HIGH is equal to 2-5 volts dc.
   C. A HIGH is equal to at least 70 percent of the power supply.
   D. A LOW is equal to or less than 10 percent of the power supply.

9. To identify the integrated circuits used in a digital system, you should:

   A. Measure the supply voltages terminal of an IC and compare the results to needed supply voltages for specific IC family.
   B. Measure a few of the signal lines.
   C. Consult a service manual.
   D. All of the above.

10. A digital family can be identified by the part number stamped on the IC. Interpret the following IC part number 74C90.

    A. It is a TTL device.
    B. It is a ECL device.
    C. It is a TTL low power device.
    D. It is a TTL compatible CMOS.

11. If you measure the signals of a digital system and found it to be a −5V, what is the digital family?

    A. TTL.
    B. ECL.
    C. CMOS.
    D. RTL.
    E. DTL.

12. If an internal IC has an open circuit, then the input gates driven will:

    A. Appear to be open.
    B. Float either HIGH or LOW to TTL or CMOS.
    C. Pull LOW for an ECL with a pull-down resistor at the input.
    D. All of the above.

13. What voltage levels exist it the 7400– series integrated circuit chips?

    A. GROUND and +15 volts.
    B. −12 volts, ground, and +12 volts.
    C. −0.9 volts and −1.7 volts.

D. GROUND and +5 volts.
E. GROUND and +15 volts.

14. An internally shorted input or output of a gate can be identified by
    A. Input remaining fixed at a HIGH.
    B. Input remaining fixed at a LOW.
    C. Output remaining at a HIGH.
    D. Output remaining at a LOW.
    E. All of the above.

15. A logical 1 state for a CMOS chip with V_{DD} at 15 volts correspond to
    A. GROUND potential.
    B. +5 volts.
    C. +12 volts.
    D. −1.7 volts.

16. When troubleshooting digital equipment that contains MOS chips you should:
    A. Not touch anyone who is working because of static electricity transfer.
    B. Discharge yourself of static electricity by touching ground.
    C. Work as normal, because ECL technology is not affected by static electricity.
    D. Follow the rules in A and B.

17. A logic pulser indicates the presence of a pulse by:
    A. A letter P being displayed.
    B. An LED staying ON.
    C. An LED staying OFF.
    D. An LED blinking ON and OFF.
    E. Either A or D.

18. When removing ICs from a socket, you should use:
    A. A small bladed screwdriver to rock out the IC.
    B. An IC remover tool.
    C. Your fingers and carefully rock it out of the socket.
    D. All of the above.

19. When desoldering an IC chip from a printed circuit board, what wattage soldering iron should be used?
    A. 25 watts.
    B. 45 watts.
    C. 65 watts.
    D. 75 watts.

20. A safety rule to remember when soldering or desoldering is to:
    A. Make sure the power is disconnected.
    B. Make sure not to use a soldering iron that does emit huge transient voltages.
    C. Make sure you use a low wattage soldering iron.
    D. All of the above.

21. When inserting an IC in a socket make sure that it is:
    A. Not offset.
    B. Oriented properly for pin 1 alignment.
    C. Not inserted wrong way (backwards).
    D. Not forced.
    E. All of the above.

Chapter 10

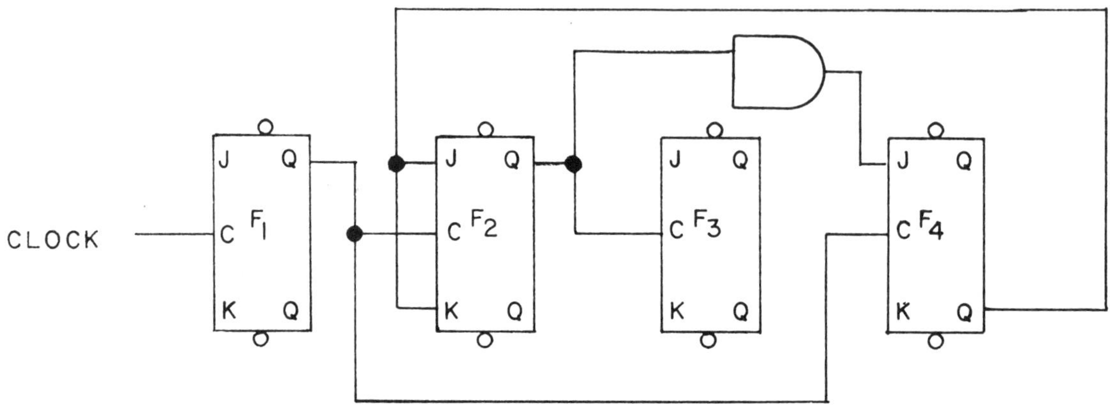

Troubleshooting Combinational and Sequential Digital Circuits

Digital logic gates are combined to perform many functions such as decoding and encoding, multiplexing and demultiplexing, and arithmetic operations. Logic gates are also configured to construct flip-flops, a device which remembers. Flip-flops are combined to create counters and registers. Let's examine these devices and how to troubleshoot them.

DECODING

A decoding function converts some type of coded information such as binary (1 or 0) into a recognizable number form such as decimal. For example, the block diagram shown in Fig. 10-1 converts a four-bit binary code into an appropriate decimal digit. A decoder is a combinational logic circuit that performs the decoding function. It produces an output that indicates the "meaning" of the input.

In Fig. 10-2 is an example of a combinational circuit that can convert a binary code into a decimal. It works as follows: if the binary code is 0000 at the inputs, then the gate 0, whose output represents a decimal zero, will be enabled LOW while all the other gate outputs will remain HIGH. If the binary code 0001 is present at the input, then gate G1 will be enabled LOW while all the other gate outputs would remain HIGH. The output of G1 would represent decimal 1. This process will continue through to binary 1001 at the inputs. Gate G9 would be enabled LOW while all the others would remain HIGH. Once again, the output of G9 would represent decimal 9. This logic circuit is called a *binary coded decimal* (BCD) decoder. Examine the truth table and you will notice that each decimal has a unique binary code and one of ten outputs are indicated by a LOW signal. The 7442 is a BCD-to-decimal decoder which utilizes the above combinational logic circuit.

Four-Line to 16 Line Decoder

A decoder similar to the 7442 is the 74154 which converts a four-bit binary number into one of 16. Since there are 16 possible outputs, the device is called one-of-16 or a four-line to 16-line decoder. The latter descriptor for the device focuses on 4 input lines and 16 output lines. A logic diagram, truth table and pinouts are shown in Fig. 10-3. Notice that there are additional inverters on the inputs. These are inserted to prevent excessive loading (drawing of large amounts of current) of the driving source(s). Also, there is a two-input AND gate which serves as a chip enable. That is, when both

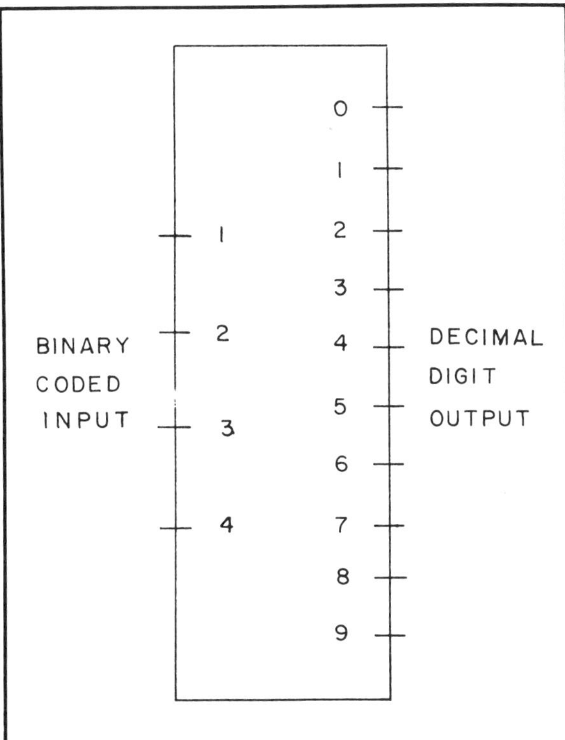

Fig. 10-1. Block diagram of binary to decimal decoder.

inputs are HIGH, then the output is HIGH; and this signal is sent to each of the other gates, thereby enabling them. If the inputs were LOW, then the output of the AND gate would be LOW. A LOW signal to all the other gates' inputs would disable them, and their outputs would be HIGH regardless of the binary inputs.

BCD to Seven Segment Display Decoder

This is a decoder constructed from logic gates to convert a binary coded decimal (BCD) into the appropriate signals which illuminates a seven-segment display. Each decimal has a unique four-bit binary code as follows:

0. 0000	5. 0101
1. 0001	6. 0110
2. 0010	7. 0111
3. 0011	8. 1000
4. 0100	9. 1001

A seven-segment display is shown in Fig. 10-4. Each segment is labeled "a" through "g" in a clockwise rotation, and each is a light emitting diode (LED). Notice the various segments which are activated to produce the appropriate decimal character. Let us examine a 7447 device which is a BCD to seven-segment decoder shown in Fig. 10-5. It has four input lines (A, B, C and D) which accept the BCD code. The logic gates convert the code to light the appropriate segments. The 7447 decoder also has a lamp test input. When this input has a LOW applied all segments will light if they are good.

The decoder also has a ripple blanking input (RBI) and a ripple blanking output (RBO). These are used to blank out any unnecessary zeros in a multidigit display. Figure 10-6 illustrates the blanking feature of leading zero suppression. Notice that the highest order value in the display is always blanked if a binary code equaling decimal zero appears at its BCD inputs and the blanking input is HIGH. Each lower order digit position is blanked if a zero code appears at the BCD inputs and the next higher display is zero as indicated by a HIGH on its blanking outputs.

There are other decoders for liquid crystal displays plus others for alphanumeric displays. Figure 10-7 illustrates an alphanumeric. These displays are manufactured in all the major digital families.

ENCODERS

Whenever you push the button on the key pad of a digital dialing telephone, a hidden circuit converts the decimal number pressed into a binary coded format. An encoder is a combinational logic device that converts a familiar symbol such as decimals, other numeric bases, or alphabetic characters into a coded output such as BCD code.

The basic logic for a decimal to BCD code and its truth table are shown in Fig. 10-8. It functions in the following way: when a HIGH appears on one of the decimal inputs, then the appropriate output line will go HIGH. For example, if decimal 5 input line is HIGH, then the BCD outputs A and C will go HIGH while B and D outputs will go LOW indicating the BCD code (0101) for decimal 5.

Another type of encoder is called a priority encoder. A priority function means that if two decimal input lines are activated at the same time, then the highest order decimal will be encoded and the others disregarded.

MULTIPLEXER/DATA SELECTOR

The process of switching data from several sources or channels into one is called multiplexing. A multiplexer is a combinational logic device which performs the multiplexing function. A multiplexer is frequently used in computer systems to connect several serial data sources onto a single computer input as shown in Fig. 10-9. The

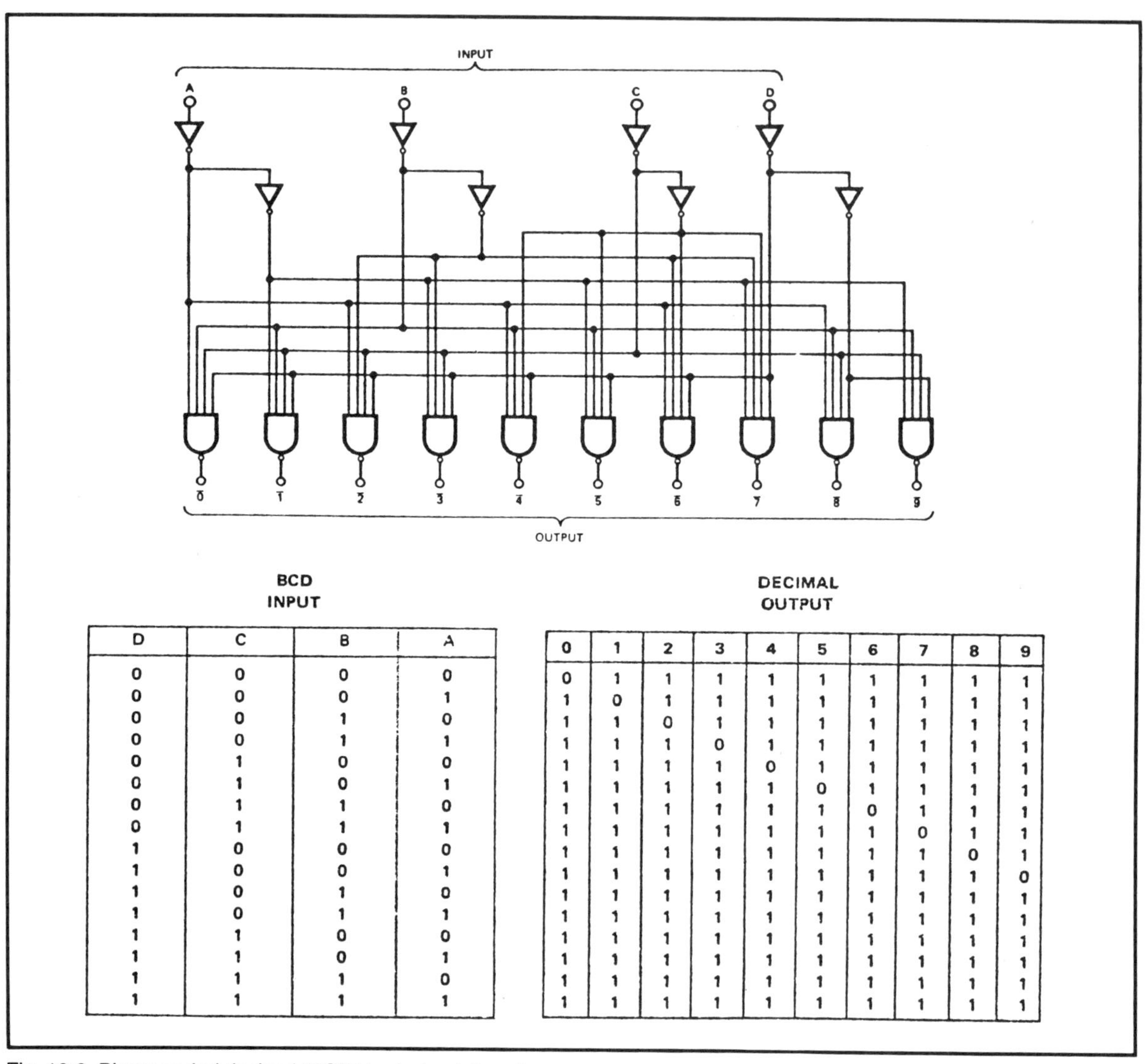

BCD INPUT

D	C	B	A
0	0	0	0
0	0	0	1
0	0	1	0
0	0	1	1
0	1	0	0
0	1	0	1
0	1	1	0
0	1	1	1
1	0	0	0
1	0	0	1
1	0	1	0
1	0	1	1
1	1	0	0
1	1	0	1
1	1	1	0
1	1	1	1

DECIMAL OUTPUT

0	1	2	3	4	5	6	7	8	9
0	1	1	1	1	1	1	1	1	1
1	0	1	1	1	1	1	1	1	1
1	1	0	1	1	1	1	1	1	1
1	1	1	0	1	1	1	1	1	1
1	1	1	1	0	1	1	1	1	1
1	1	1	1	1	0	1	1	1	1
1	1	1	1	1	1	0	1	1	1
1	1	1	1	1	1	1	0	1	1
1	1	1	1	1	1	1	1	0	1
1	1	1	1	1	1	1	1	1	0
1	1	1	1	1	1	1	1	1	1
1	1	1	1	1	1	1	1	1	1
1	1	1	1	1	1	1	1	1	1
1	1	1	1	1	1	1	1	1	1
1	1	1	1	1	1	1	1	1	1
1	1	1	1	1	1	1	1	1	1

Fig. 10-2. Binary coded decimal (BCD) to decimal decoder.

multiplexer acts like a rotary switch. The address inputs S0 and S1 are digital inputs which determine the position of the switch. The address lines are usually controlled by the computer.

Let's take a closer look at a multiplexer. Figure 10-10 shows a 74153, which is a dual 4-line to 1-line multiplexer, and its truth table. The address input lines control which of the four input signals will be transmitted to the output. The strobe input line is used for controlling the output as well as which of the two multiplexers is activated. Examine the truth table; if the address inputs A and B have LOW signals on each, the data input IC0 has a HIGH signal and the strobe or enable line 1G is activated LOW, then the data at input C0 would be transmitted to the output 1Y. On the other hand, if the address inputs A and B have a HIGH signal applied to each, the data input 2C3 has a LOW applied and the enable line 2G is activated LOW, then the data at input 2C3 will be transmitted to the output 2Y.

A multiplexer is sometimes called a data selector and is frequently used in microcomputers systems to route data from one of several sources to the computer's memory. Also the multiplexer is used in some multicharacter displays as shown.

LOGIC DIAGRAM

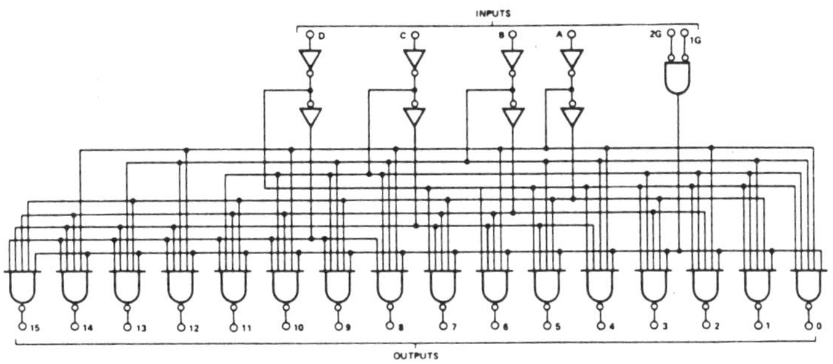

INPUTS

D C B A 2G 1G

15 14 13 12 11 10 9 8 7 6 5 4 3 2 1 0

OUTPUTS

TRUTH TABLE

INPUTS						OUTPUTS															
G1	G2	D	C	B	A	0	1	2	3	4	5	6	7	8	9	10	11	12	13	14	15
L	L	L	L	L	L	L	H	H	H	H	H	H	H	H	H	H	H	H	H	H	H
L	L	L	L	L	H	H	L	H	H	H	H	H	H	H	H	H	H	H	H	H	H
L	L	L	L	H	L	H	H	L	H	H	H	H	H	H	H	H	H	H	H	H	H
L	L	L	L	H	H	H	H	H	L	H	H	H	H	H	H	H	H	H	H	H	H
L	L	L	H	L	L	H	H	H	H	L	H	H	H	H	H	H	H	H	H	H	H
L	L	L	H	L	H	H	H	H	H	H	L	H	H	H	H	H	H	H	H	H	H
L	L	L	H	H	L	H	H	H	H	H	H	L	H	H	H	H	H	H	H	H	H
L	L	L	H	H	H	H	H	H	H	H	H	H	L	H	H	H	H	H	H	H	H
L	L	H	L	L	L	H	H	H	H	H	H	H	H	L	H	H	H	H	H	H	H
L	L	H	L	L	H	H	H	H	H	H	H	H	H	H	L	H	H	H	H	H	H
L	L	H	L	H	L	H	H	H	H	H	H	H	H	H	H	L	H	H	H	H	H
L	L	H	L	H	H	H	H	H	H	H	H	H	H	H	H	H	L	H	H	H	H
L	L	H	H	L	L	H	H	H	H	H	H	H	H	H	H	H	H	L	H	H	H
L	L	H	H	L	H	H	H	H	H	H	H	H	H	H	H	H	H	H	L	H	H
L	L	H	H	H	L	H	H	H	H	H	H	H	H	H	H	H	H	H	H	L	H
L	L	H	H	H	H	H	H	H	H	H	H	H	H	H	H	H	H	H	H	H	L
L	H	X	X	X	X	H	H	H	H	H	H	H	H	H	H	H	H	H	H	H	H
H	L	X	X	X	X	H	H	H	H	H	H	H	H	H	H	H	H	H	H	H	H
H	H	X	X	X	X	H	H	H	H	H	H	H	H	H	H	H	H	H	H	H	H

H = High, L = Low, X = Irrelevant

Fig. 10-3. Four-line to 16-line decoder.

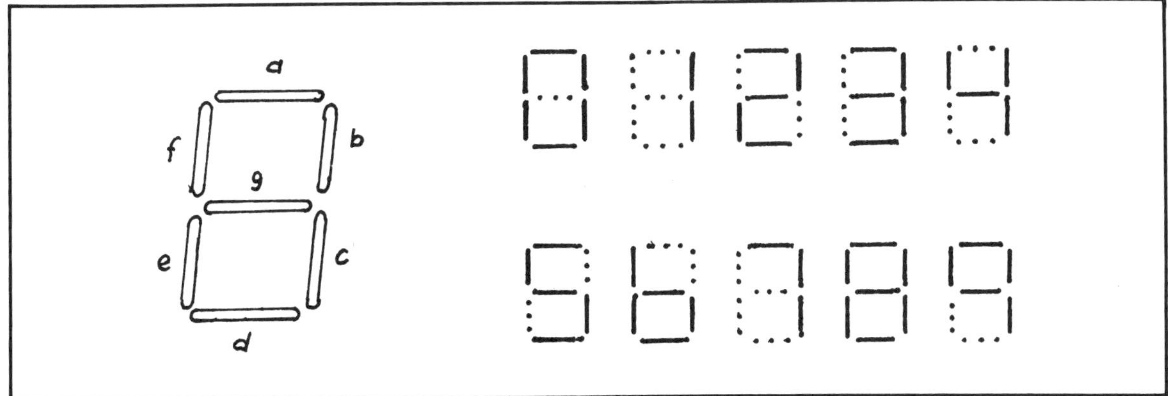

Fig. 10-4. Seven-segment display format showing arrangement of segments.

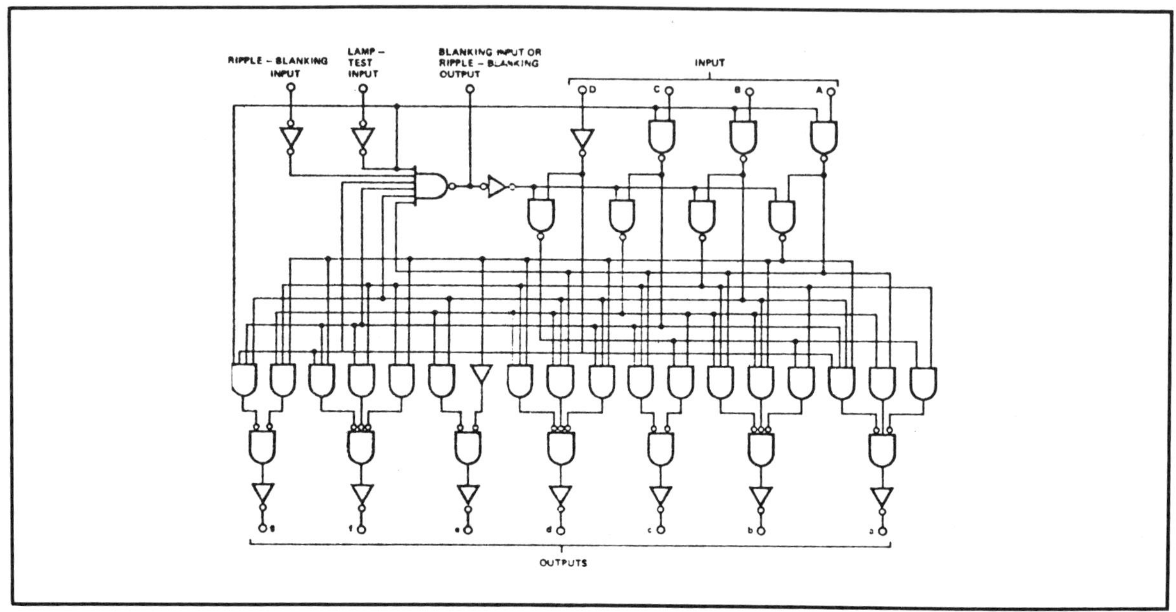

Fig. 10-5. BCD to seven-segment display decoder.

DEMULTIPLEXER

The function of a demultiplexer is to separate a multiplexed signal onto separate output channels. Figure 10-11 shows a single input data line which is switched to one of the four output channels. Also shown is a simplified communication system with a multiplexer on the left and a demultiplexer on the right. It works this way: the address input of both the multiplexer and demultiplexer are the same, binary value (00) S1 of both devices are switched on, the N is transmitted, the address inputs of both devices are incremented by one (01), S2 on both devices are switched on, the E is transmitted, and the process continues.

Figure 10-12 shows a dual 2-line to 4-line demultiplexer with its truth table and pinout. One of the dual multiplexers could be used to implement the above com-

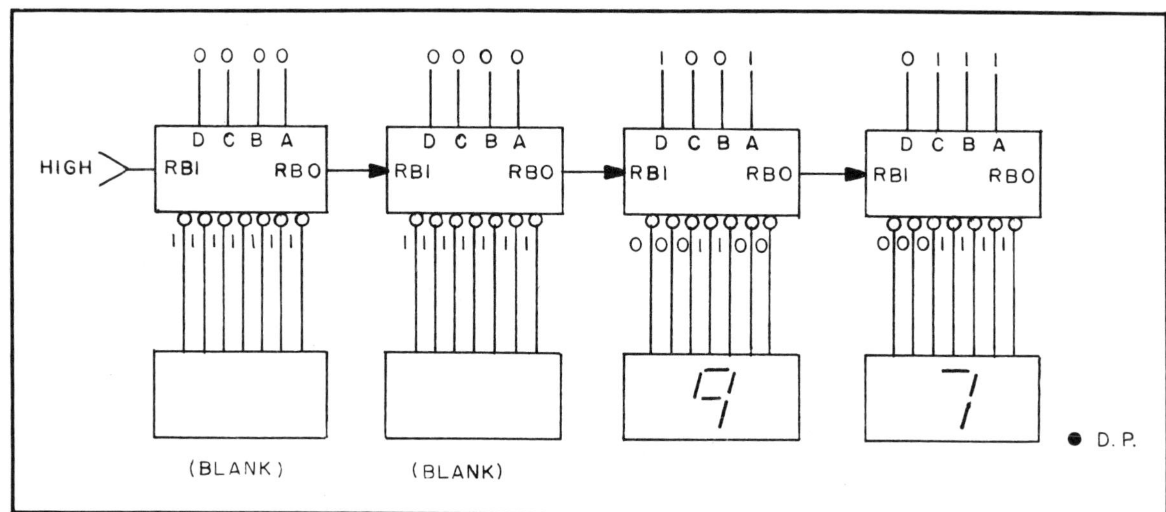

Fig. 10-6. Illustration of leading edge zero suppression display when the BCD inputs are zero and the (RBI) is HIGH.

Fig. 10-7. Alphanumeric dot matrix display. (courtesy Hewlett-Packard Company).

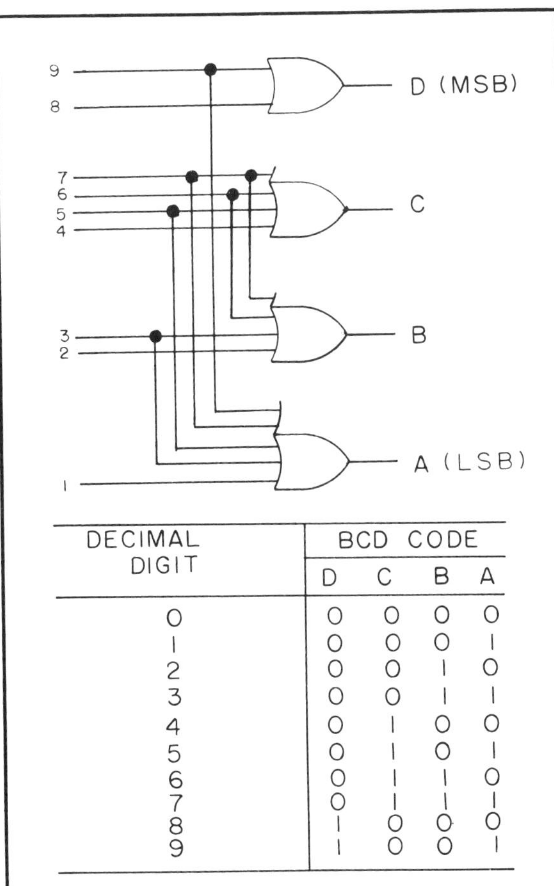

DECIMAL DIGIT	BCD CODE			
	D	C	B	A
0	0	0	0	0
1	0	0	0	1
2	0	0	1	0
3	0	0	1	1
4	0	1	0	0
5	0	1	0	1
6	0	1	1	0
7	0	1	1	1
8	1	0	0	0
9	1	0	0	1

Fig. 10-8. Decimal-to-binary coded decimal (BCD) decoder.

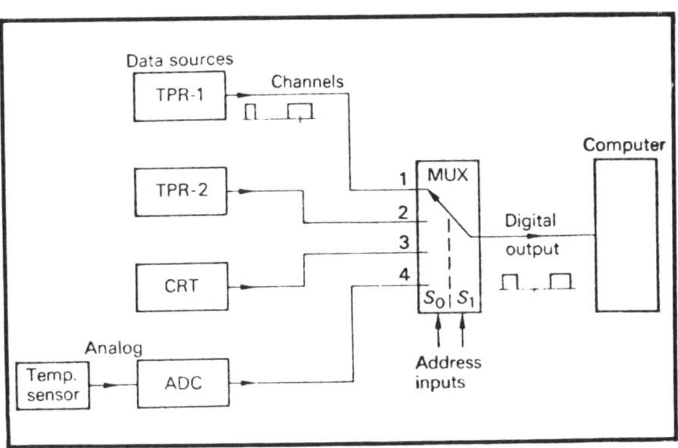

Fig. 10-9. Multiplexing four input channels into a computer.

munication system demultiplexer. Examine the logic diagram. In implementing the above communication demultiplexer, the transmitted data comes in on data line 1C. The four outputs to which receiving devices are attached are 1Y0, 1Y1, 1Y2, and 1Y3. The select line A and B are like the address lines of the multiplexer and determine which output is selected based upon the appropriate signal level (H or L) as shown in the truth table.

BINARY NUMBER SYSTEM

Counting in decimal numbers is second nature to everyone. However, not everyone is familiar with the nature of a number scheme. It is very important to know how the binary numbering system is used in digital equipment so that you, as a troubleshooter, can interpret the outputs of devices such as registers and counters.

Basically, a number is a group of symbols representing a quantity. Each symbol in the group is called a digit. Each digit is unique in that it is different in appearance and value. The decimal system that we use has 10 digits that are 0, 1, 2, 3, 4, 5, 6, 7, 8, 9. The system's name is derived from the number of digits. For example, base 10 means there are 10 digits.

It is fascinating to note that with only 10 digits, any quantity can be represented. When you count greater than 9, it is necessary to assign a *positional* value; that is, allow the digits to represent a greater power of the base number by placing them in a higher order position. Examine the base 10 number of "7777." Each "7" is in a different position and therefore represents a different power. Look how the number "7777" may be broken down:

$$
\begin{array}{llll}
7000 & +\ 700 & +\ 70 & +\ 7 \\
(7\times1000) & +\ (7\times100) & +\ (7\times10) & +\ (7\times1) \\
7\times10^3 & +\ 7\times10^2 & +\ 7\times10^1 & +\ 7\times10^0
\end{array}
$$

LOGIC DIAGRAM

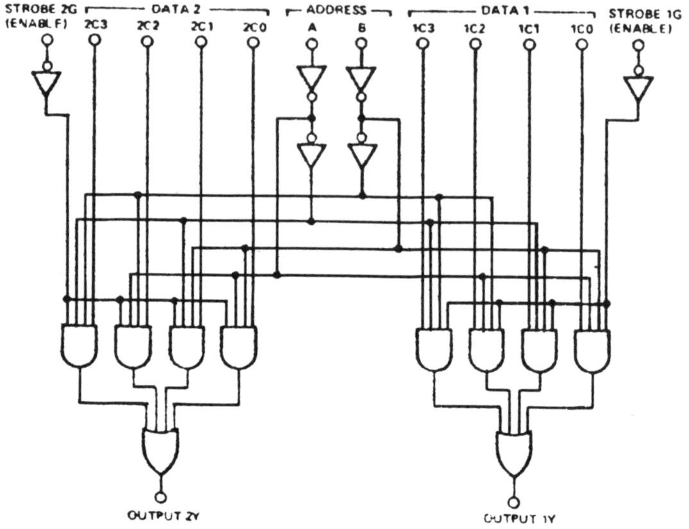

STROBE 2G (ENABLE) — DATA 2: 2C3 2C2 2C1 2C0 — ADDRESS: A B — DATA 1: 1C3 1C2 1C1 1C0 — STROBE 1G (ENABLE)

OUTPUT 2Y   OUTPUT 1Y

ADDRESS INPUTS		DATA INPUTS				STROBE	OUTPUT
B	A	C0	C1	C2	C3	G	Y
X	X	X	X	X	X	H	L
L	L	L	X	X	X	L	L
L	L	H	X	X	X	L	H
L	H	X	L	X	X	L	L
L	H	X	H	X	X	L	H
H	L	X	X	L	X	L	L
H	L	X	X	H	X	L	H
H	H	X	X	X	L	L	L
H	H	X	X	X	H	L	H

Address inputs A and B are common to both sections. H = high level, L = low level, X = irrelevant.

Fig. 10-10. Dual 4-line to 1-line multiplexer and truth table.

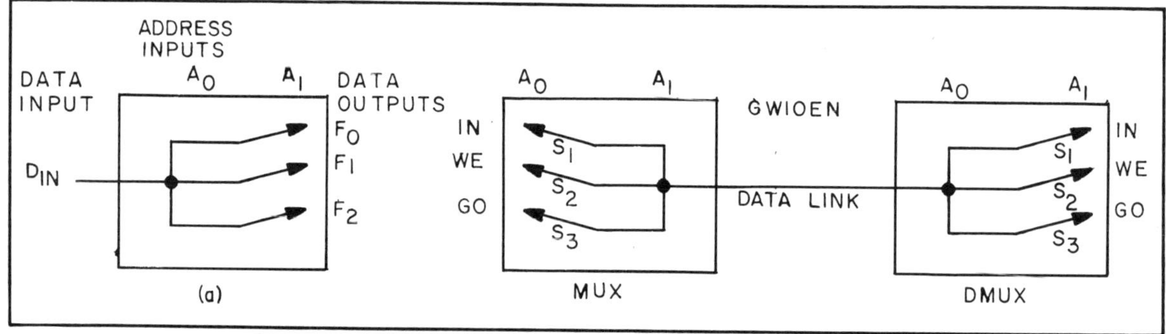

Fig. 10-11. A demultiplexer and its use in a multiplexed system.

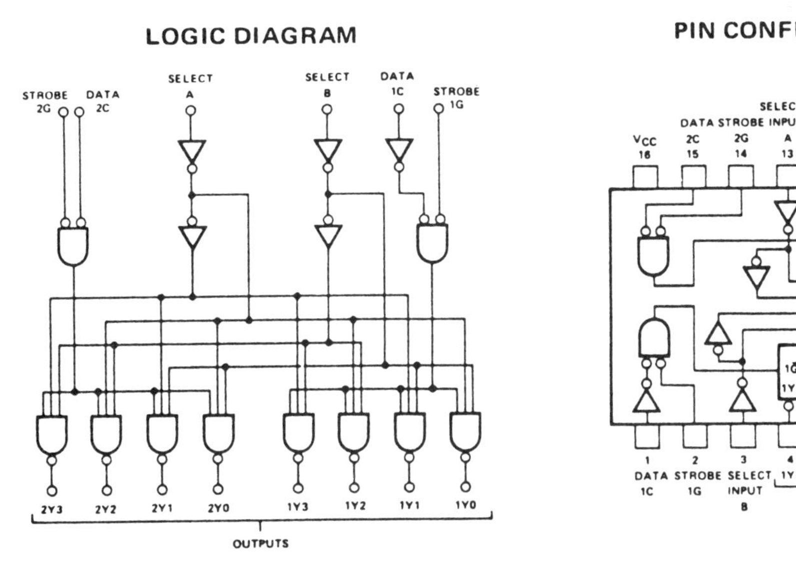

TRUTH TABLES

INPUTS				OUTPUTS			
SELECT		STROBE	DATA				
B	A	1G	1C	1Y0	1Y1	1Y2	1Y3
X	X	H	X	H	H	H	H
L	L	L	H	L	H	H	H
L	H	L	H	H	L	H	H
H	L	L	H	H	H	L	H
H	H	L	H	H	H	H	L
X	X	X	L	H	H	H	H

Fig. 10-12. Dual 2-line to 4-line demultiplexer.

A closer examination would reveal that the numbers are arranged in "powers of 10." It is important to note that any number taken to the zero power is 1. Therefore, $10^0=1$ as would $7^0=1$ or $2^0=1$.

There are many other numbering systems, but one of particular importance to us in working with digital electronics is the base 2 system, or binary system. The base 2 system has two symbols "0" and "1." Even though no other symbols are allowed, any quantity may be represented by them by the use of positional values. Let's analyze the binary number 1101.

$$
\begin{aligned}
1101 &= (1\times2^3) + (1\times2^2) + (0\times2^1) + (1\times1^0)\\
&= (1\times8) + (1\times4) + (1\times2) + (1\times1)\\
&= 8 + 4 + 0 + 1\\
&= 13 \text{ in base } 10
\end{aligned}
$$

Notice that it takes more digits to express a number in binary than in the decimal system.

Converting a Decimal Number to Binary

To convert a decimal number to binary number, divide the decimal number by base 2 until the quotient equals zero. For example, convert decimal number 14 to binary. You begin by dividing 14 by 2 which gives a quotient of 7 and a remainder of 0 then divide 7 by 2 giving a quotient of 3 and a remainder of 1. You continue until the quotient equals 0. The remainders form the binary number.

	Remainders
$14 \div 2 = 7$	0 (least significant bit)
$7 \div 2 = 3$	1
$3 \div 2 = 1$	1
$1 \div 2 = 0$	1

The first remainder is the least significant bit (LSB), while the last remainder is the most significant bit (MSB).

FLIP-FLOPS

There are essentially three basic logic elements: the OR, AND and inverter. However, a fourth one called a flip-flop may be added. It belongs to a class of electronic circuits called multivibrators and is the bistable multivibrator. The primary function of this circuit is to store or remember the binary digits of "1" or "0" for a definite period of time. The flip-flop is differentiated from previously discussed logic elements in that it has ability to retain either logic level at its output after input conditions have been removed. The flip-flop is constructed by combining basic gates. The basic digital elements of gates, inverters, and flip-flops are used to construct many more complex logic circuits such as counters and registers as well as arithmetic operations.

There are four types of flip-flops: RS-latch, strobed RS latch, D-type and J-K type. Flip-flops are available in integrated circuit (IC) form from all the basic digital families.

RS Latch

The simplest flip-flop available for storing a logical "1" or a "0" is the R-S latch. Figure 10-13 shows the symbol for the R-S latch. This device can be constructed from NAND or NOR gates. For example, the NAND gate is used for analysis of the R-S latch.

Since the flip-flop initial condition can be set or reset, this analysis will begin with the assumption that the flip-flop will be reset and represent a binary 0. Examine the latch made from NAND gates and its truth table shown in Fig. 10-14. Since the flip-flop is reset, the output of Q is LOW and feeds the input of G2, and the output of $\overline{Q}$ is HIGH and feeds the input of G1. This is a latched condition and will remain in this condition until

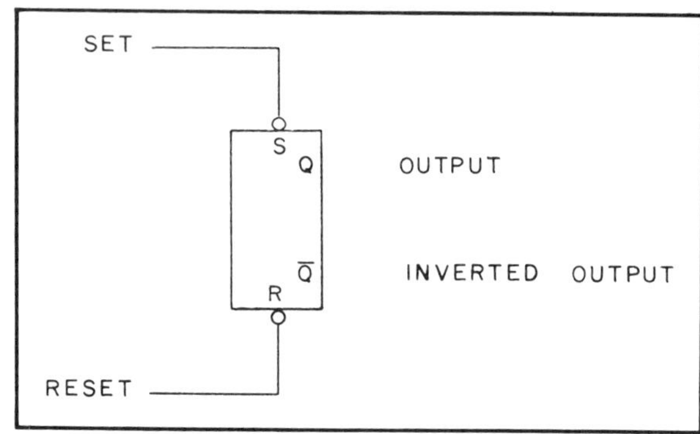

Fig. 10-13. Reset/set (RS) latch schematic symbol.

the appropriate input signals at the set or reset alters the state.

If the set and reset signals are both LOW at the same time, then G1 will have a HIGH and LOW input forcing its output HIGH. The same input conditions exist for G2, and its output is also forced HIGH. Since both outputs are HIGH at the same time, you don't know if the flip-flop contains a logic 1 or a logic 0; therefore, it is called an indeterminate state.

The designers of systems avoid the conditions which create the indeterminate state. If the set input has a LOW and the reset input has a HIGH, then G1 will have the appropriate input condition to force the output Q HIGH, while G2 will have both inputs HIGH and the output Q will go LOW. These are the appropriate conditions for setting the R-S flip-flop, and it is said to contain a logical 1. However, if the set input is HIGH and the reset input is LOW, then the R-S flip-flop is reset and contains a logical 0. Should both the set and reset input have a HIGH signal, then no change in the output condition will occur. Since the initial condition of the R-S flip-flop was reset, it will remain reset. The truth table summarizes the action of the R-S latch.

Strobed R-S Flip-Flop

The strobed R-S flip-flop, its truth table and timing diagram are shown in Fig. 10-15. Once again it is assumed that the initial state is reset. Notice that the truth table is inverted as compared to the R-S latch. Upon examining the truth table you will notice the following: 1) if R and S are different, Q and $\overline{Q}$ follow the HIGH, and 2) if both R and S are LOW, Q never changes; it remembers the previous state.The R-S latched component behaves as a normal R-S latch. The gates G3 and G4 pass the set or reset input signal only when a clock or strobe pulse goes from a LOW to HIGH signal level. The timing diagram in

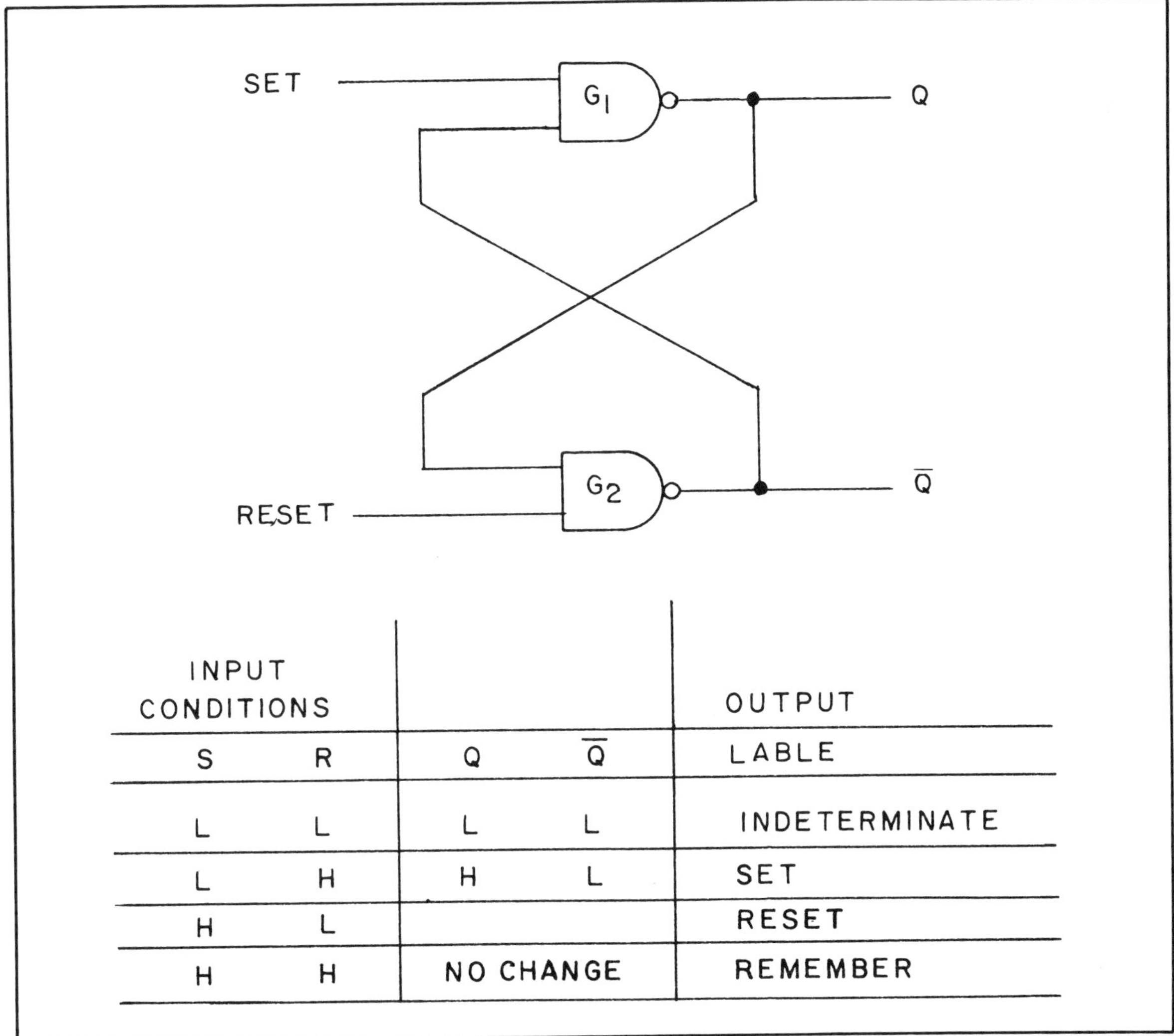

INPUT CONDITIONS				OUTPUT
S	R	Q	$\overline{Q}$	LABLE
L	L	L	L	INDETERMINATE
L	H	H	L	SET
H	L			RESET
H	H	NO CHANGE		REMEMBER

Fig. 10-14. R-S latch circuit and truth table.

Fig. 10-15 illustrates this point. Notice that only when a clock or strobe pulse occurs do the steering gates pass the input signal.

J-K Type Flip-Flops

While the R-S flip-flop works well for a temporary storage device, it is not sufficiently versatile to be used in constructing complex memory systems, counters or resisters. A variation of the R-S flip-flop which retains the latch or memory capabilities of the R-S flip-flop is the J-K flip-flop. Figure 10-16 shows a schematic, truth table and a timing diagram. Notice there are set and reset inputs, which behave the same as the R-S type, as do the $\overline{Q}$ and Q outputs. The J and K inputs are control inputs, while the C input is the clock. Notice the bubble on the clock input. It signifies that a HIGH to LOW clock transition is required before appropriate action is taken.

Examine the truth table and you will notice the following: 1) if the J and K are both LOW, Q never changes, 2) if J and K are different, Q follows whatever signal is at J, and 3) if J and K are both HIGH, Q changes state on every clock pulse. The timing diagram in Fig. 10-16 illustrates these points. Notice that the triggering action occurs at the trailing edge of the clock pulse.

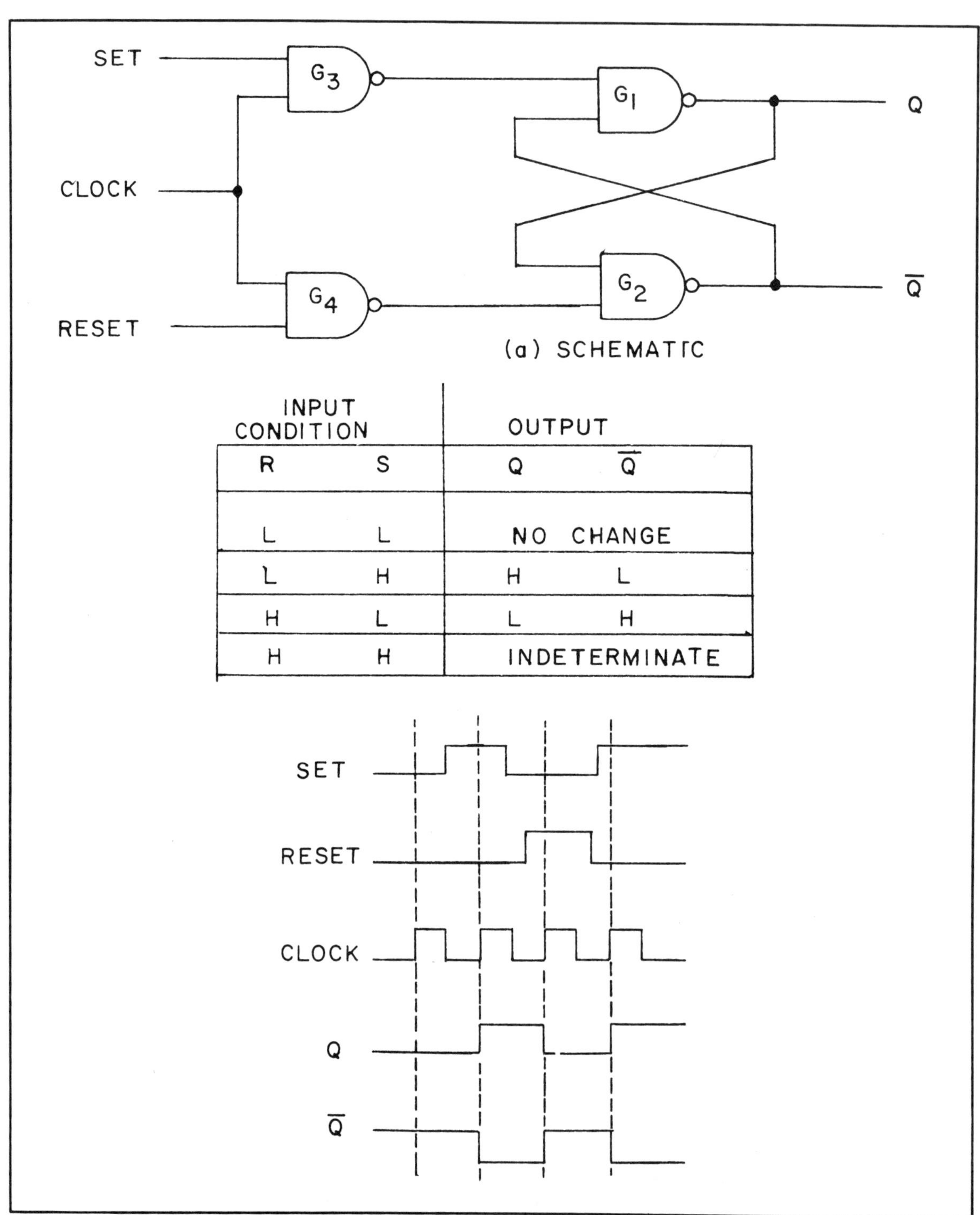

INPUT CONDITION		OUTPUT	
R	S	Q	$\overline{Q}$
L	L	NO CHANGE	
L	H	H	L
H	L	L	H
H	H	INDETERMINATE	

Fig. 10-15. Strobed R-S flip-flop circuit, truth table, and timing diagram.

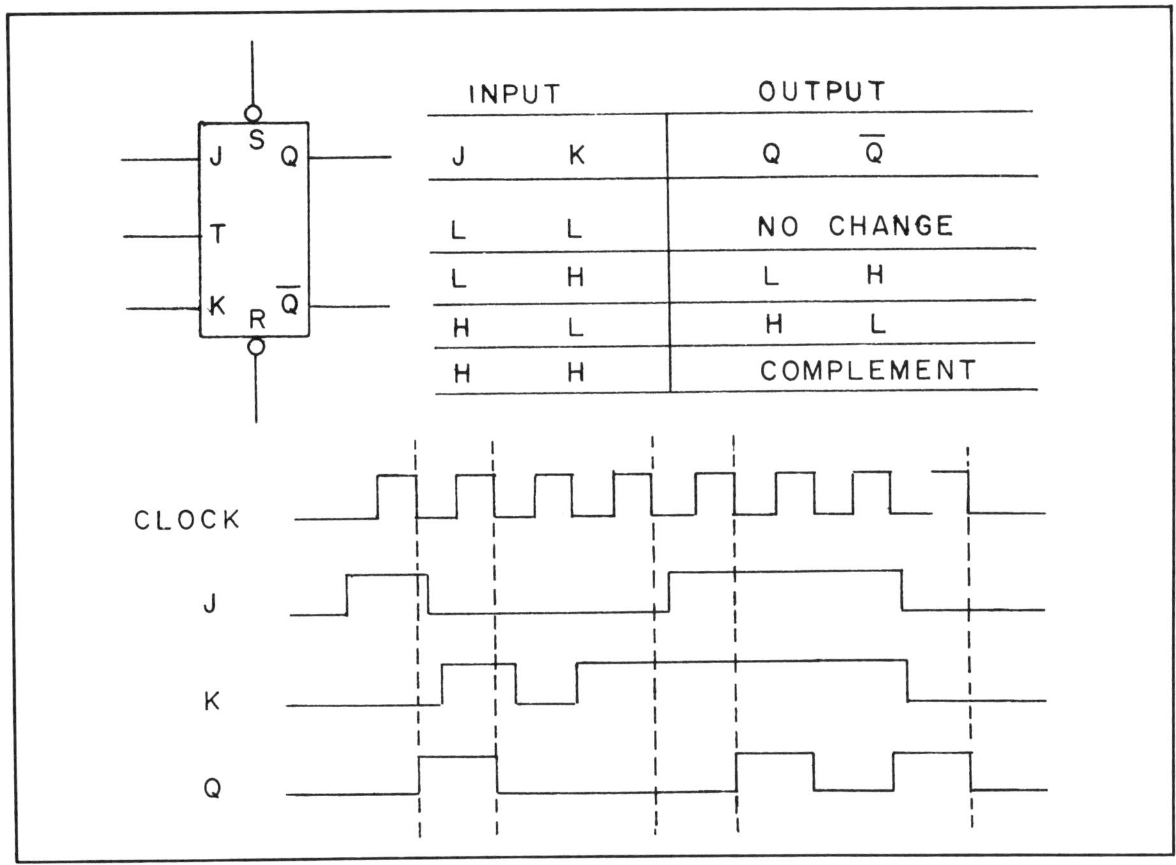

Fig. 10-16. J-K flip-flop schematic symbol, truth table, and timing diagram.

D-Type Flip-Flop

The D-type flip-flop is very useful in cases where a single data bit (1 or 0) needs to be stored. Its schematic symbol, truth table and timing diagram are shown in Fig. 10-17. Notice that there is only one input in addition to the clock. This is called the data input and labeled D. Its operation is very simple. If there is a HIGH on the D input and there is a clock pulse, then the flip-flop will be set and a logical 1 will be stored. Now, if there is a LOW on the D input and there is a clock pulse, then the flip-flop will be reset and a logical 0 will be stored. Another way to describe the action of a D type flip-flop is that the Q output assumes the state of the D input when the clock pulse occurs. Notice that the D-type triggers on the leading edge of the clock pulse.

Buffer/Storage

Very often the transfer of data from one subsystem to another must be latched temporarily. A common IC configuration is a quad, hex or octal D-type flip-flop. A logic diagram of a quadruple bistable D latch is shown in Fig. 10-18. Notice that there is a bubble on the clock input to each flip-flop. This means that a trailing edge of a clock pulse will trigger it. Thus the data on the input line will be latched in the buffer or storage resister on the trailing edge of the clock pulse.

COUNTERS

Counters are sequential circuits that perform tasks ranging from simple counting to controlling the sequence of several events. Some basic types of TTL counters are shown in Table 10-1. Notice that some counters are presettable; that is, a value can be loaded into the counter and then count up or down from the preset value. Also identified is whether the clock triggers the flip-flop on a rising edge (positive) or trailing edge (negative). The

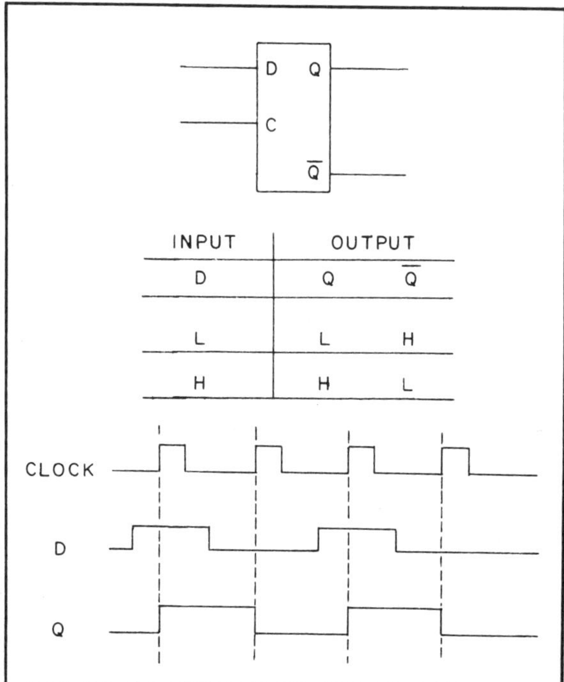

Fig. 10-17. D-type flip-flop schematic symbol, truth table and timing diagram.

modulus of a counter is the number of count states that a counter uses and is indicated in the counter table.

Ripple-Counter

Earlier we noted that the J-K flip-flop was very versatile and that holding the J and K input HIGH while clocking it would cause it to toggle or change states. This behavior is employed in constructing a binary counter. Figure 10-19 shows a simplified binary counter and a timing diagram.

It is assumed that all flip-flops are initially reset. The first clock pulse to FF-1 changes the output to a HIGH. The second clock pulse changes FF-1 output to a LOW. As FF-1's output Q changes from a HIGH to a LOW, it serves as a clock input to FF-2 and its output Q goes from LOW to HIGH. The third clock pulse will set FF-1 and FF-2 will remain set. The fourth clock pulse will cause FF-1 to reset and FF-2 to reset. And as FF-2 resets, it triggers and sets FF-3. This process continues for fifteen counts and will have all the flip-flops set as shown in the truth table. A timing diagram illustrates the pulsing action of the ripple counter.

It should be pointed out that the truth table shows FF-1 on the right, whereas in the counter circuit FF-1 is on the left. Both are conventions. The truth table always has the least significant bit on the right, just as we have the units position on the right when we count in the decimal system.

This type of counter is called a ripple counter because the clock sort of "ripples" down through the flip-flop. It is also an asynchronous counter because the clock pulse affects only the first flip-flop. The other flip-flops are triggered by the previous flip-flop.

Decade Counter

A binary counter can be used to construct a decade counter by resetting the counter after it counts from 0 to 9 and then reset to 0 at the next count. Figure 10-20 shows a schematic of a decade counter.

The flip-flops, FF-1, FF-2 and FF-3 will count from 1 to decimal 7 on the first seven counts. With the counters containing 7 (1110), the outputs of FF-2 and FF-3 feed and enable the AND gate, causing the J input of FF-4 to go HIGH. With the J input HIGH on FF-4, the next clock

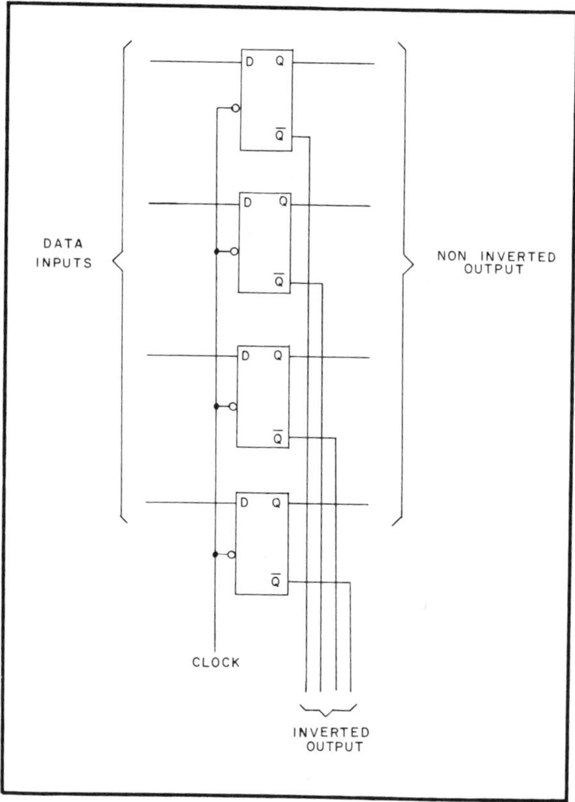

Fig. 10-18. Simplified quad D-type latch circuits.

Table 10-1. TTL MSI COUNTERS.

Type	Mod.	Count Direction	Synchronous Max. Clock Rate MHz	Ripple Max. Clock Rate MHz	Configuration	Clock	Unit Cascadable
Non-Presettable							
7490	10	up		18	2×5	neg. edge	no
7492	12	up		18	2×6	neg. edge	no
7493	16	up		18	2×8	neg. edge	no
Presettable							
74160	10	up	32		1×10	pos. edge	yes
74161	16	up	32		1×16	pos. edge	yes
74190	10	up/down	25		1×10	pos. edge	yes
74191	16	up/down	25		1×16	pos. edge	yes
74192	10	up/down	32		1×10	pos. edge	yes
8280	10	up		35	2×5	neg. edge	no
74177	16	up		35	2×8	neg. edge	no
8283	12	up		35	2×6	neg. edge	no

pulse complements FF-4 and resets FF-1, FF-2 and FF-3. The counter reads 0001 or decimal 8.

On the ninth clock pulse, FF-1 complements, and the counter reads 1001, or decimal 9. However, the AND gate is no longer enabled; therefore a LOW appears on the J input to FF-4. As a result, FF-4 is reset on the next clock pulse. Thus on the tenth clock pulse, FF-1 and FF-4 both reset, restoring the counter to 0000. A feedback line from FF-4 to FF-2 prevents FF-2 from responding to the HIGH to LOW transition at its clock input as the counter reset occurs.

Synchronous Binary Counter

The synchronous binary counter shown in Fig. 10-21 is made from four rising edge JK flip-flops. Notice that each flip-flop clock is connected to a single external clock so that each is clocked at the same time. Not all the inputs are tied HIGH as in the asynchronous counter, but some are controlled with logic gates.

The simplified timing diagram shows that FF-1 is toggled on the rising edge because of the inverter inserted at the clock input. The first clock pulse caused FF-1 to be set since both J and K inputs are connected permanently HIGH. The Q of FF-1 feeds a HIGH signal to the inputs of J and K of FF-2. The second clock pulse sets FF-2 because J and K inputs were HIGH and resets FF-1. The Q of FF-1 and FF-2 feed the AND gate G1. Only the Q of FF-2 is HIGH; therefore the AND gate G1 is disabled, and its output will send a LOW to the J and K inputs of FF-3. The third clock pulse will not affect either the FF-2 or FF-3 state because the J and K inputs are LOW, but FF-1 will be set. The count contains 0011 or decimal 3.

The fourth clock pulse will affect FF-1, 2 and 3. Since the AND gate G1 has both inputs HIGH, it is enabled and feeds a HIGH signal on the J and K inputs of FF-3. The results of the fourth clock pulse will cause FF-3 to set, and FF-1 and FF-2 will be reset. The content of the counter will be 0100 or decimal 4. Rather than review each count, let's examine how FF-4 is toggled.

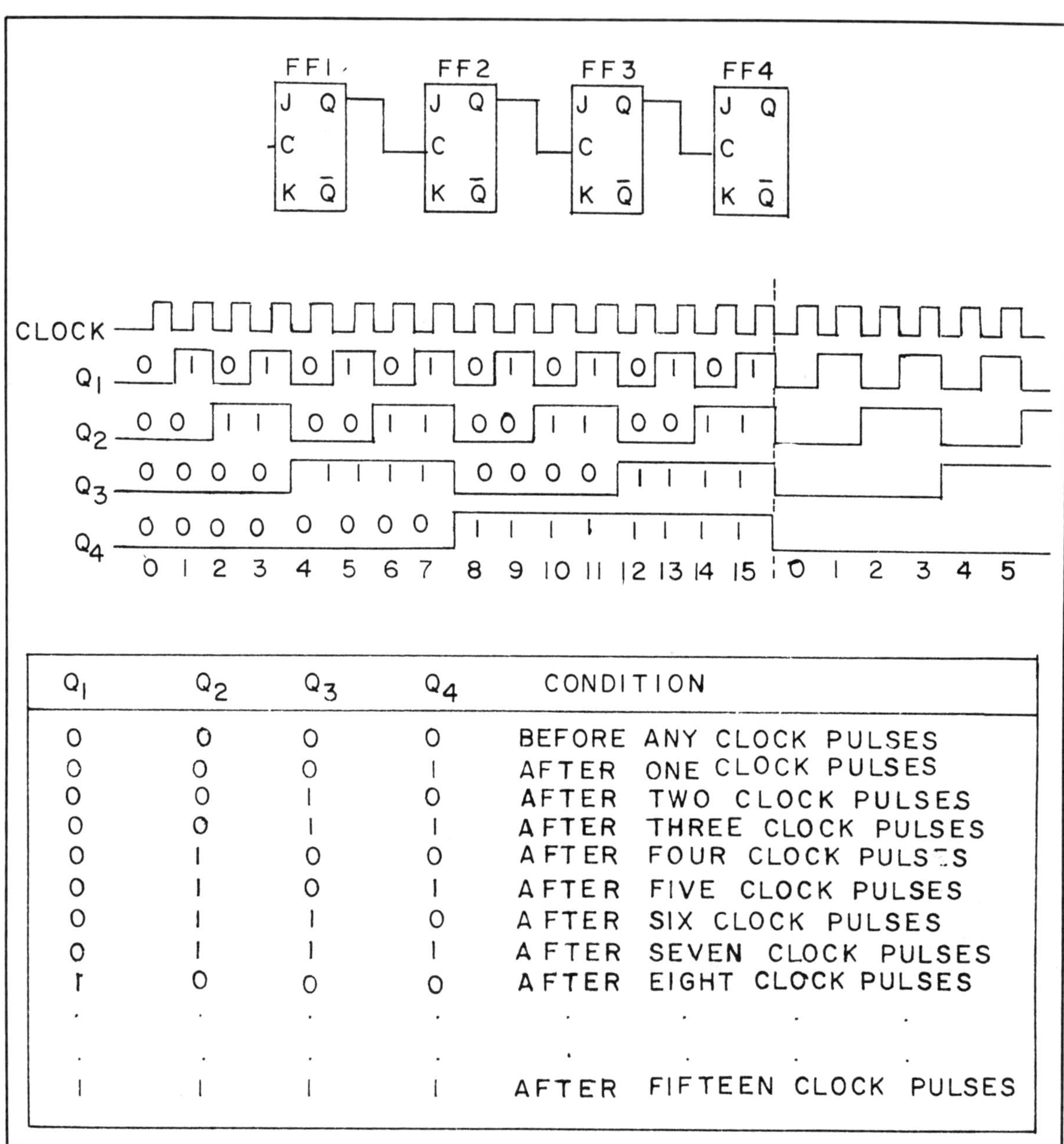

Q1	Q2	Q3	Q4	CONDITION
0	0	0	0	BEFORE ANY CLOCK PULSES
0	0	0	1	AFTER ONE CLOCK PULSES
0	0	1	0	AFTER TWO CLOCK PULSES
0	0	1	1	AFTER THREE CLOCK PULSES
0	1	0	0	AFTER FOUR CLOCK PULSES
0	1	0	1	AFTER FIVE CLOCK PULSES
0	1	1	0	AFTER SIX CLOCK PULSES
0	1	1	1	AFTER SEVEN CLOCK PULSES
1	0	0	0	AFTER EIGHT CLOCK PULSES
.	.	.	.	. . . .
.	.	.	.	. . . .
1	1	1	1	AFTER FIFTEEN CLOCK PULSES

Fig. 10-19. Binary ripple counter, timing diagram and clock condition.

At count seven the counter contains 0111. The outputs of the first three flip-flops are fed to the input of the AND gate G2. Since all inputs to G2 are HIGH, the gate is enabled HIGH and sends a HIGH to the J and K inputs of FF-4. On the eight count the clock will trigger FF-4 because the J and K inputs are HIGH and FF-1, FF-2 and FF-3 will be reset. The counter will contain 1000 or decimal 8.

There are many other types of counters. Some are up counters; others are only down counters, while some are a combination as shown in Table 10-1. Logic gates are used to control the direction of counting.

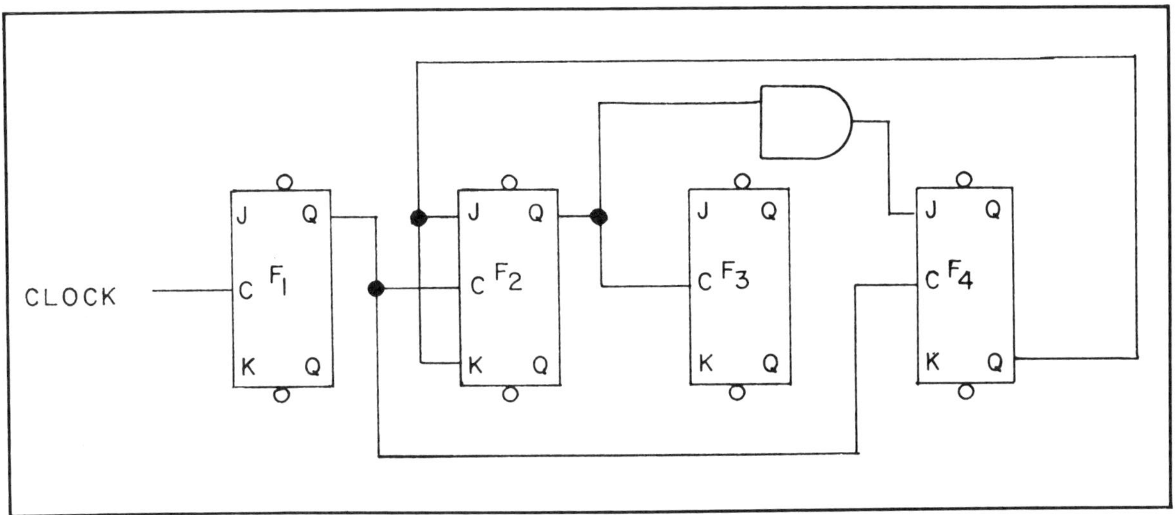

Fig. 10-20. Decade counter, counts 0 through 9, then resets to 0 and repeats the count sequence.

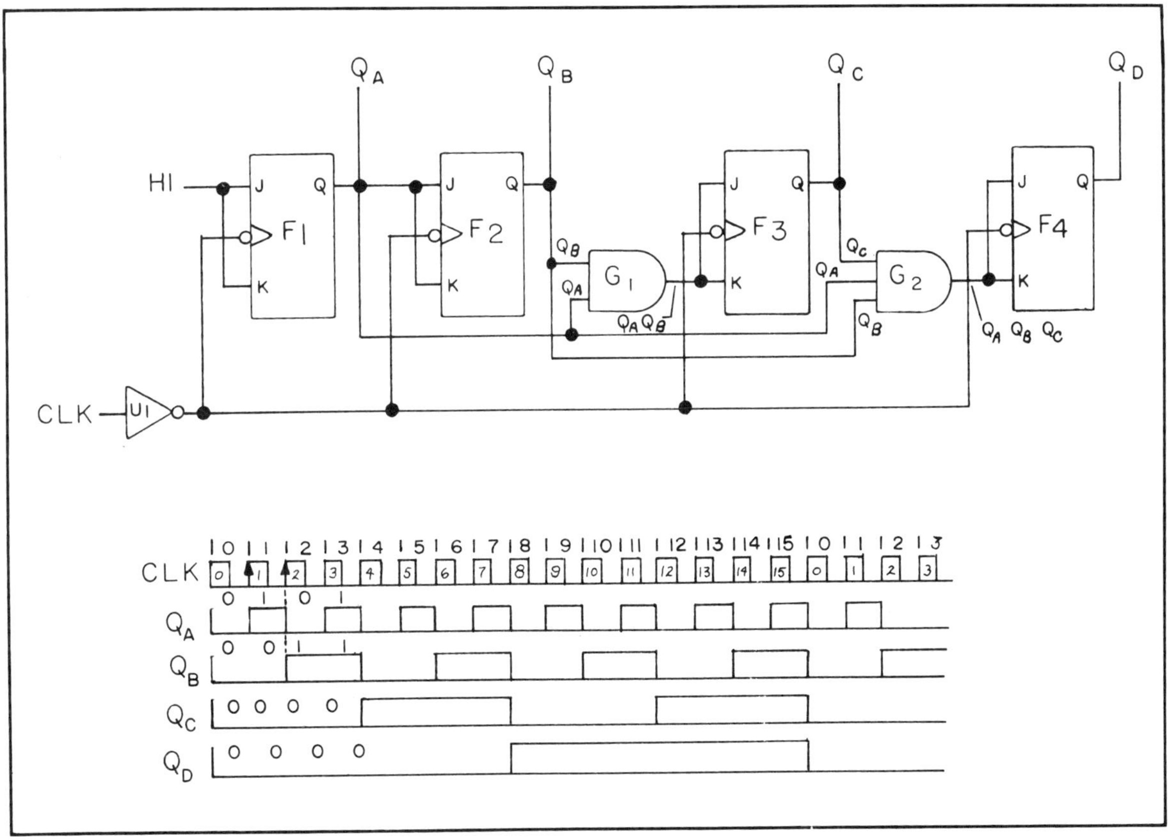

QA
QB
QC
QD
HI
F1
F2
G1
F3
G2
F4
CLK
U1

Fig. 10-21. Synchronous 4-bit binary counter and timing diagram.

SHIFT REGISTER

A register consists of a number of interconnected flip-flops and is used to store information. By adding appropriate logic gates to a basic storage register, it can perform a variety of functions such as serial and parallel conversion as well as counting or introducing a time delay.

A shift register is a special kind of register which performs the above-mentioned functions. Table 10-2 lists the common types of shift registers. Notice the characteristics of the shift registers, such as length or number of flip-flops, whether the outputs are parallel, the direction of shifting and if it has a parallel load and a clear or reset capability.

A basic shift register and its timing diagram are shown in Fig. 10-22. Notice that each flip-flop can be cleared by applying a LOW to the clear line (CLR) and then returning it to HIGH. If the serial input had a signal applied which was but two pulses as shown in the timing diagram, then two pulses would be inputted with two clock pulses.

When the serial code is not present, a LOW signal is applied to the J input while the K input is HIGH because of the inverter. Since the flip-flops are J-K types, whenever the above condition is present, the flip-flop is reset. And whenever the S input is HIGH and the R input is LOW and a clock pulse is applied, the flip-flop is set. Notice that the outputs of one flip-flop are connected to the input of another except for the first and last ones. Examine their state prior to each clock pulse as shown in the timing diagram, and you will see how those two pulses were entered and why just two pulses were shifted into the registers. Once the data is shifted in, then the data can be sent out serially or in parallel.

CLOCKS

We have been talking about clocks, and in a digital system a clock is a pulse generator or oscillator that produces a continuous pulse train. In a synchronous digital system, the continuous pulse train is produced by a clock which is usually crystal controlled for a high degree of stability. However, clocks are made from logic gates and use a register and capacitor for feedback to continue the oscillation as illustrated in Fig. 10-23.

PULSE CHARACTERISTICS

In all of the timing diagrams, the clock pulses have been shown as ideal. However, in reality the pulses have a rise time and a fall time as shown in Fig. 10-24. If a pulse is repeated at regular intervals, the time interval from the leading edge of one pulse to the leading edge of another is called the period (T) or the pulse repetition time (prt). The pulse frequency (prf) is the reciprocal of the pulse repetition time. Mathematically, this is expressed as follows:

$$\text{prf} = \frac{1}{\text{prt}}$$

Table 10-2. SHIFT REGISTER.

Length (Stages)	Type Number	TTL or C-MOS	Parallel Outputs	Direction	Parallel Load	Clear
8	7491	TTL	no	right	no	no
4	7494	TTL	no	right	preset only	yes
4	7495	TTL	yes	right/left	synchronous	no
4	74C95	C-MOS	yes	right/left	synchronous	no
5	7596	TTL	yes	right	preset only	yes
8	74164	TTL	yes	right	no	yes
8	74C164	C-MOS	yes	right	no	yes
8	74165	TTL	no	right	asynchronous	yes
8	74C165	C-MOS	no	right	asynchronous	yes
8	74166	TTL	no	right	synchronous	yes
4	74194	TTL	yes	right/left	synchronous	yes
4	74195	TTL	yes	right	synchronous	yes
4	74C195	C-MOS	yes	right	synchronous	yes

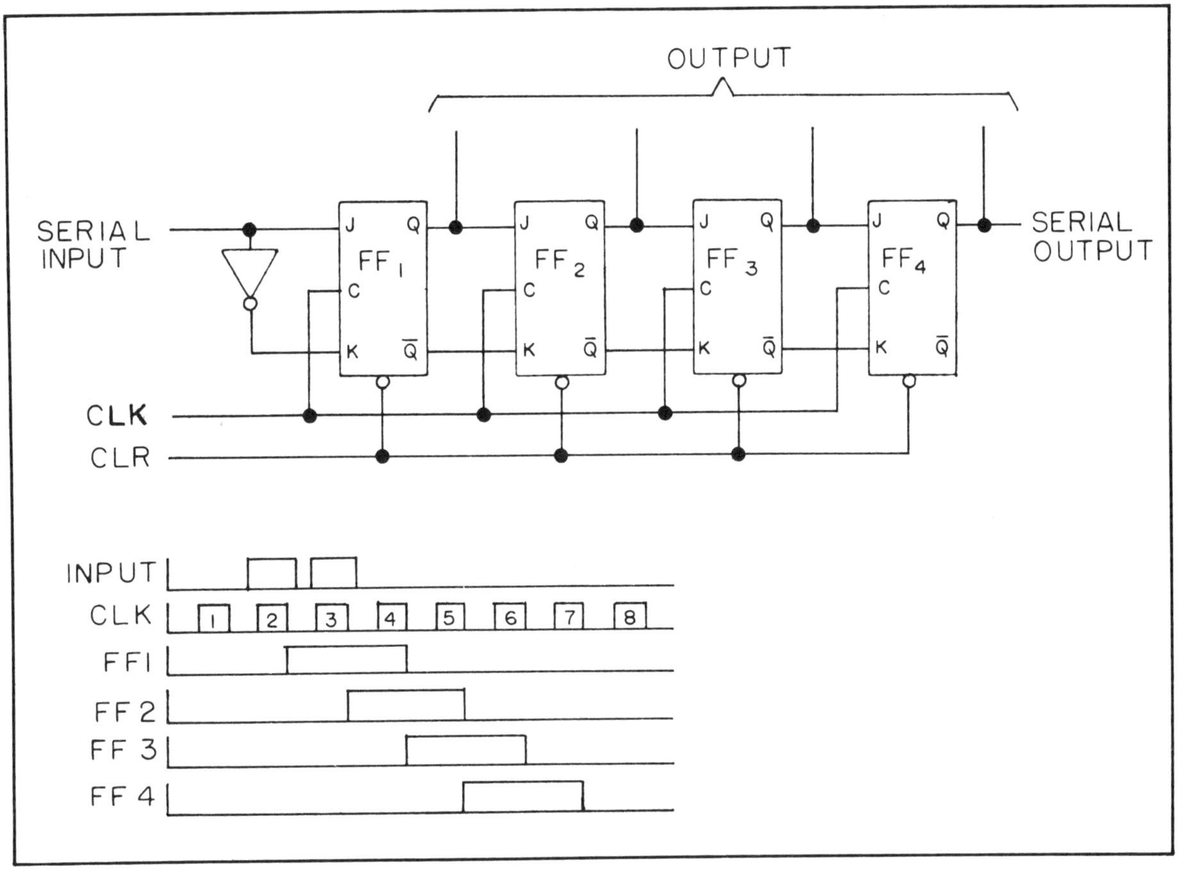

Fig. 10-22. Basic shift register and timing diagram.

The duty cycle (dc) of a pulse refers to the ratio of on time period (t_p) to the pulse repetition. Mathematically expressed as follows:

$$dc = \frac{t_p}{prt}$$

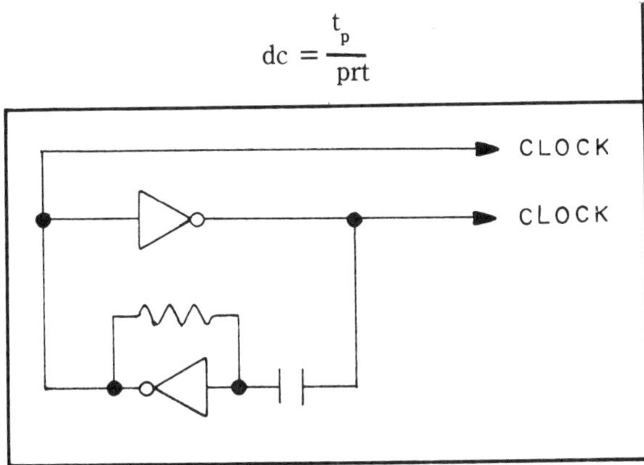

Fig. 10-23. A logic gate clock circuit.

A square wave refers to a clock pulse that is HIGH 50 percent of the time and LOW the other 50 percent of the pulse repetition time. If a pulse has a duty cycle of 0.2, then the pulse is HIGH during just two-tenths of each cycle. If the duty cycle is greater than 10 percent but less than 50 percent, then the pulse train is commonly called a rectangular wave. The terminology of the pulse is important when examining the characteristics of the clock signals at different points in the digital system using an oscilloscope.

TROUBLESHOOTING COMBINATIONAL AND SEQUENTIAL CIRCUITS

Finding failures in combinational and sequential circuits can be approached similarly to other circuits. You should begin by using your senses. Look for obvious problems such as loose connectors, or ICs not firmly seated.

Another technique is to touch the ICs while the

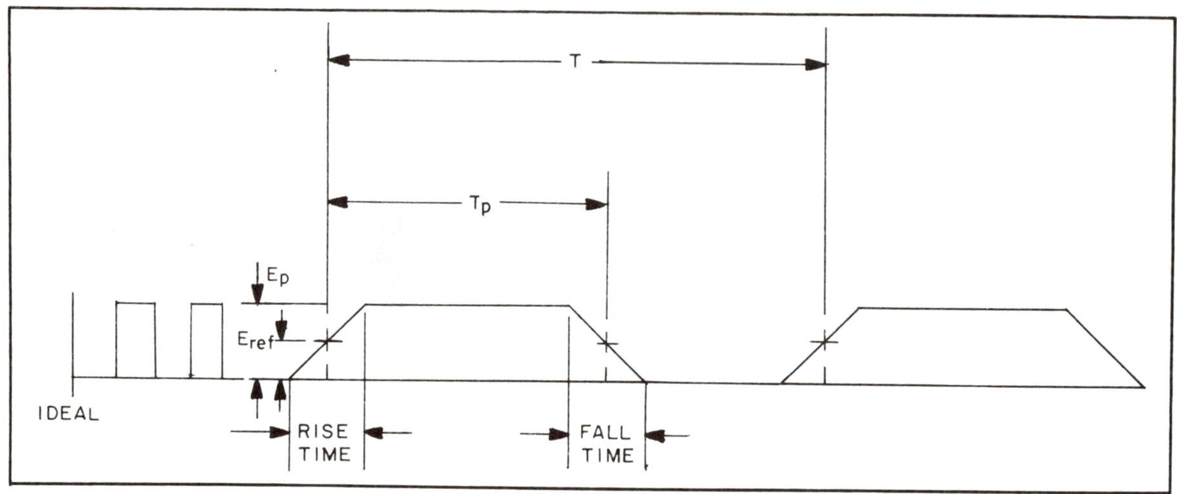

Fig. 10-24. Pulse characteristics showing an ideal pulse and the rise and fall time of a typical pulse.

circuit is operating to determine if any IC is too hot. A hot IC is suggestive of a shorted IC. It should be noted that large scale ICs, 40 pins, do operate warmly as compared to small scale ICs, 16-20 pins, which are cooler. A warm to hot small IC should be considered suspect. However, if a large IC operates coolly instead of warmly it may not be receiving supply voltage, or an internal open could exist rendering it inoperative.

Checking Displays

By observing an alpha or numeric display output, an easy check can be made of its circuitry. For example, if a seven-segment display does not light when a lamp test is made, then the problem is a segment failure and the LED seven-segment display should be replaced. However, if all segments light, then check the decoder for failure. Another method of testing a display failure would be to interchange two LED displays to see if the problem moves. Making voltage checks of the LED segments is another test that could be made. Typically a segment drops 1.6 volts. Also monitor the input signal to the display; if they are all right, then check the inputs to the decoder. Each display technology has its own unique characteristics that will become evident with experience.

Testing a Decoder

To test a suspect decoder or decoder/driver for proper operation, you will need to consult a data manual for the pinouts and the truth table. Clip a logic monitor to the chip. With a logic pulser, input the various codes and observe the logic monitor for the output signals. Now compare the observed action with the truth table. Should the truth table not match, then replace the decoder.

If the decoder/driver is used to activate a display, then pulse the inputs to the decoder and observe the display for proper decoding.

Some decoders have an inhibit line which is used to construct a larger decoding network. Should the particular decoder being checked have an inhibit line which is not used for expansion, make certain that it is connected to a permanent HIGH. Also make certain that the particular chip has power. Measure the supply voltage at the Vcc pin to the ground pin. This measurement technique will assure a complete circuit.

Testing Latches and Registers

To troubleshoot a defective register, you must analyze its operation. For example, if the register is used for temporary storage between a processing unit and the input/output indicator, then the register may be a bistable latch. The information is presented at its input, then depending on whether the latch is level or trigger-active clock, the information is transferred to the outputs of the latches. Thus, to check a latch register, you will need to verify that the appropriate input data is present, then observe the clock input for proper level or trigger action, and verify that the input data was transferred to the output. If the output information does not match the input, the register is defective. Another signal line which should be checked is the clear line. If this line has an open, it would "float" either HIGH or LOW. Should the clear line be open and pulled LOW because of electrical noise the latch register would clear unintentionally. Don't forget to check that the chip is receiving proper supply voltage and is securely seated in its socket.

The troubleshooting process is more complex if the register is a shift register. First, a knowledge of the shift register's characteristics is important. For instance, is the register serial-in and serial-out or parallel-in and serial-out or other combinations of data transfers? Also, it is important to know the direction of the data transfer. If you know what the input characteristics are and have verified their presence, pulse the shift register and observe each state to determine if there is a defective state. If the shift register operates at a high frequency and the clock signal cannot be interrupted, then use a multitrace oscilloscope or Logic Comparator to verify the proper operation of the shift register. Make sure you check preset and clear signal input lines for proper operation, as well as power supply voltage.

Testing Multiplexer and Demultiplexer

A multiplexer or data selector has four major signals to monitor for testing the device. The signal lines are the inputs, address, strobe and output. You should identify the multiplexer type, such as a dual 4-line to 1-line, and acquire the pinouts. Clip a logic monitor on the chip and observe the patterns. Examine the address lines to determine which data input line was selected and its signal level, then watch for the strobe pulse and see if the input data is transferred to the output line. This analysis must be repeated for each active input line. To accomplish the step-by-step analysis, attempt to interrupt the system clock and pulse it.

The demultiplexer sometimes is used as a decoder. As a demultiplexer, the serial input data is applied to the data input line and the address line selects the appropriate output. Therefore, to test a demultiplexer, you must observe the address selected and the data level on the data input line and verify that it was transmitted to the appropriate output line.

If the multiplexer or demultiplexer appears to operate intermittently, you may wish to use the piggyback technique. The piggyback technique parallels ICs by placing an identical good IC on top of the suspect IC as shown in Fig. 10-25. Make certain that the piggyback IC is properly oriented for pin 1. Hopefully the intermittent behavior will disappear. If it does, the suspect chip is defective. If the chip is soldered, remove the chip as shown in Chapter 9. Before soldering a new chip in the circuit, you may wish to verify if the chip was defective and that it was the only problem. Figure 10-26 shows how to deform a wire wrap socket and insert it in the circuit with a new chip. Operate the circuit and note if the problem has been eliminated.

Fig. 10-25. Illustration of the piggyback technique of placing an identical IC on top of a suspect one.

Fig. 10-26. Illustration of deforming a wire wrap IC socket and insertion into the desolder suspect IC for testing diagnosis.

Checking Counters

Counters are comparatively easy to troubleshoot because the outputs are eventually displayed. By observing the display response, you can infer a great deal about the counter operation.

A good way of testing a counter is to get control of the clock, slow it down, and monitor the counter output for proper functioning. If this cannot be done, then a Logic Comparator or triggered oscilloscope should be used to dynamically test the counter.

Erratic Operation

Erratic operation of a functional block in a digital system can be caused by either mechanical or electrical defects. When this problem appears, signal tracing procedures using a logic probe or oscilloscope will localize the fault. Possible causes for erratic operation could be an IC not seated in its socket, corroded or damaged contacts on the module, or a cold-soldered connection. Check the digital module for these potential mechanical problems. Also, a high ripple on the power supply lines or fluctuating power supply voltage can be a cause for erratic operation. An oscilloscope should be used to observe the power supply line ripple or voltage fluctuations.

Another source of erratic operation is when a pulse has less than effective threshold to trigger a gate of flip-flops. It is similar to having no trigger pulse at all. Thus, when a digital system behaves erratically or inconsistent in response, check for both missing pulses and for pulses with low amplitudes. The missing or low amplitude pulse can be due to leakage of the signal between printed circuit board connectors. Check for dirt, dust or grime and clean with solvent and a brush. A marginal IC in the pulse circuit channel could reduce the amplitude or lose it.

Testing Clocks

The clock in a system can vary in frequency for various reasons such as temperature or aging or components. To check a clock for proper frequency, an oscilloscope or a frequency counter is used. Figure 10-27 shows a frequency counter being used to verify the clock frequency in a digital system.

Should the clock be unstable or in error, check for a defective crystal. If the clock is constructed from logic gates, chances are the capacitor or resistor is open, shorted or changed in value. The resistor or capacitor must be replaced.

The primary difference between troubleshooting a combinational circuit and a sequential circuit is that a combinational circuit is placed in the defective state; begin troubleshooting the circuit from that reference. On the other hand, when troubleshooting a sequential circuit, the fault is found by placing it into the state which just preceeds the defective state. By analyzing the inputs prior to the clock input, you can determine why the circuit goes to the defective state.

Fig. 10-27. Using a counter/timer for measuring and verifying clock frequency (courtesy Global Specialties, Inc.).

There are some sequential circuits which don't lend themselves to the previous method of troubleshooting. These circuits are designed to operate at high clock frequencies. The clock cannot be interrupted to monitor the signal processed at various points at a leisurely pace. In this situation, force the inputs into a state that will cause the circuit to repeatedly pass through the wrong state. By using a triggered oscilloscope you can monitor or trace the signals at flip-flop input with reference to the clock frequency and determine the cause of the defective state.

SELF EXAMINATION

Select the best answer:

1. The term "to enable" a gate is the opposite of which of the following terms?
   A. To "unlock" a gate.
   B. To "inhibit" a gate.
   C. To "disable" a gate.
   D. To "close" a gate.
   E. All of the above.
2. A positive clock pulse represents a transition from logical 1 to logical 0 and occurs on the:
   A. Rising edge.
   B. Negative leading edge.
   C. Positive trailing edge.
   D. Positive leading edge.
   E. Negative trailing edge.
3. The binary number 11011 represents which of the following:
   A. 16.
   B. 24.
   C. 27.
   D. 32.
4. A four-decade counter can count from
   A. 0000 to 10000.
   B. 0000 to 9999.
   C. 0001 to 10000.
   D. 0000 to 9998.
5. The largest binary number that can exist in binary coded decimal (BCD) is
   A. DCBA = 1111.
   B. 1000.
   C. 1100.
   D. 1001.
   E. 0111.
6. The BCD code 0111 = DCBA is equal to
   A. Decimal 5.
   B. Decimal 6.
   C. Decimal 7.
   D. Decimal 8.
   E. None of the above.
7. A multiplexer is
   A. A device for data acquisition.
   B. Analogous to a rotary switch.
   C. A device that transfers data from a single line to one of many output lines.
   D. A triggering mechanism for an encoder.
8. A demultiplexer is similar to
   A. A device that transfers data from many input lines to a single output line.
   B. Data selector.
   C. Decoder.
   D. Encoder.
   E. None of the above.
9. In the clear condition, a JK flip-flop will set when:
   A. The J input is HIGH and the leading edge of the clock pulse occurs.
   B. The J and K inputs are LOW and the trailing edge of the clock pulse occurs.
   C. The J input is HIGH and the trailing edge of the clock pulse occurs.
   D. Both J and K are HIGH and the leading clock pulse occurs.
10. Four JK flip-flops are arranged to form a 4-bit counter. If all four flip-flops are in their LOW state before the first clock pulse, how many pulses are needed to return the flip-flops to their LOW state?
    A. 8.
    B. 9.
    C. 15.
    D. 16.
    E. 32.
11. A six stage up counter provides a count from zero to:
    A. 177.
    B. 25.
    C. 64.
    D. 77.
12. A typical flip-flop can have the following signal and control lines:
    A. Count, J, K, clear, and Q output.
    B. Clock, count, Q output, and clear.
    C. Strobe, Q output, and count.
    D. Clock, J, K, Q output, clear, and preset.
13. A flip-flop is a:
    A. One-state device.
    B. Two-state device.
    C. Three-state device.
    D. Either a one-state or two-state device, depending upon the logic state at the preset.
14. Which of the following list is actually a flip-flop?
    A. D flip-flop.
    B. J-K flip-flop with preset and clear.
    C. RS flip-flop.
    D. Master slave J-K flip-flop.
    E. All of the above.

15. If J=0 and K=1 on a J-K flip-flop, then after the clock pulse
    A. Q remains at its logic state.
    B. Switches back and forth, complements.
    C. Q can change to or stay at logical 0.
    D. Q can change to or stay at logical 1.
16. The preset and clear input to a flip-flop
    A. Don't take precedence to the clock, but do take precedence to the J-K inputs.
    B. Don't take precedence to the J-K inputs, but do take precedence to the clock input.
    C. Don't take precedence to the D-input, but do take precedence to the clock input.
    D. Take precedence over all inputs.
17. A typical shift register consists of
    A. Several flip-flops arranged in parallel and clocked by a common clock.
    B. Several flip-flops arranged in series and clocked by a common clock.
    C. Several counters arranged in sequence.
    D. None of the above.
18. A shift register can
    A. Store both serial and parallel data.
    B. Delay serial data.
    C. Convert parallel data to serial data.
    D. Convert serial data to parallel data.
    E. All of the above, but depends upon the specific shift register used.
19. Which of the following best describes an Asynchronous Counter?
    A. Counter in which all stages simultaneously shift.
    B. Counter stages shift in ripple fashion.
    C. Counter shift does not require clock pulses.
    D. Up-counter only.
20. Counters can be either
    A. Asynchronous or synchronous.
    B. Presetable or non-presetable.
    C. Gate or strobed.
    D. Cascadable or non-cascadable.
    E. All of the above.
21. If a counter receives a clock pulse but does not count, then a possible cause for the fault is:
    A. The clear line is open and resets the counter.
    B. The parallel load lines are open and constantly load ones into the counter.
    C. The up or down count could be stuck LOW.
    D. All of the above.
22. Erratic operation can be isolated by:
    A. Heating the component with a hot blower and note performance of the circuit.
    B. Cooling or freezing the component.
    C. Piggybacking IC.
    D. All of the above.
23. A method of checking display is to:
    A. Interchange two LED displays and see if the problem moves.
    B. Check the voltages across the segments and compare with typical values.
    C. Monitor the input signals to the display and compare with typical values.
    D. All of the above.
24. A low amplitude pulse can be due to:
    A. Signal leakage between printed circuit board connectors.
    B. A marginal IC in the pulse circuit channel.
    C. An open.
    D. All of the above.
    E. Only A and B.
25. If a counter is locked in a constant HIGH, then the problem might be
    A. Slow clock pulse.
    B. The load or preset line is enabled.
    C. The clear line is enabled.
    D. All of the above.

Chapter 11

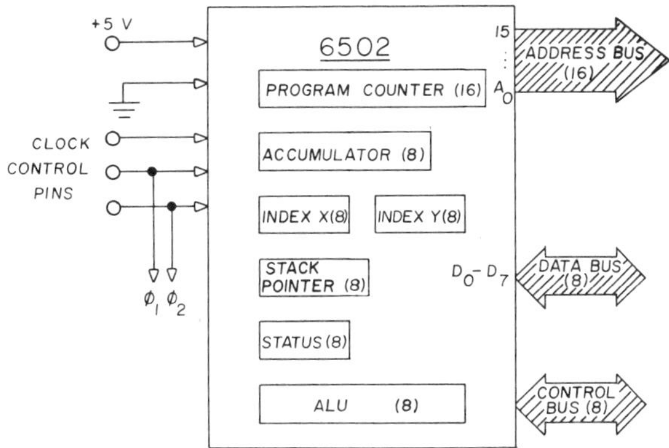

Troubleshooting Microcomputer Systems

As integrated circuit technology evolves, greater numbers of gates are added to a silicon chip. Consequently, more digital logic functions that once were made from discrete transistor circuits and small scale integration are added to a single chip creating large-scale integration. One significant large-scale integrated circuit is the microprocessor. The microprocessor is a very important component of every new digital system design. By taking a microprocessor and adding a few other specialized integrated circuit chips, a computer is created.

DIGITAL COMPUTER BASICS

A microcomputer system consists of five major units as shown in Fig. 11-1. They are the input, output, memory, arithmetic logic (ALU), and the control units. The control unit sequences the flow of instructions and data between the other units. The ALU performs arithmetic and logical operations, and the memory unit stores the instructions and data which constitute a program. Sometimes the ALU and control unit are combined in one package called the central processing unit (CPU).

MICROPROCESSOR AND MICROCOMPUTER

The microprocessor, which is fabricated on a silicon chip less than ¼ inch on a side, is the central processing unit. It differs from the microcomputer because it does not have the memory and input/output (I/O) units. A microcomputer contains all five major units and is fabricated on a single silicon chip. Figure 11-2 shows a picture of Texas Instruments 4-bit microcomputer. Microcomputer systems are generally found in small dedicated industrial controllers, while the microprocessor is central to the design of very powerful general purpose microcomputers.

MICROCOMPUTER SYSTEMS

A basic personal microcomputer system is shown in Fig. 11-3. It has a monitor, a keyboard input, a central processing unit and memory, disk drives and a printer. The microcomputer system can be analyzed in terms of system hardware and system software.

System Hardware

The system hardware also includes input peripheral devices such as: keyboard, paper punch reader, card readers, or analog to digital converters as well as high and low level switch inputs. The output peripheral devices include such devices as the monitor, printer, plot-

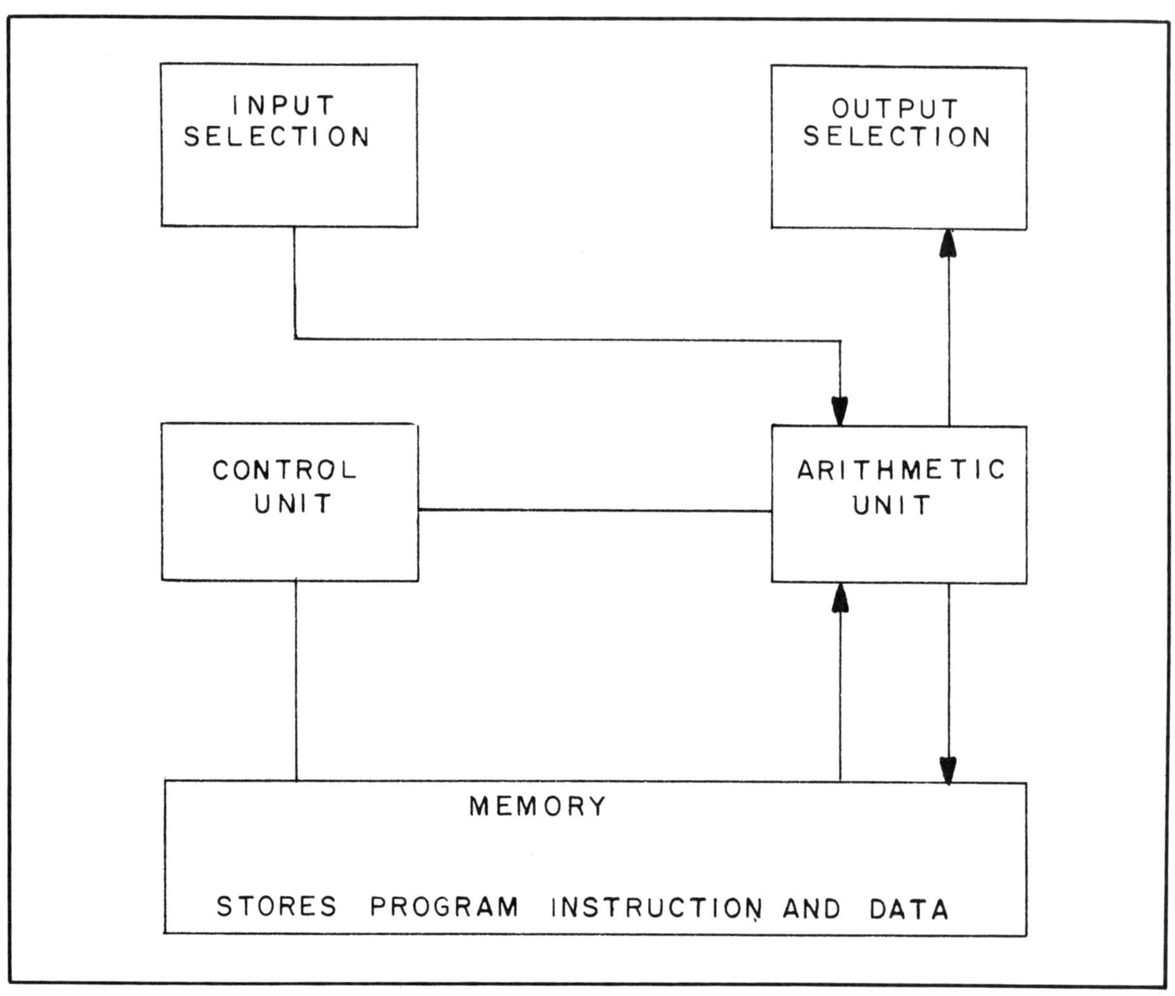

Fig. 11-1. Five major units of a computer.

ter, and digital-to-analog converters. A disk drive constitutes a combination I/O device that is used for storage.

The main hardware of a microcomputer consists of a microprocessor and the main memory. To adapt and transmit information between the microprocessor and peripheral devices, special circuits called interfaces are used. Some of the interfacing circuits are built in on the mother board as shown in Fig. 11-4. The slots on the mother board, are for expansion of the basic system. In the slots, special interface cards or modules are inserted with the appropriate interfacing circuits as shown in Fig. 11-5. In some cases, a memory chip with a permanent program called firmware is inserted to aid the transmission of information.

A rack-mounted microcomputer system, which is composed of many module cards, is shown in Fig. 11-6. The modular card includes a microprocessor mother board and bus-driven integrated circuits with the memory unit on separate module cards for expansion. Other interfacing cards or modules can be inserted in a particular slot and connected to input or output devices. The backplane holds the edge connectors and is wired to complete a communication path between the various modules.

Disk Operating System

There are various disk operating systems (DOS) such as Apple's DOS 3.3 or the more popular C/PM by Digital Research. The DOS is a program which manages storage on a floppy or hard diskette and works hand in hand with high-level languages such as BASIC and its interpreter (translator). The language interpreter or translator converts typed words and abbreviations into the machine language code (binary numbers) for the processor. The interpreter is a language which starts the program running and translates the program for continuous operation.

Whenever DOS stores information such as a program, data or a text report on a disk, it is stored in files and is given a file name. An image of a file can be a steel file cabinet and each drawer has a file name. The file names are stored on the disk in a special track or tracks. This special track area is called the catalog or directory and is the table of contents of the disk, much like the table of contents of a book. Each type of disk operating system has commands which are similar among various manufacturers as well as unique commands.

When you buy a new diskette, it is blank; nothing is recorded on it, and it is similar to a blank tape for a tape recorder. Before information can be stored on the disk, it

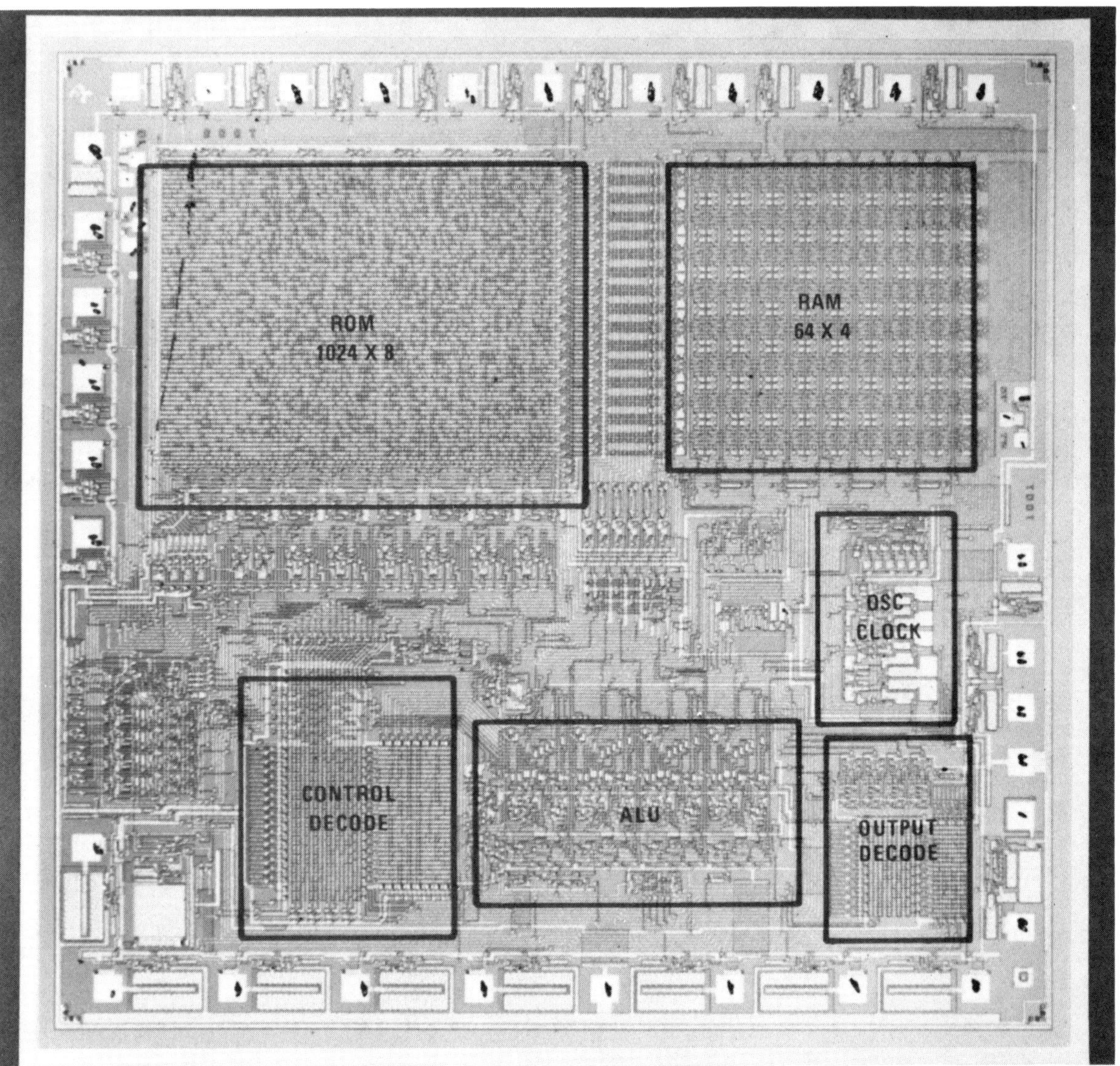

Fig. 11-2. Texas Instruments' 4-bit microprocessor chip (courtesy Texas Instruments).

Fig. 11-3. Basic personal computer systems which include the microprocessor, monitor, disk drive and printer (courtesy Hewlett-Packard Company).

must be prepared by initializing or formatting. This process stores a DOS program which permits the storage and retrieval of information.

System Software

The system software is composed of a group of sequential instructions called a *program* which the central processing unit executes or carries out. The programs are stored in a read only memory (ROM) integrated circuit called a monitor program that is located on the mother board. The monitor program has the ability to start the microcomputer running, as well as other tasks, when you turn it on. The monitor program varies from one microcomputer system to another. The programs are called firmware because they are programmed by the manufacturer at the factory and cannot be changed.

Most programs are loaded into the computer main memory from a magnetic disk. The main memory is called random access memory and is labeled RAM. The monitor program, along with the disk operating system, accomplishes the movement of information and is referred to as system software.

Languages

Information is what you type on the keyboard, what you see on a video screen, or the output that is viewed on a printer. This information or communication takes place at three different levels. Figure 11-7 illustrates the language levels. The lowest level is the machine language which is composed of a string of binary numbers. How these binary numbers are interpreted and what actions are taken by the microprocessor depends upon the type of microprocessor. Because of the long numbers which result from the binary number system, they are very awkward to remember and use efficiently. Therefore, to overcome this problem, the *hexadecimal* (HEX) scheme is employed as a type of shorthand.

Hexadecimal is a shorthand form of representing a binary number. Using binary numbers by people is very cumbersome because they are long and consist only of 1's and 0's. On the other hand, binary numbers are the only type that can be used by the computer. All the machine level instructions for the microcomputer can be represented in a hexadecimal code. An example of how the code is formed for an eight-bit binary word is shown in Fig. 11-8.

Let's say you want to convert decimal 255 to a hexadecimal number. First, you would convert the decimal number to a binary number by applying the repeated division method. Then arrange the binary number into groups of four bits beginning from the right. If the last group has less than four bits, then add the leading zeros. The last step is simply to convert the four-bit groups to a hexadecimal. Notice that the highest number that can be represented by a four-bit binary group is decimal 15. However, the four-bit binary group needs to be converted into only a single symbol or character. For this reason, hexadecimal uses six letters A through F and the decimal characters 0 through 9.

The next level of languages is the assembler lan-

Fig. 11-4. Mother board of an Apple //e showing the custom LSI devices, the 8-bit 6502 microprocessor, eight 64K random access memory chips, seven expansion slots across the rear, and a special purpose video output at lower left (courtesy Apple Computer, Inc.).

Fig. 11-5. Mother board of an Apple II+ computer and special interface cards.

Fig. 11-6. Rack mounted microcomputer system which includes the microprocessor, memory and other process, and communication cards in the rack.

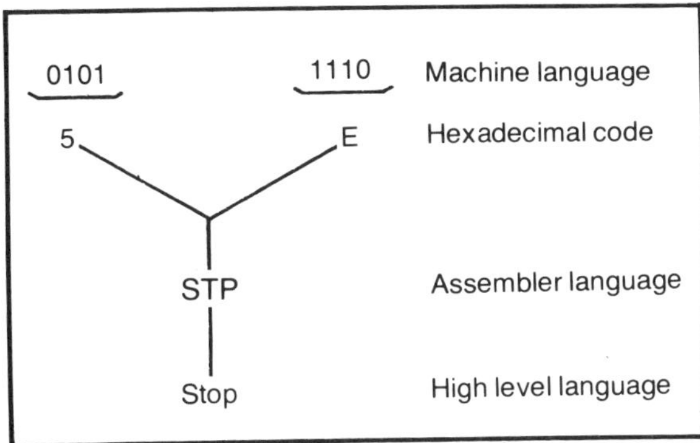

Fig. 11-7. Computer language levels from machine to high level.

guage. The assembler consists of a symbolic form of abbreviations called "mnemonics." These mnenonics are a step closer to English and are, therefore, easier for us to remember and use in writing programs. The word "mnemonic" is derived from the Greek which means "to aid the memory."

The high-level languages are the easiest for us to use. In high-level language such as BASIC, Pascal or FORTRAN, the commands or instructions are generally short English words. In general, the easier it is for us to communicate with a computer in a high-level language, the more sophisticated the interpreter program must be to translate the high-level language instructions into machine language instructions.

ASSEMBLER

Of the three languages, the assembler requires knowledge and understanding of the operation of a microprocessor. It is at this language level that a troubleshooter needs to operate. The troubleshooter may write or use assembler programs that cause the microprocessor to generate appropriate timing and control signals to observe the operation of the computer system.

Let's examine the assembler language more closely. Every microprocessor has a set of instructions which are unique to it. The instructional set can be grouped into categories such as data transfer which moves data between registers, memory, and input/output devices. Other instructional groups are arithmetic, logical, branch (conditional and unconditional) and machine control which controls, interrupts and stacks.

Every instruction has a unique operation code (op code) and a mnemonic symbology. Thus AD 00 C0 in HEX machine code would represent LDA C0 00 and mean "Fetch and load the accumulator with the content at

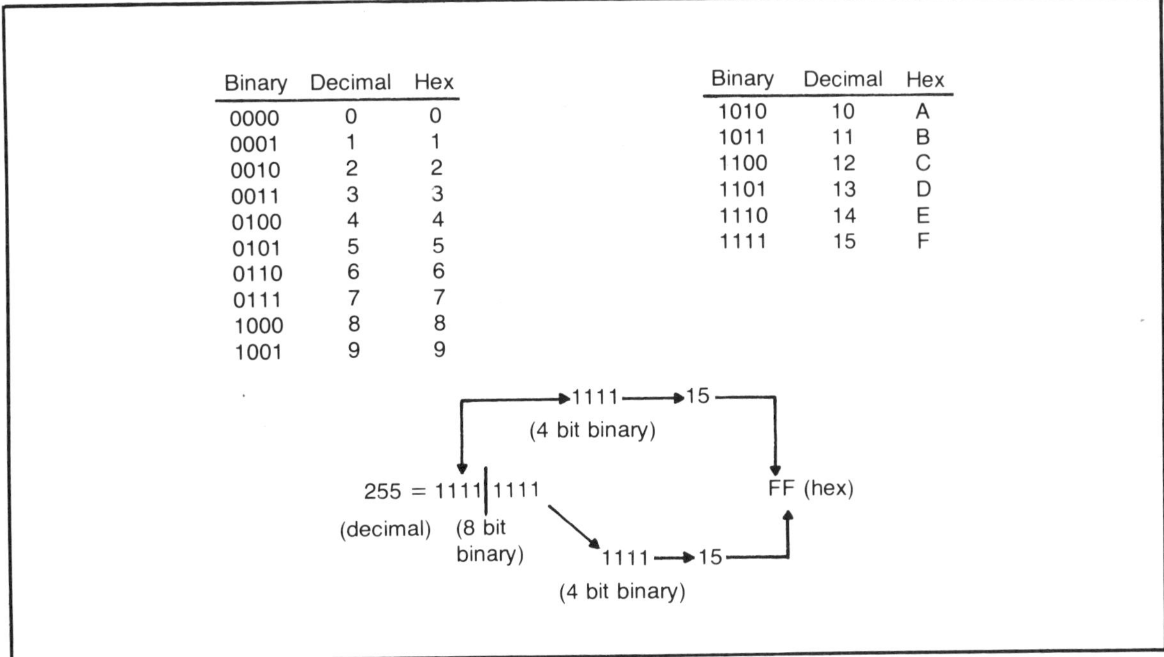

Binary	Decimal	Hex
0000	0	0
0001	1	1
0010	2	2
0011	3	3
0100	4	4
0101	5	5
0110	6	6
0111	7	7
1000	8	8
1001	9	9

Binary	Decimal	Hex
1010	10	A
1011	11	B
1100	12	C
1101	13	D
1110	14	E
1111	15	F

Fig. 11-8. Comparison of binary, decimal and hexadecimal number plus a decimal to hexadecimal conversion.

memory address C000." If you programmed this statement in assembler language, the keys LDA C0 00 would be pressed on a keyboard. The assembler language program would generate 10101101 00000000 1100000 as three 8-bit *bytes* in machine language necessary for the processor to interpret and store these instructions in memory. A byte is a group of eight binary bits. The assembler would likely generate a print-out for you in HEX code. A completed assembly program would look like this:

OPERATION CODE	MNEMONIC	OPERAND FIELD
AD 00 C0	LDA	$C000
8D F7 07	STA	$07F7
00	BRK	

This program would fetch the content at location C000 and put it in the accumulator, store the accumulator content at memory location 07F7 and then halt. The dollar sign denotes for the assembler that the code is HEX. In order for the computer to interpret the instructions, the assembler would generate the following machine codes:

AD = (10101101)
00 = (00000000)
C0 = (11000000)
8D = (10001101)
F7 = (11110111)
07 = (00000111)
00 = (00000000)

MICROCOMPUTER ARCHITECTURE

The architecture of a microprocessor refers to how a particular semiconductor manufacturer has designed its central processing unit and instruction set. The microprocessor performs many different functions, including:

- —generating appropriate timing and control signals for all components of the microprocessor
- —fetching instructions and data from memory
- —decoding instructions
- —transferring data to and from I/O devices
- —performing arithmetic and logic operations as required by the instruction
- —responding to RESET and INTERRUPTS as requested by I/O devices.

Without any knowledge of a microprocessor's architecture, it is very difficult to troubleshoot a system. Typically, the microprocessor is a 40-pin integrated circuit and is only a single component in a microcomputer system.

The Three Busses

Three busses exist, which are a set of wires or printed circuit conductors that carry all the information and control signals involved in a computer system. Specificially, these busses are: 1) an address bus having 16 lines A0-A15, 2) a bidirectional data bus with eight data lines designated D0-D7, 3) and a control bus containing a varying number of lines such as read/write (R/W), ready (RDY), interrupt request (IRQ), non-maskable interrupt (NMI), reset (RES) and synchronization (SYCH). A pin-out of the 6502 microprocessor is shown in Fig. 11-9. The internal architecture of the 6502 microprocessor is illustrated in Fig. 11-10.

The address bus of 16 lines of a microprocessor can generate 2^{16}, which equals 65,536 different possible addresses. Each of these addresses corresponds to one memory location or one I/O device.

When the microprocessor is commanded to READ or WRITE to a specific memory location or I/O device, the appropriate address code is placed on the 16 lines (A0-A15). The address bits are sent to a decoder which selects the appropriate memory location or device by generating a chip enable signal.

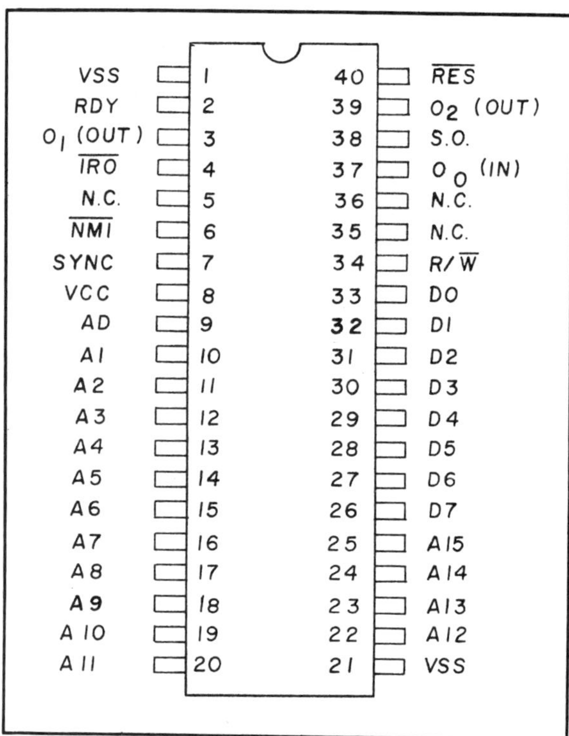

Fig. 11-9. 6502 microprocessor pin configuration.

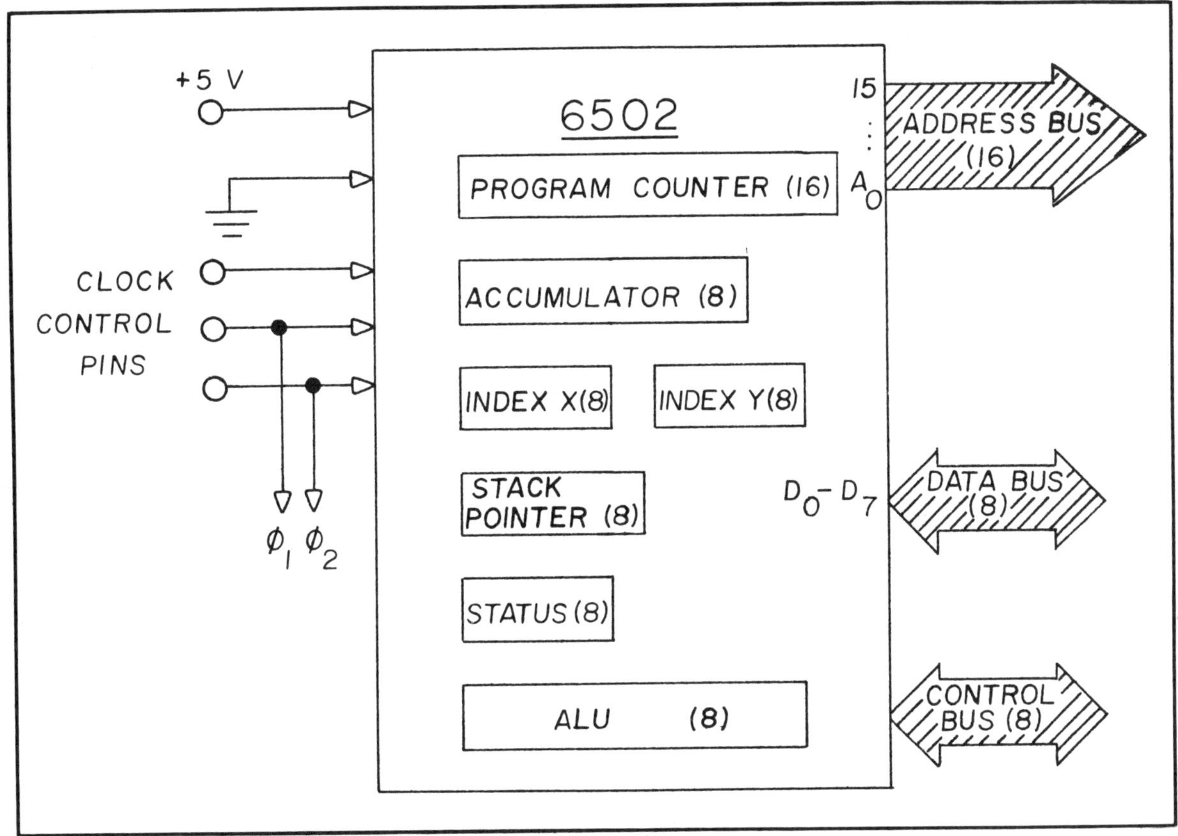

Fig. 11-10. Simplified internal architecture of the 6502 microprocessor.

The data bus is bidirectional because data can flow to or from the microprocessor on the same eight data lines (D0-D7). During a READ command, the data bus carries inputs; while during a WRITE command, the data bus carries outputs. In all cases the data transmitted is a group of 8-bits because this microprocessor handles 8-bit data words, thus making it an 8-bit processor. Newer microprocessors can handle 32-bit data words, thereby making them 32-bit processors.

The control bus generates or receives signals used to synchronize the activities of the internal and external elements of a microcomputer system. Some of these control signals, such as read/write R/W, are sent from the microprocessor to other external devices telling them what operation is currently being performed by the microprocessor. The I/O device can send signals to the microprocessor requesting service by way of the interrupt input line (INT). Also, the reset input line (RES) could be driven LOW which causes the microprocessor to reset to a particular location and start executing instructions from that point.

A general microcomputer system is shown in Fig. 11-11. The microprocessor contains the ALU and control circuitry to execute a sequence of instructions called a program which is stored in RAM and ROM. The microprocessor continually executes READ and WRITE commands as the program is executed.

Described below are the steps taken for a READ cycle:

1. The microprocessor generates a logical 1 on the R/W line, thus initiating a READ cycle; this signal is sent to all memory chips or I/O device.
2. At the same time, the microprocessor places the appropriate 16-bit address code onto the address bus (A0-A15) to select the specific memory location or I/O device from which it wishes to receive data.
3. The selected memory location or I/O device places an 8-bit word on the data bus (D0-D7).

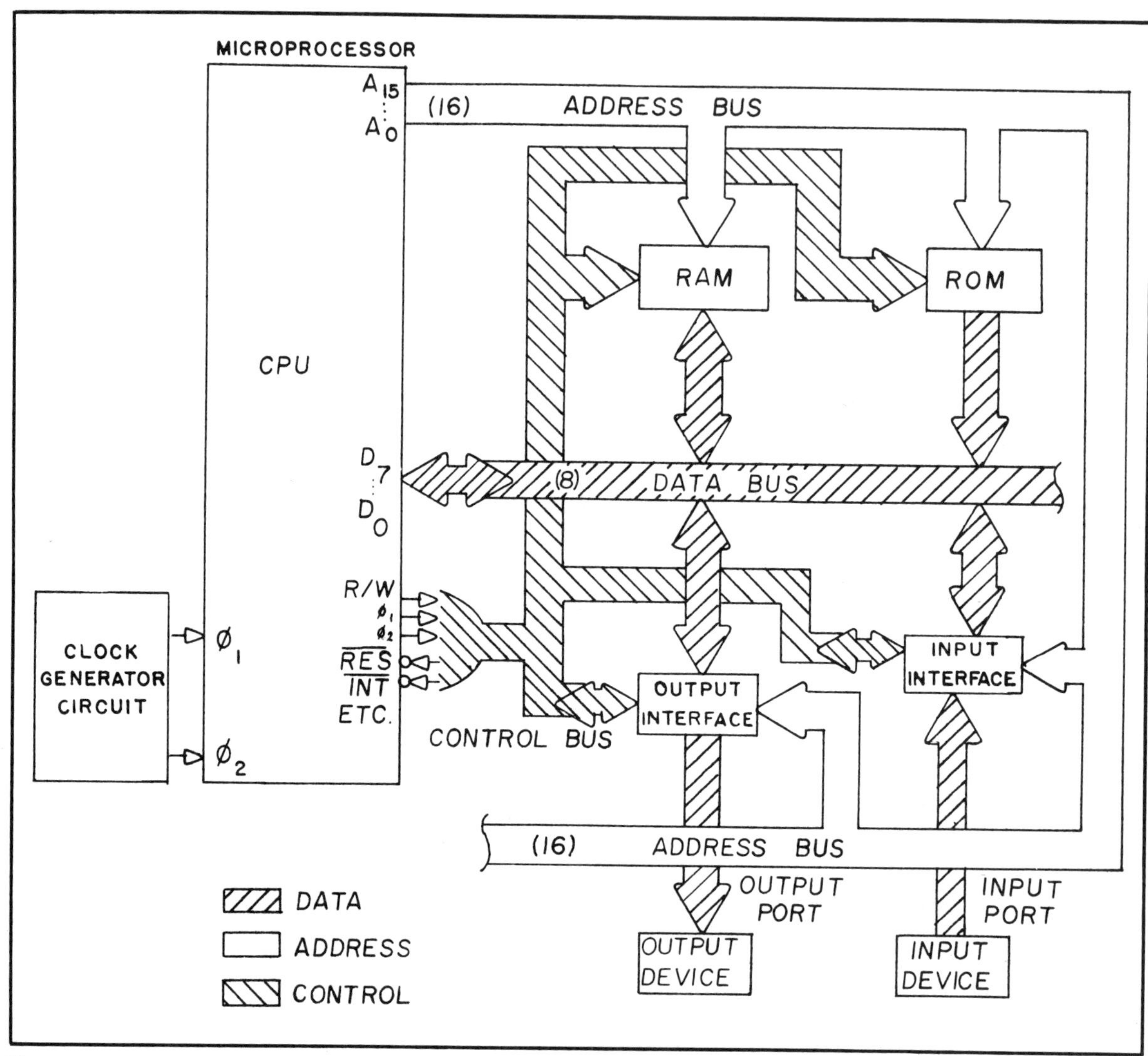

Fig. 11-11. General 8-bit microcomputer system including the microprocessor, memory and busses.

All memory locations or I/O devices not selected will have no effect on the data bus because their tristate outputs are disabled and in high Z state (in effect, electrically disconnected from the bus).

4. The microprocessor receives the 8-bit data word from the data bus (D0-D7). The data word is then latched in one of the microprocessor registers such as the accumulator.

A timing diagram of a READ cycle is shown in Fig. 11-12. Notice that phase 1 (ϕ1) initiates the READ cycle at reference 1. The R line and appropriate address lines then go HIGH. When phase 2 (ϕ2) clock cycle transitions from HIGH to LOW, the data on the data bus lines are latched in the appropriate microprocessor register.

In order for the microprocessor to WRITE, the following steps occur:

1. The microprocessor generates a logical 0 on the R/W line, thus initiating a WRITE cycle; this signal is sent to all memory chips and I/O devices.
2. At the same time the microprocessor places the

appropriate 16-bit address code onto the address bus.

3. The microprocessor then places an 8-bit data word on the data bus (D0-D7). These words are now acting as output data lines. The 8-bit data word comes from a specific internal register such as the accumulator.

4. The selected memory location or I/O device accepts the data from the data bus lines.

The WRITE cycle is illustrated in a timing diagram in Fig. 11-13. Once again, the leading edge of the phase 1 ($\phi 1$) clock cycle initiates the R/W signal (logical 0) and the address bus lines at reference 1. On the leading edge

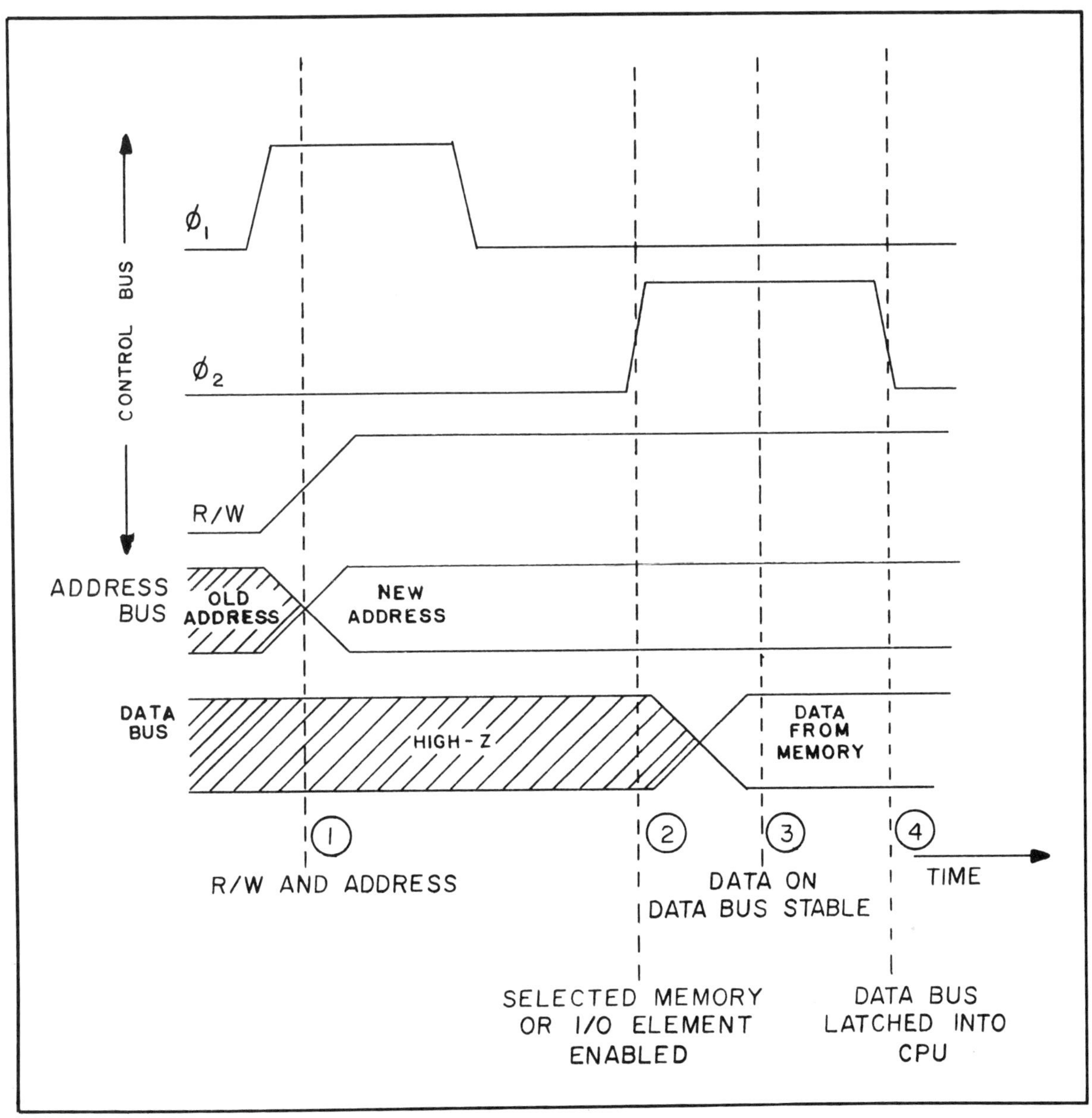

Fig. 11-12. Timing diagram for a READ cycle.

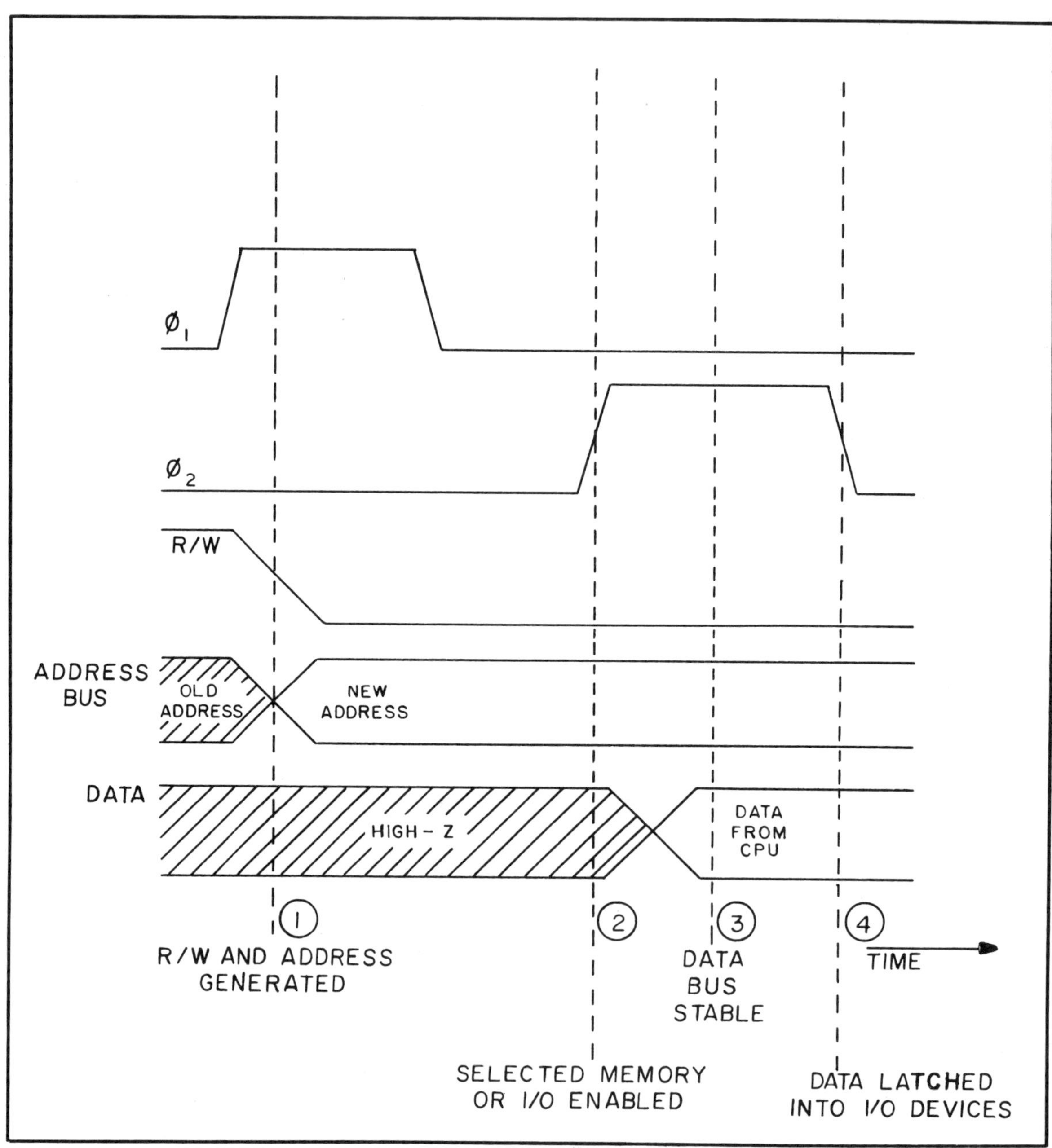

Fig. 11-13. Timing diagram for a WRITE cycle.

of the phase 2 ($\phi2$) clock pulse, the selected memory location or I/O device is enabled. At reference 3, the data bus line becomes stable. And on the trailing edge of the phase 2 ($\phi2$) clock pulse, the data on the data bus is written or latched into the appropriate memory location or I/O device.

The READ and WRITE cycles entail most of the microprocessor's activities that take place outside the actual microprocessor. Internally, the microprocessor performs arithmetic, logical and data transfers between the internal registers. The preceding steps in a READ or WRITE cycle are typical. However, each specific microprocessor may have some variations.

PRIME MEMORY

There are two major memory types used in microcomputer systems. One is the read only memory (ROM) from which content of a specific memory can only be retrieved or read. The other is the random access memory (RAM) in which the content of a specific location can be stored (write) as well as retieved (read). This type is better described as a read/write memory. A characteristic memory is that if the stored information is retained when the power to the computer system is shut off, the storage is called nonvolatile. A ROM memory has this characteristic. In contrast, memory which loses its stored content when the power is shut off is called volatile; a RAM has this characteristic.

Random Access Memories

There are two types of MOS fabrication of RAM semiconductor memories. One is called static RAM because power supply voltage must be constantly applied to the appropriate terminals. Figure 11-14 illustrates a static MOS RAM memory cell that forms a MOS flip-flop. The other type of memory is called dynamic. In construction of the memory cell, a capacitor of small value is inserted between the gate and the substrate as shown in Fig. 11-5. A charge can be temporarily stored in the capacitor, and the power supply can be shut pff most of the time by a switching circuit controlled by a clock pulse train. The charged capacitors, representing a logical 1, must be refreshed about 500 times per second. However, this technique permits an increase in the number of cells which can be placed on a single chip and is cheaper to construct because of simpler cell design. It has the disadvantage of needing to refresh the memory. Critical timing must be accomplished without interfering with other computer operations. These dynamic memory cells are more prone to "soft fail," which means the program in memory is lost or changed because of electrical noise transients.

Memory cells are often arranged in an X/Y coordinate grid. A specific cell is written to or read from, based

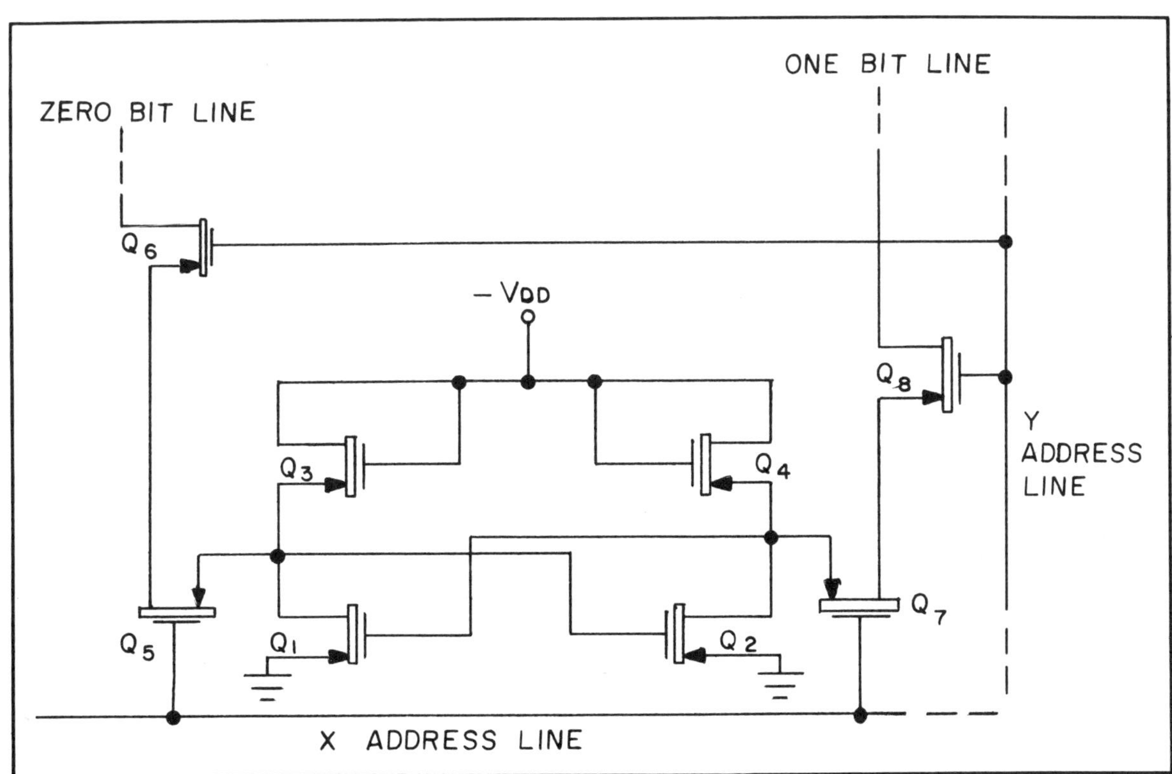

Fig. 11-14. Static MOS RAM memory cell.

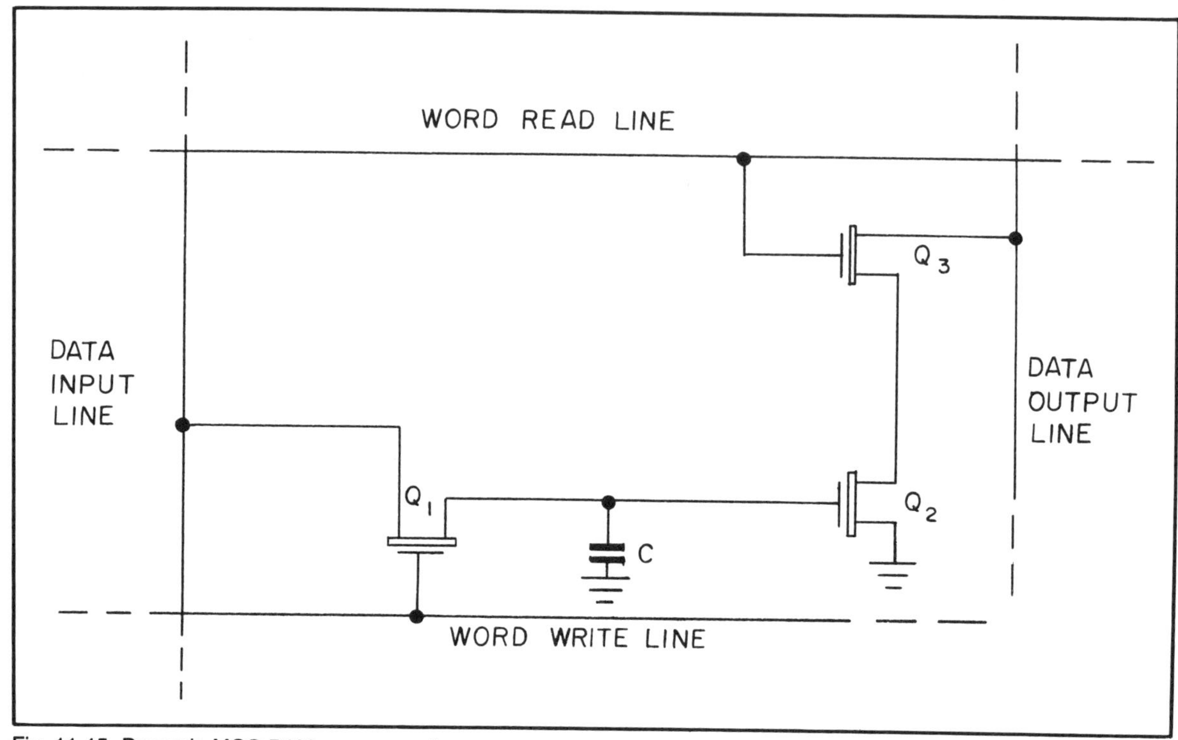

Fig. 11-15. Dynamic MOS RAM memory cell.

on the specific selection and activation of the row (X) and column (Y) lines. A decoder built on the chip is used to select one line at a time on the X (row) axis, and another decoder is used to select the Y (column) axis. The decoder inputs are connected to the address line of the microprocessor. When the appropriate address lines are activated (logical 1 or 0), the specific memory location is accessed.

There are two types of memory organization; namely, a bit-organization and word-organization. In a bit-organization shown in Fig. 11-16, you will notice that the address line when appropriately activated will select only one cell. In contrast to a word-organization, a given memory address will select several cells at the same time as shown in Fig. 11-17. Study the address line and observe that it has 010, which is decoded and the cells for the word (8-bit) are available to be read or written.

Semiconductor RAM is also fabricated from bipolar devices (TTL). A TTL RAM is shown in Fig. 11-18, and the transistors form a flip-flop.

Read Only Memories

There are two general types of ROM memories in use. One is fabricated using MOS technology, while the other uses bipolar technology. ROM memories are sometimes mask programmed by the semiconductor manufacturers to the customer's specification. An example is illustrated in Fig. 11-19. A line is masked open during the fabrication, creating an unchangeable logical 1. Another type of ROM which is programmable by the user only once is called a programmable ROM (PROM). A program directs hardware to apply sufficient current to flow through a cell in which a logical 1 is to be stored, causing a fuse link to open. The content cannot be changed. This principle is illustrated in Fig. 11-20.

There are ROM memories which can be reprogrammed a number of times. Figure 11-21 shows the general construction of a ROM cell which can be programmed. Notice the gate is suspended in an insulated material. To store a logical 1 in a cell, high voltage is applied and a charge is placed on the gate. This charge is retained for a long time, many years. However, the charge can be released by exposing the chip to ultraviolet light; thus the ROM can be reprogrammed. This type of ROM is called EPROM (erasable programmable ROM). Figure 11-22 shows the masked ROM and two different sizes of EPROM. Notice the window on the EPROMs.

The ultraviolet light is focused on this window for erasing the content. Often a gum label is placed over the window after it is programmed.

Memory Implement

A memory chip has address lines, data input and output lines, a read/write line, and a chip select (CS) or chip enable (CE) line. The address lines are used to select a memory cell or word. The chip select or enable is used to select a memory chip within a memory consisting of many chips. A read or write line or the combination on one line permits the reading or writing of data. The data lines are used to transmit information between the memory and the microprocessor.

Larger memory systems can be implemented by combining several memory chips as illustrated in Fig. 11-23. Each chip can store 64K 1-bit words. A 16-bit address line can select any bit. The data lines are individually connected to one of the eight chips creating the 8-bit data word. All chip select or enable lines are connected to a single line, and there is a single read/write line. Thus, when a memory location is addressed for a read function, all the chips are enabled simultaneously, followed by a read signal to all the chips. The content of

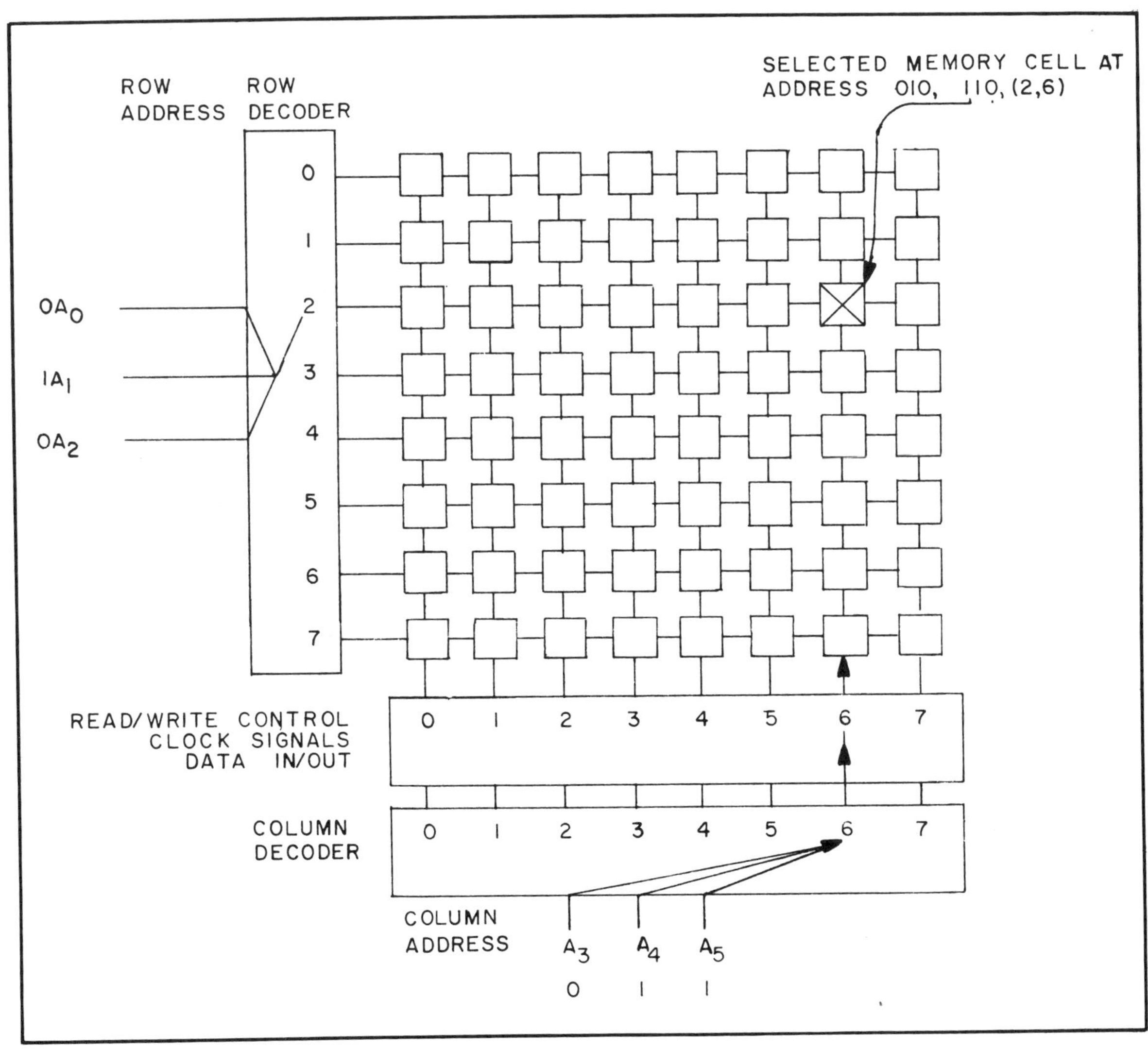

Fig. 11-16. Bit-organization memory.

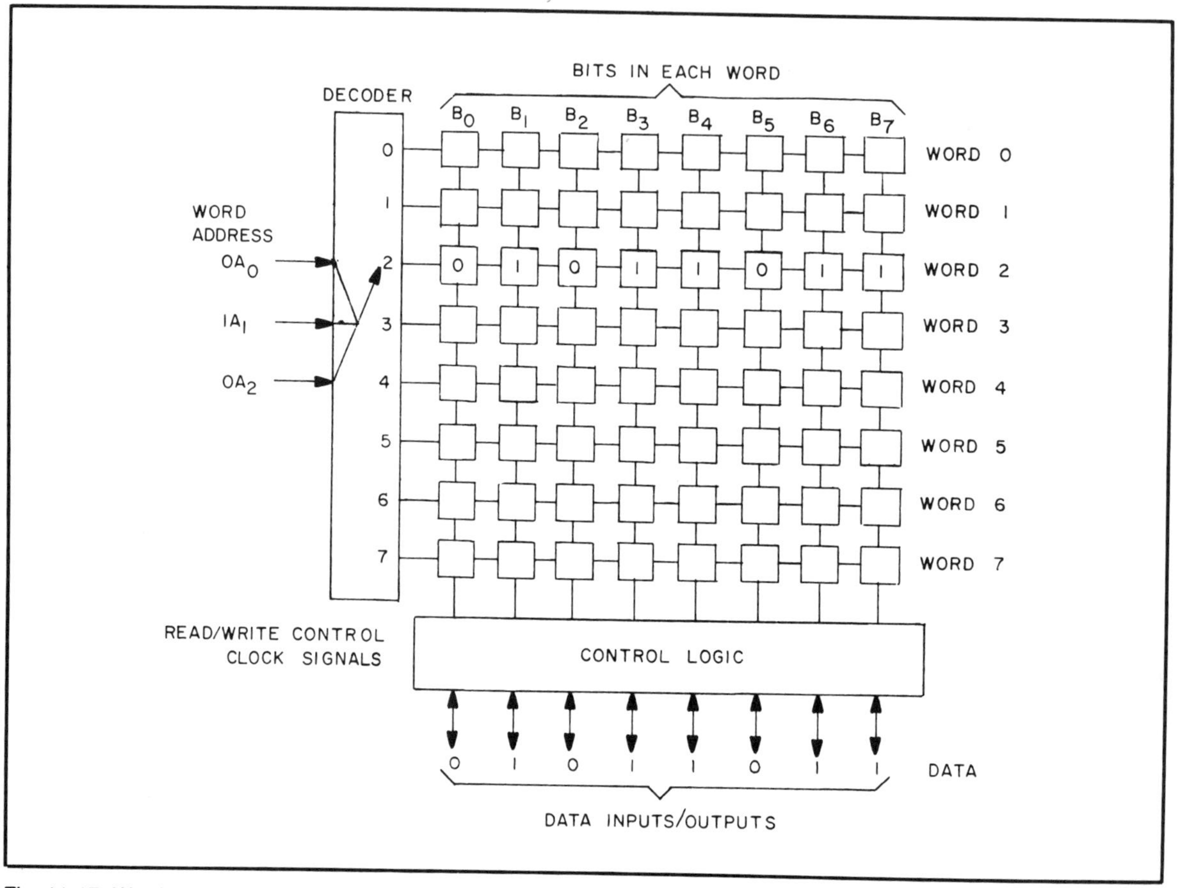

Fig. 11-17. Word organization memory.

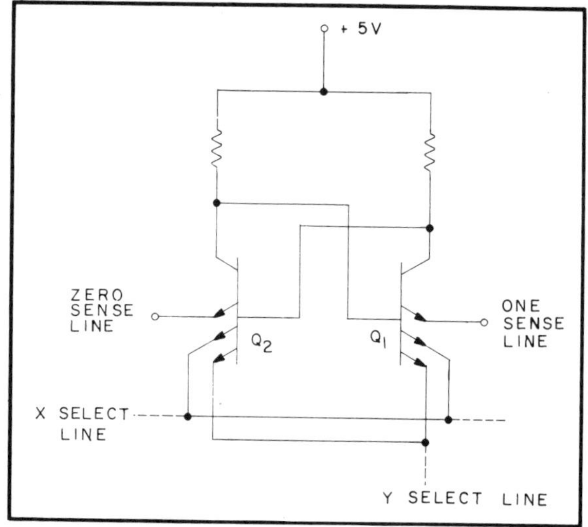

Fig. 11-18. Bipolar RAM cell.

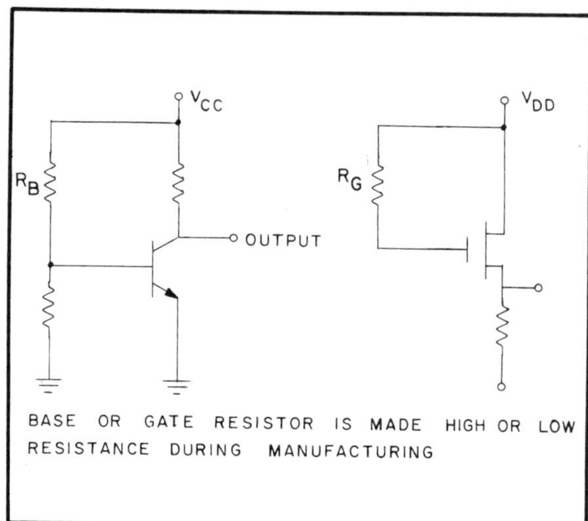

Fig. 11-19. Masked ROM memory cells.

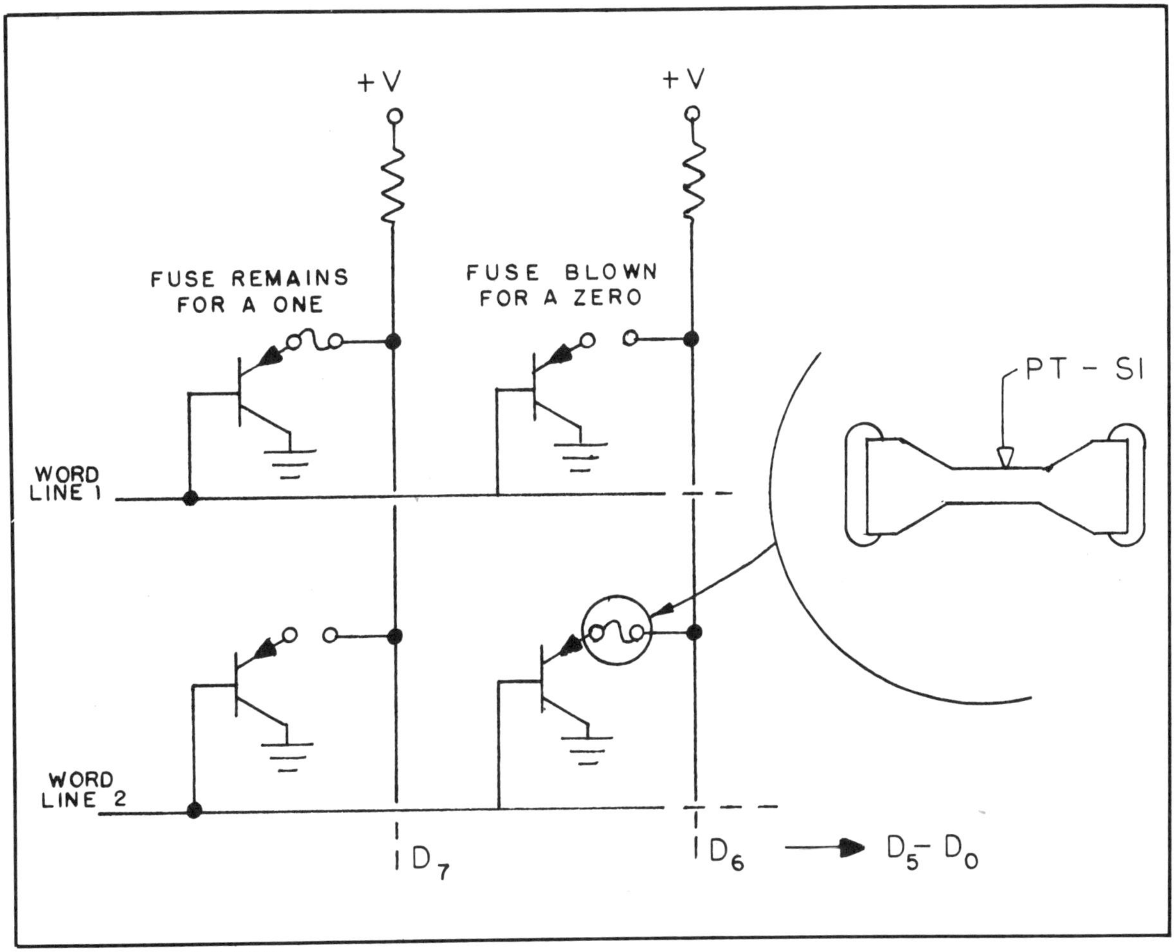

Fig. 11-20. Fused link ROM.

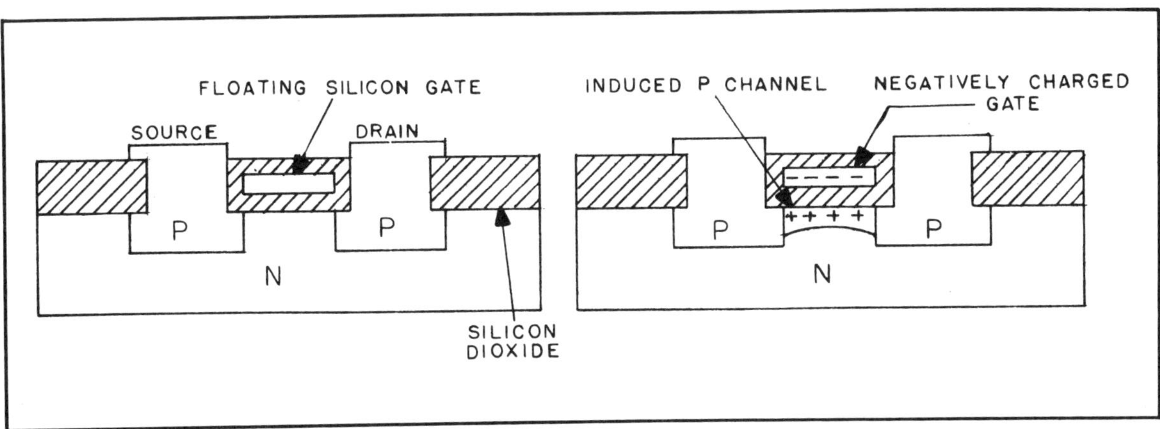

Fig. 11-21. Erasable PROM sectional view which shows how a charged gate creates a conductive P channel to store a logical 1.

Fig. 11-22. Masked ROM and two sizes of EPROM.

the addressed location is then transferred to the data lines.

Another more complicated technique used to combine memory chips to produce a larger memory system is shown in Fig. 11-24. Notice that two inputs of a 3-line to 8-line decoder are connected to the address lines. The code on these address lines permits the decoder to select any one of the four 256 × 8-bit memory chips. The decoder chip select or enable is activated using an address line such as 15 or phase 2 of a clock cycle. By using more address lines and more memory chips, the memory capacity can be increased.

Timing Diagram for Memory

Each memory product has a slightly different timing diagram, due to memory design, which can be very important for proper operation. Figure 11-25 illustrates some typical memory timing diagrams. Notice the general sequence of events: first, the address input must be present and stable; then the chip select or enable signal is activated, followed by a read or write signal, after which data is transmitted.

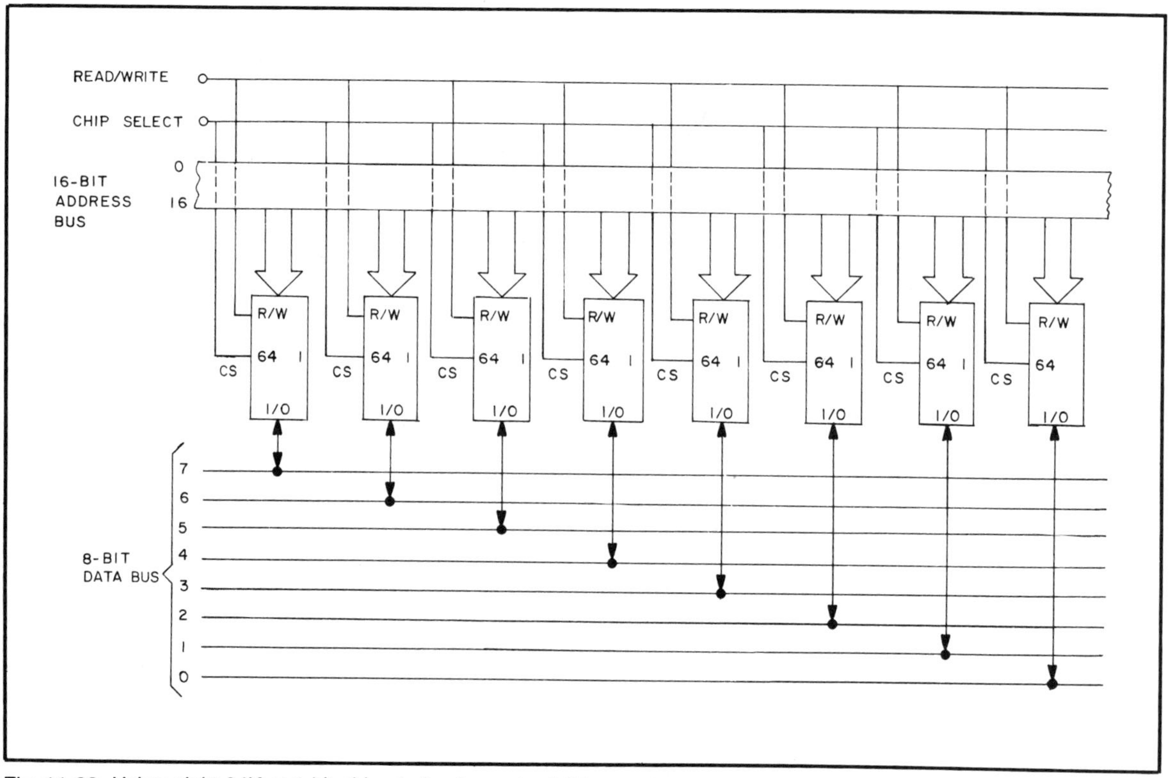

Fig. 11-23. Using eight 64K × 1-bit chips to implement a 64K × 8-bit memory.

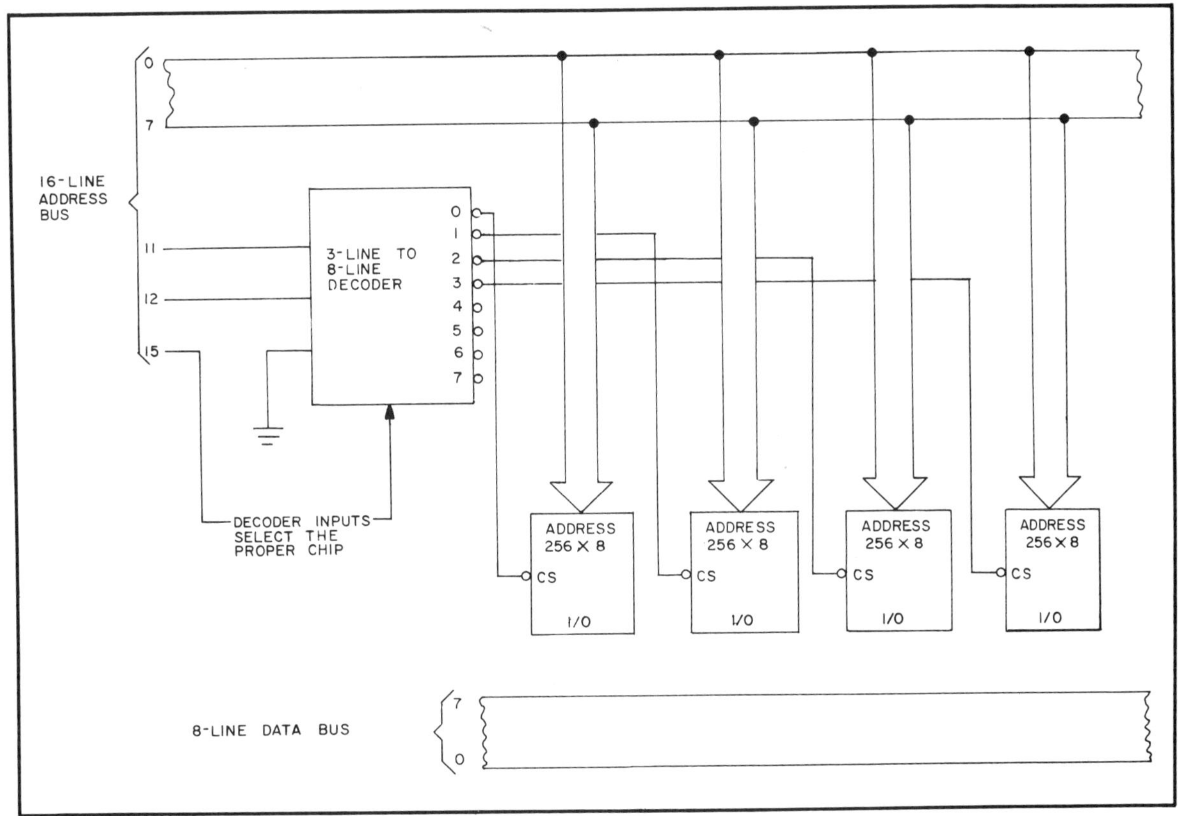

Fig. 11-24. Using four 256 × 8-bit chips to implement a 1K × 8-bit memory.

SECONDARY AND BACKUP MEMORY

Secondary and backup memory takes the contents of the RAM memory and stores it, acting as a vast data warehouse such as a library. A library can store a great number of books; and a person can select one book from the shelf, read it and return it. Also, an author can write a new book and store it on the bookshelf. The secondary and backup memories all store information in a similar manner with the exception of the bubble memory. Secondary storage is achieved by using a thin layer of iron oxide deposited on floppy or hard disks or tape for support.

The medium becomes useful for memory because two distinct magnetic patterns can be created. Figure 11-26 shows the different magnetic patterns of an unmagnetized segment and magnetized segments. The hysteresis property of a magnetic material guarantees that the pattern will remain locked for a considerable period of time. Figure 11-27 illustrates a reading (sensing) and writing (setting) of magnetic domains employing a conventional inductive recording head and a thin-film head.

In order to write, a sufficient current must be forced through a loop to align the magnetic domains creating a strong magnetic field. To store a logical 1 or 0, the current flow direction is changed. When reading, a coil of wire on the magnetic head senses the changing magnetic patterns. These changes in magnetic patterns induce a voltage in the coil, resulting in a voltage wave form that is decoded by logic circuitry to recreate the serial patterns of "ones" and "zeros."

Backup memory devices commonly associated with microcomputer systems are floppy-disk, cassette recorder or paper tape. The latter is used in industrial environments.

Floppy Disk Drives and Servicing

On the outside, a floppy disk looks much like a 45-rpm record enclosed in a plastic jacket. The disk is made of Mylar and coated with iron oxide. Unlike a 45-rpm record which has one track that spirals from the outside toward the center, the floppy disk has a large

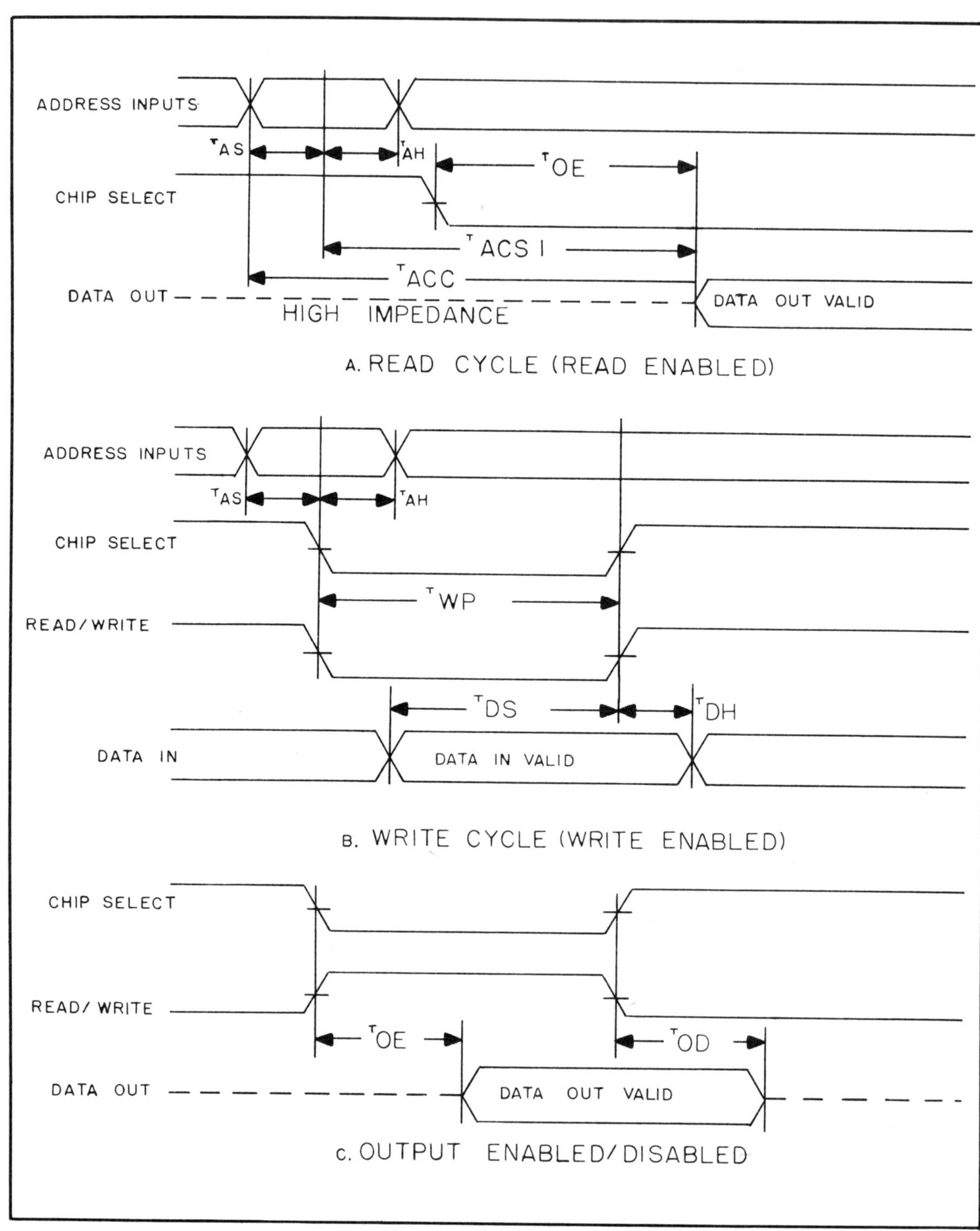

Fig. 11-25. Typical memory timing diagrams.

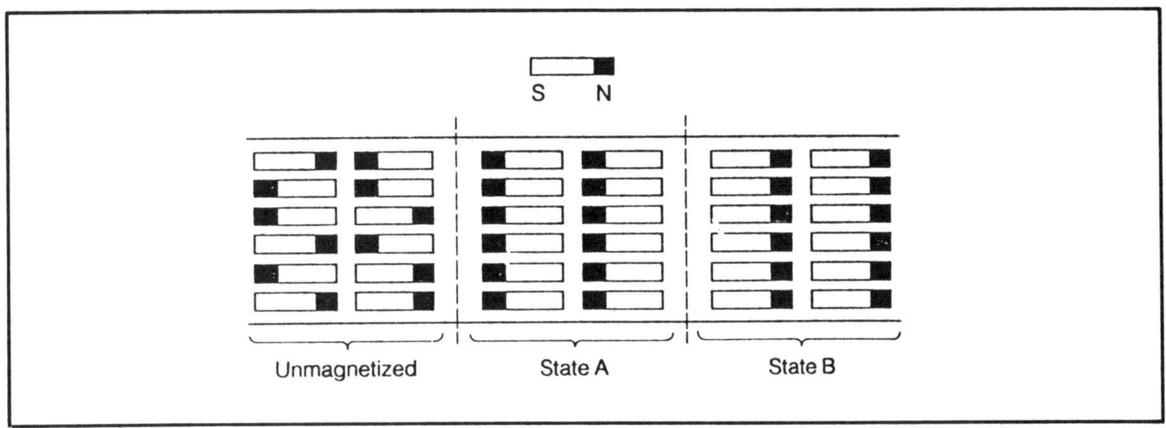

Fig. 11-26. Magnetic storage patterns for disk or tape.

number of independent concentric circles as shown in Fig. 11-28. The 8-inch floppy disk has tracks numbering from 00 through 76. The popular 5¼ floppy for personal computers has only 35 tracks. The diskette is organized into sectors, while the 5¼ inch typically has 16 sectors. Data is written to or read from the part of the track in each sector. The data are stored on the central track/sector. When addressing a specific file, the drive mechanism spins the diskette at 360 rpm using either a dc servomotor or an ac synchronous motor. These motors run continuously and are controlled to operate at precise speed for accurate information transfer. Disk speeds should be checked periodically for correct speed.

The read/write head is moved from track to track by means of a load screw driven by a small stepping motor. Every time the stepping motor is pulsed, its rotor turns at a fixed proportion of a revolution. This rotor action turns the lead screw a fixed distance which positions the read/write head. Figure 11-30 illustrates a drive mechanism for a floppy disk drive. The analog card which controls the head positioning and read/write activity is shown in Fig. 11-31. It should be noted that the read/

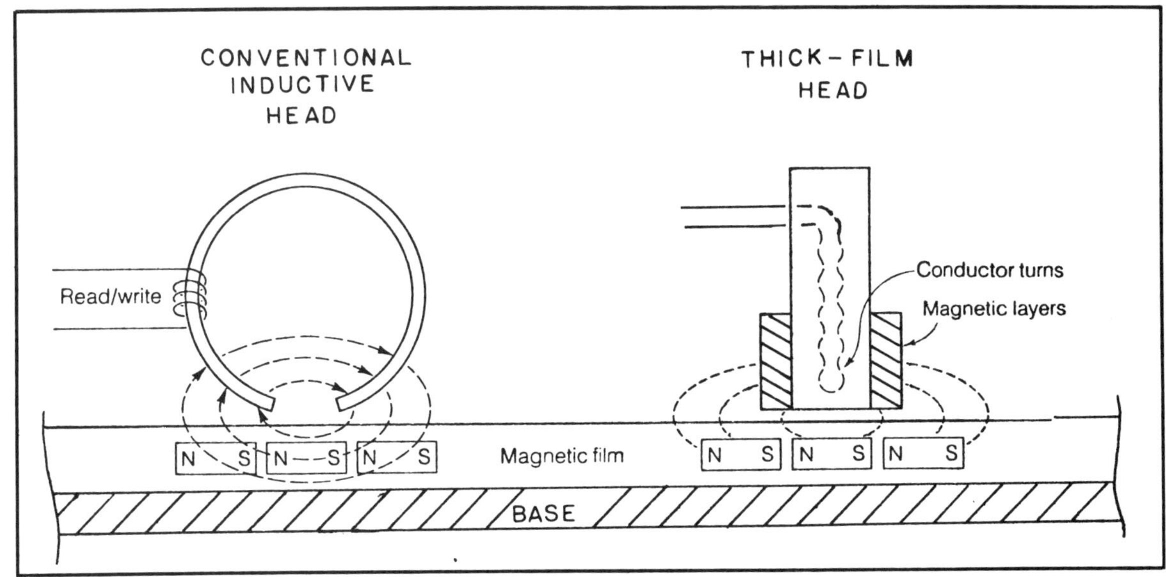

Fig. 11-27. Reading (sensing) and writing (setting) the magnetic domains using conventional inductive read and thick-film head.

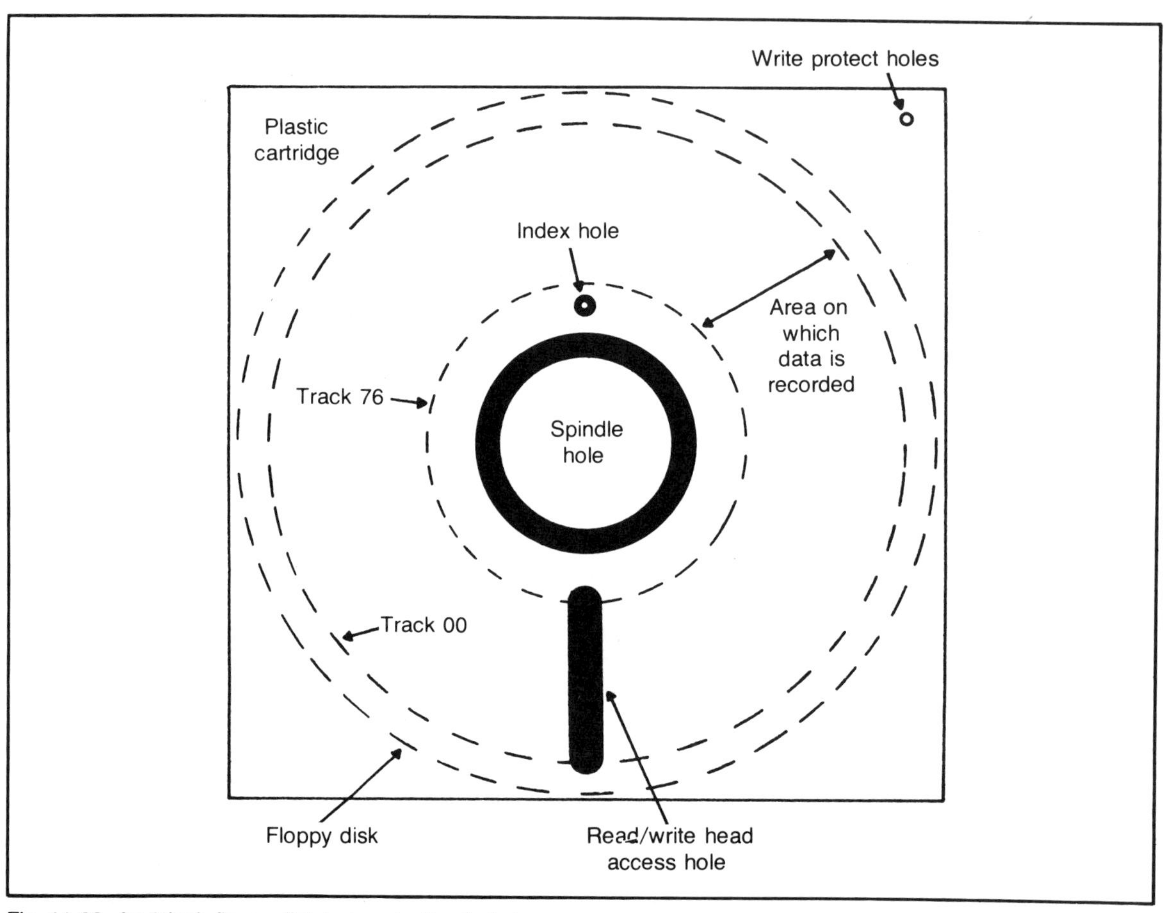

Fig. 11-28. An 8-inch floppy disk in a protective jacket.

write head actually touches the disk which does wear out the disk surface. As a consequence, the head gets covered with iron oxide and must be cleaned periodically. The iron oxide build-up will impare the electromagnetic field for data transfer. Therefore, the heads need to be cleaned periodically. Diskette manufacturers provide a special head cleaning kit. It contains a special jacket which holds a disposable disk. You insert the cleaning disk into your disk and turn it on for 30-60 seconds. A special cleaning solvent in the disk cleans the head evenly over its entire surface.

When a floppy disk drive will not load (read into the RAM memory) or it destroys the content of some of the tracks on a diskette, you should examine the analog card inside the disk drive. There are generally two ICs on the analog card which could be at fault. One is a tristate chip used to connect the disk drive to the microprocessor bus. The other is the stepping motor control chip. To identify these chips, refer to the service manual for the disk drive, or use the manufacturer's codes stamped on the IC and consult data books. Get the pinout from the data book, clip on a logic monitor and verify its proper operation.

Audio Cassettes and Recorders and Servicing

Audio cassettes are so extensively used in everyday activities that it is hardly necessary to describe one. However, certain principles are of importance. For one, the radio cassette records and retrieves information in a serial mode. Also, the cassette recorder is an asynchronous device, which means that the sender and receiver are not operated at the same speed.

There is a difference between digital recording of information on magnetic tape and using an audio cassette recorder. A digital recorder uses pulses of currents sent to its read/write head that saturates the magnetic tapes in

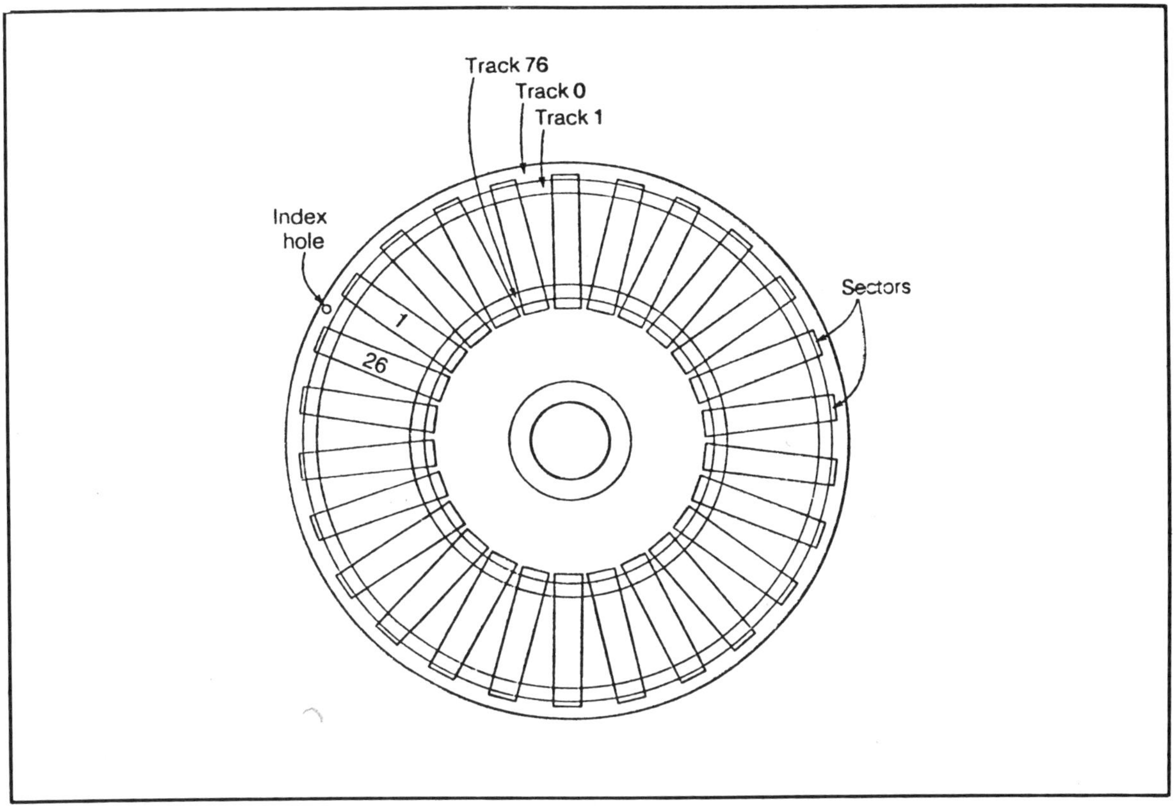

Track 76
Track 0
Track 1
Index hole
Sectors
1
26

Fig. 11-29. Floppy disk sector arrangement.

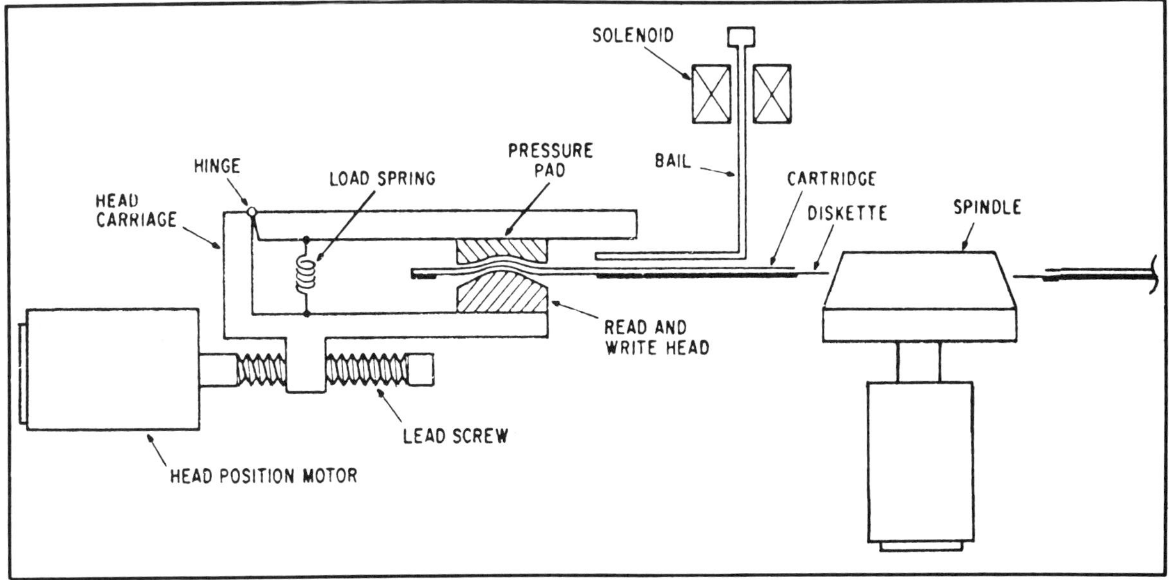

Fig. 11-30. Floppy disk drive mechanism for positioning the head.

Fig. 11-31. The analog card which controlls the read/write and head positioning.

one direction or the other for a logical 1 or a logical 0. In an audio cassette recorder, digital square waves are not directly applied to the read/write head. A universal asynchronous receiver and transmitter (UART) converts a parallel code from the microprocessor into a serial output which controls a 4800 hertz modulator. Figure 11-32 illustrates a simplified write interface. The serial digital output signals are transformed into an audio signal of 2400 hertz for a logical 1 and a 1200 hertz signal for a logical 0.

To read from an audio cassette recorder, the signal is passed through an amplifier which triggers a retriggerable one-shot. The one-shot is an IC that has only one state, normally LOW. When the one-shot is triggered, the output goes HIGH for a time period determined by a resistor and capacitor time constant. The one-shot takes about 555 microseconds to time out. If the audio signal is 2400 hertz, the one-shot is constantly being retriggered and remains HIGH and sets the flip-flop which is interpreted as a logical 1. If the incoming audio frequency is 1200 hertz, the one-shot has time to time out, resets the flip-flop to a LOW state, and is interpreted as a logical zero. In order to strobe the incoming bits, the UART must receive a clock signal. The clock signal for the

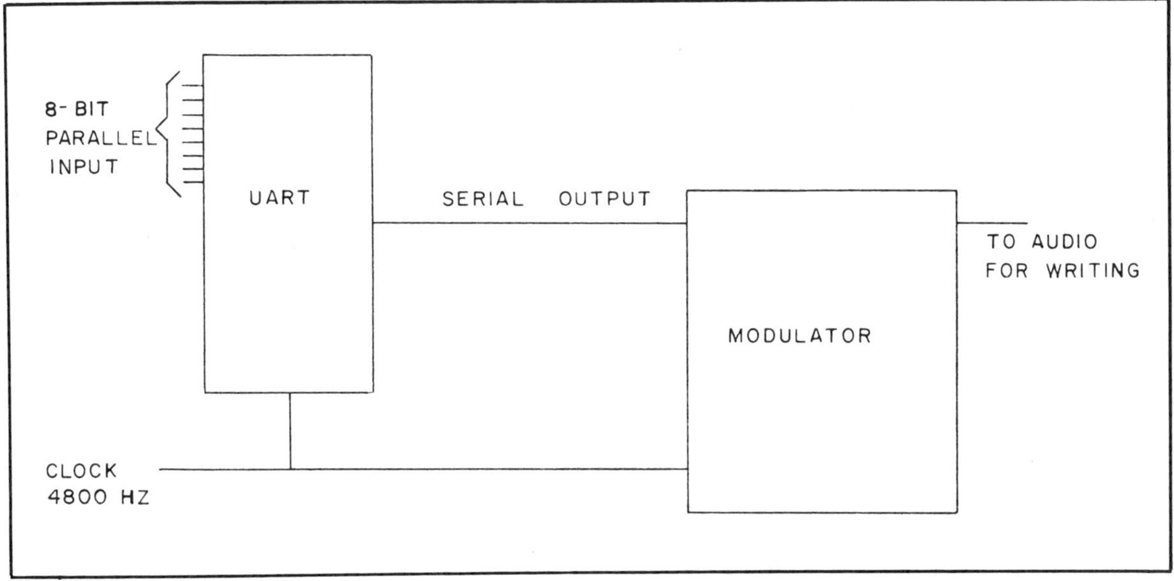

Fig. 11-32. Simplified WRITE interface for an audio cassette.

UART is derived from the audio cassette drive by monitoring the audio signal. A simplified circuit for an audio cassette read is shown in Fig. 11-33.

Generally the audio cassette is highly reliable as a backup memory system. To keep an audio cassette operating properly, you should clean the read/write head as recommended by the manufacturer.

Paper Tape Reader/Punch and Servicing

Paper tape is used as an economical backup memory storage method. Paper tape readers are available in low- to high-speed versions that read from 10-1000 characters per second. The punches will punch from 10-150 characters per second.

The paper tape width varies from ½ to 1 inch, having

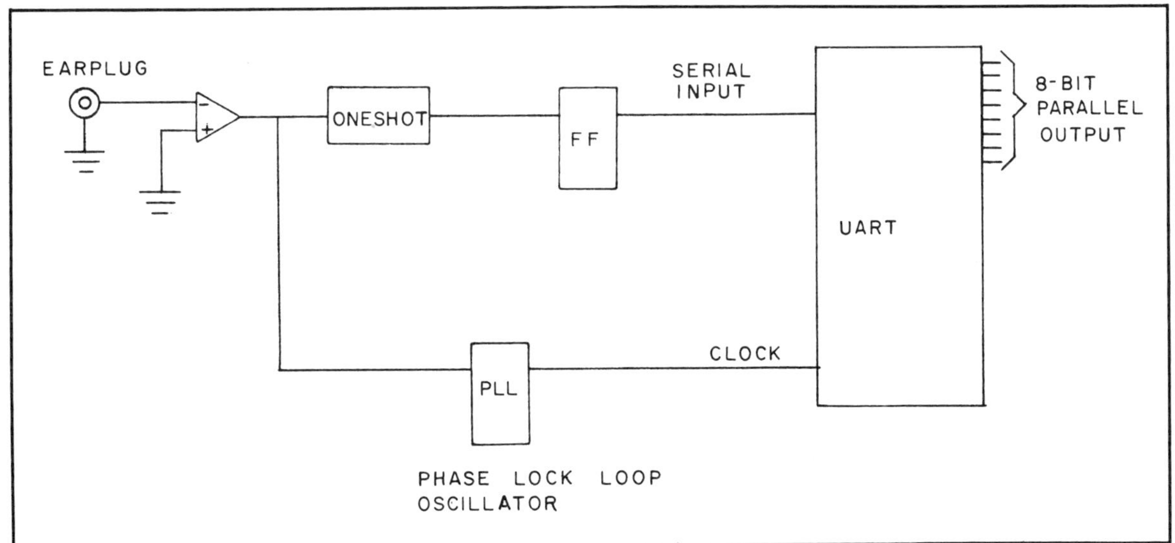

Fig. 11-33. Simplified READ interface for an audio cassette.

five- to eight-bit characters, respectively. The bits are represented by hole perforations in the tape at each bit site. Usually a hole represents a logical 1, while the absence of a hole represents a 0. Figure 11-34 shows a commonly used eight-bit paper tape. Many codes are used to represent data on the tape, such as hexadecimal, binary and the American Standard Code for Information Interchange (ASCII).

The reader for a five-bit paper tape is shown in Fig. 11-35. Power is supplied to all LEDs as the light sources. As the paper tape passes, a hole permits light to pass through to the base of a phototransistor and switches it on. Where there is no light falling on the phototransistor, it is switched off. The output of the phototransistor is decoded according to the communication code used.

The paper tape punch receives a binary code which is decoded, and the appropriate punch is activated.

To service a paper tape reader/punch, consult the appropriate service manual. However, should the reader lose a bit or read erratically, check to see if dirt is interfering with the LED light source or the phototransistor sensor. Also, force the LEDs on and check the voltage drop across them to verify their proper action. You should also check the phototransistors for proper operation by measuring their outputs when the light source is activated and deactivated.

If the reader fails to operate, check the cable for loose connections or breaks. Also, make certain that the tape is properly loaded.

DATA COMMUNICATION

Communication between the microprocessor and peripheral devices such as disk drives, printers or modems is an integral part of the system. There are two methods of data transfer: serial and parallel. A serial method has only two wires and is used for long-distance communication. Parallel transmission, on the other hand, requires as many bits for data plus a few more for control signals. Parallel transmission is used for short distances and high data rates. Figure 11-36 illustrates serial and parallel transmission.

Asynchronous and Synchronous Communication

In a serial or parallel communication system, data can be transmitted asynchronously or synchronously. In an asynchronous communication system, the data transmitted and received uses a "handshaking" scheme for transmission because the sender and receiver are not operated at the same speed. Handshaking means that the transmitting device must send a pulse that says "data is ready" and the receiver device sends a pulse "acknowledge." The handshaking signals are required for each byte (8-bits) or word transmitted between devices. How-

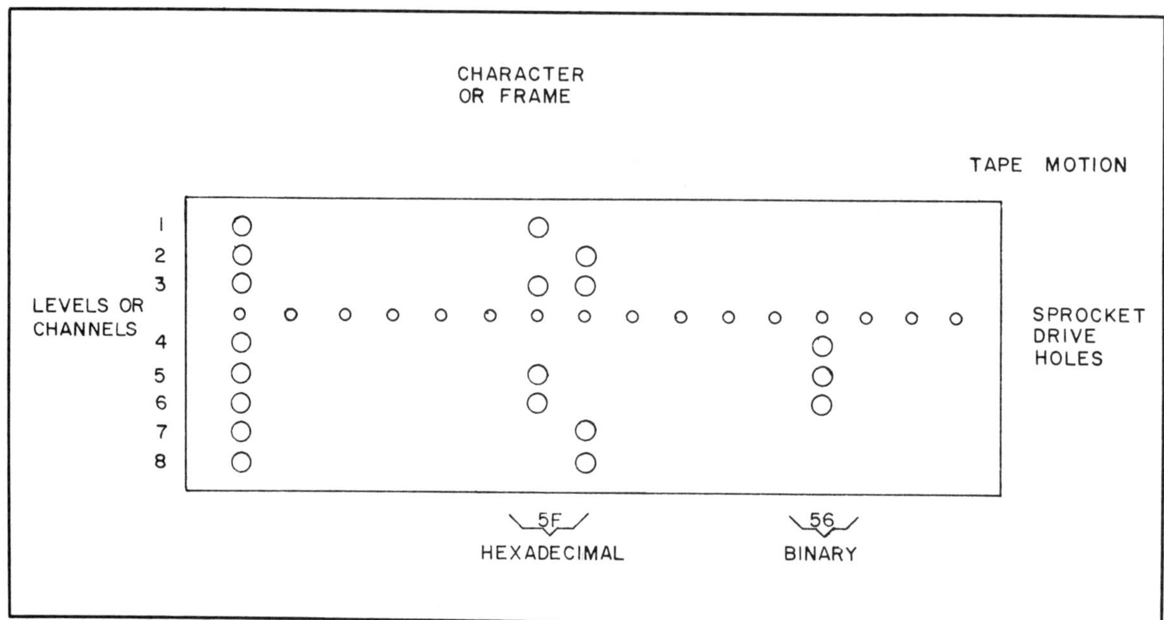

Fig. 11-34. Eight-bit paper tape and code illustration.

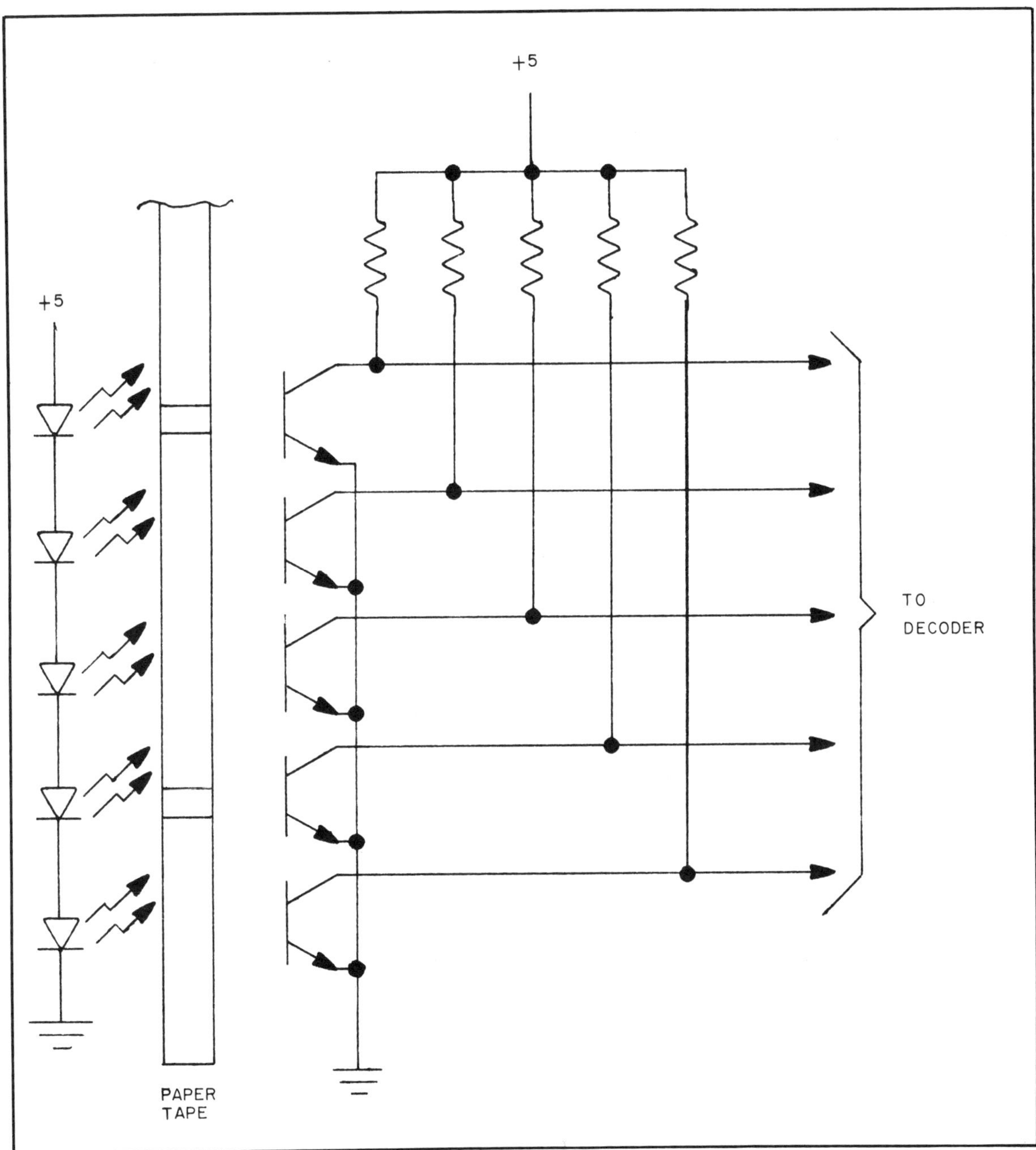

Fig. 11-35. Simplified optical paper tape reader.

ever, in a synchronous system, the data is transmitted in synchronization with a timing signal or control pulse. In a parallel synchronous system, a strobe or clock pulse is transmitted in parallel with the data. The control signal notifies the receiving device that data are ready for input. A serial asynchronous transmission for teletype is shown in Fig. 11-37. One pulse is used to designate the start, followed by a seven-bit code, and ending with two stop pulses.

Codes

To communicate, we need not only numbers, but also letters and other symbols. A set of codes which represent numbers and alphabetic characters are called alphanumeric codes. One popular code is the American Standard for Information Interchange (ASCII) which is a seven-bit code. Decimal digits are represented in BCD format preceded by 011. The letters of the alphabet and other symbols and instructions are represented by other codes as shown in Table 11-1. For example, the letter A is represented by the code 1000001, the letter Z by 1011010, and a line feed instruction by 0001010.

When you strike a key, it is encoded into an ASCII code and stored in RAM memory. Also, when you send data to the printer or store data on a floppy disk, the same code is used. However, for IBM equipment the Extended Binary-Coded Decimal Interchange Code (EBCDIC), a 9-bit code, is the standard language.

Bus and Communication Standards

To make order out of a world of intercommunication networks, standards have been adopted. Some of these

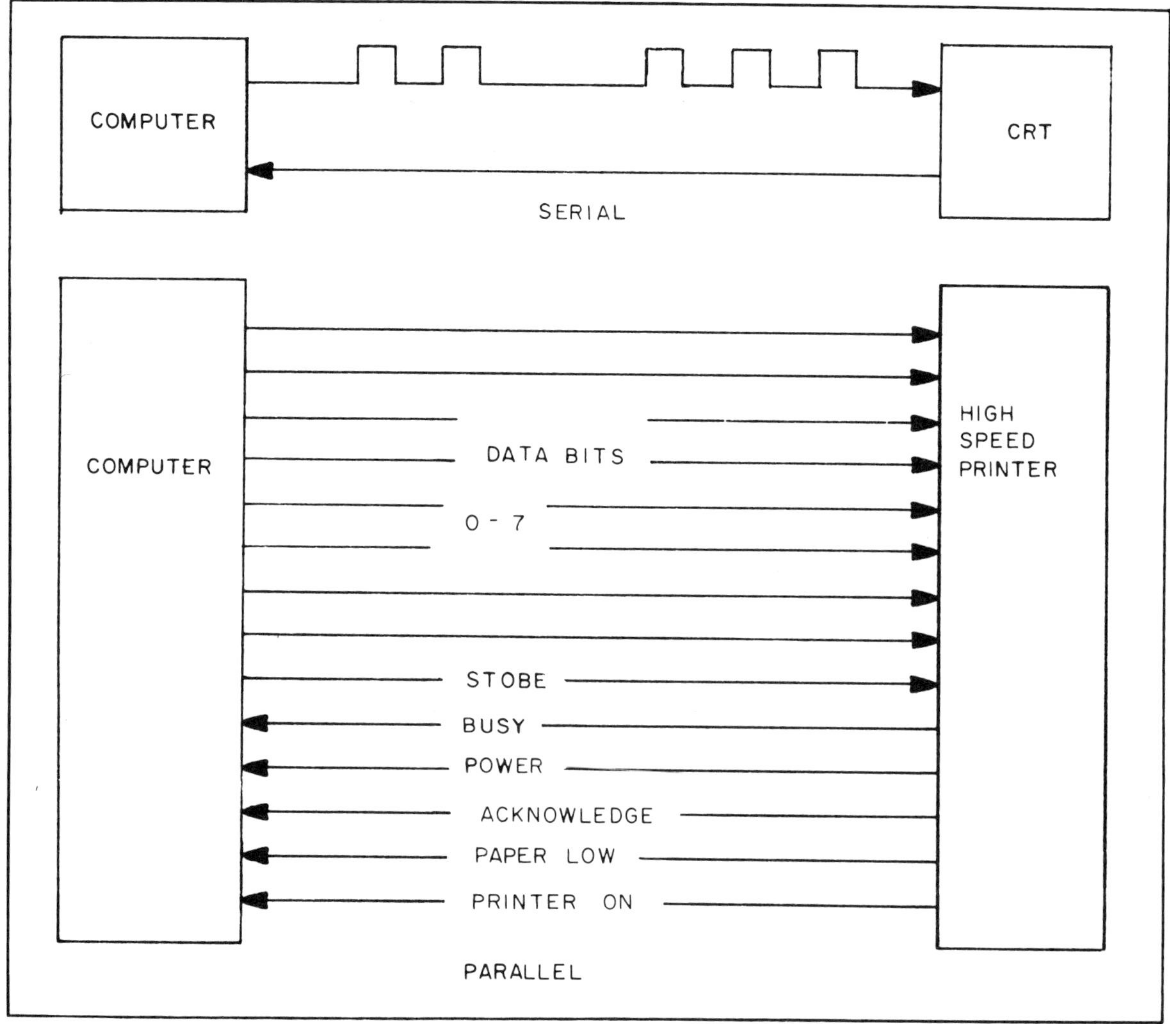

Fig. 11-36. Serial and parallel transmission.

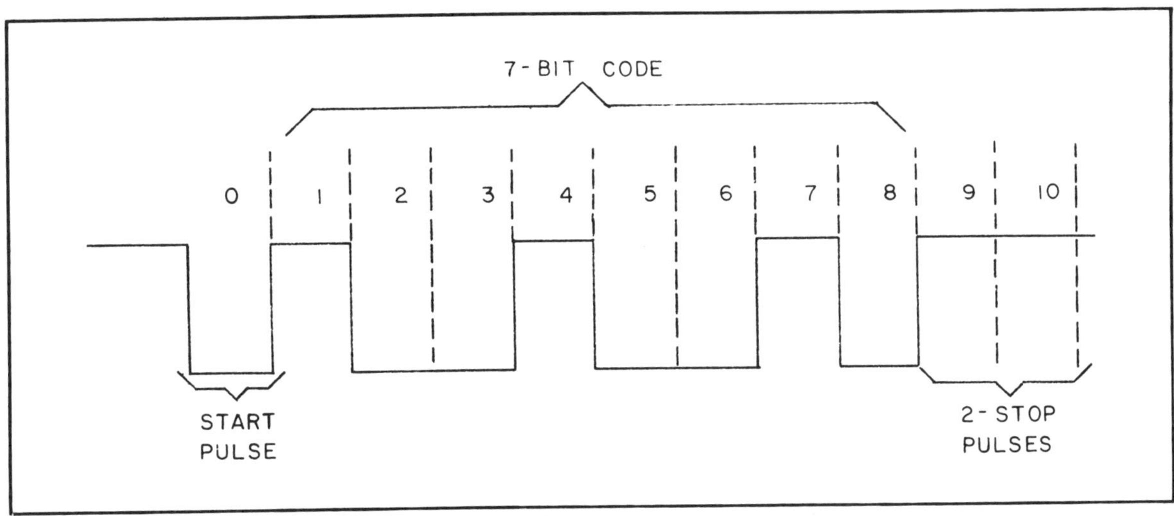

Fig. 11-37. Teletype asynchronous serial-transmission format.

standards have evolved because of successful use and adopted by various manufacturers. These standards can generally be categorized by distance, rate of transmission and serial or parallel transmission.

The serial standard is RS-232C and was one of the first standards developed. Logic level is not defined in terms of the standard of +5 volts but defines a space, or logical 0, as a voltage level from +3 to +15 and a mark, or logical 1, as a voltage level from −3 to −15. The standard further defines the 25 pin connector as shown in Fig. 11-38. More advanced serial standards are RS-422 and 423; however, the widespread acceptance of RS-232C has decreased their popularity.

The parallel standard IEEE 488 (also known as the General Purpose Interface Bus—or GPIB) is a bus standard developed by Hewlett-Packard. It is specifically designed to tie together peripherals and instrumentation systems under the control of a microprocessor. The bus is composed of 24 lines, of which eight are data/address/control bus lines, five are bus interface management lines, three are for handshaking, while the remaining lines are used for grounding and shielding. Figure 11-39 illustrates the bus structure. The GPIB connector pin assignments are shown in Fig. 11-40.

DATA CONVERTERS AND SERVICING

Digital computers require input signals in a binary format, and they can generate only digital signals for outputs. Unfortunately, most signals external to the computer operate as analog; that is, varying voltage levels. For a computer to input an analog signal, a sensor is used to detect the signal's magnitude and change it into an electrical equivalent. An example is a microphone which detects sound and changes it into a voltage that is analogous to the sound.

Another device which sits between the sensor and the computer is a converter. Its function is to convert the analog signal voltage into a binary representation, and it

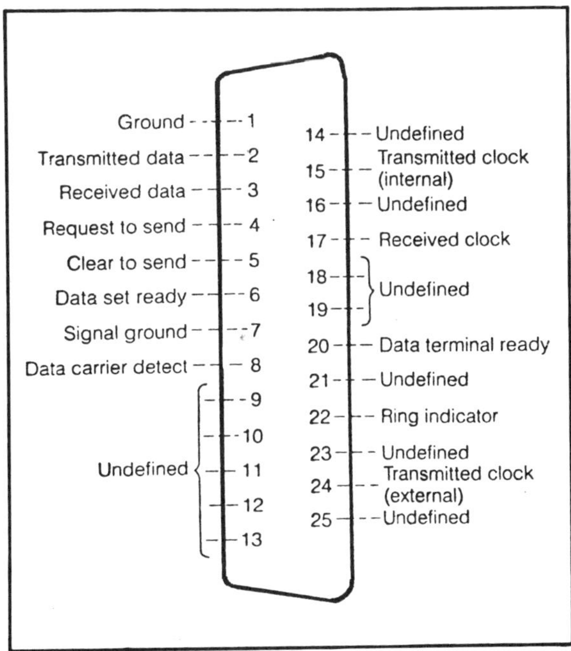

Fig. 11-38. RS-232C pin assignment for a connector.

Table 11-1. ASCII CODE SET.

Bits 765 / 4321	000	001	010	011	100	101	110	111
0000	NUL	DLE	SP	0	@	P	\	p
0001	SOH	DC1	!	1	A	Q	a	q
0010	STX	DC2	"	2	B	R	b	r
0011	ETX	DC3	#	3	C	S	c	s
0100	EOT	DC4	$	4	D	T	d	t
0101	ENQ	NAK	%	5	E	U	e	u
0110	ACK	SYN	&	6	F	V	f	v
0111	BEL	ETB	'	7	G	W	g	w
1000	BS	CAN	(	8	H	X	h	x
1001	HT	EM	)	9	I	Y	i	y
1010	LF	SUB	*	:	J	Z	j	z
1011	VT	ESC	+	;	K	[	k	{
1100	FF	FS	,	<	L	\	l	¦
1101	CR	GS	–	=	M	]	m	}
1110	SO	RS	.	>	N	⌒	n	~
1111	SI	US	/	?	O	—	o	DEL

NUL	Null
SOH	Start of Heading
STX	Start of Text
ETX	End of Text
EQT	End of Transmission
ENQ	Enquiry
ACK	Acknowledge
BEL	Bell (audible signal)
BS	Backspace
HT	Horizontal Tabulation (punched card skip)
LF	Line Feed
VT	Vertical Tabulation
FF	Form Feed
CR	Carriage Return
SO	Shift Out
SI	Shift In
DLE	Data Link Escape
DC1	Device Control 1
DC2	Device Control 2
DC3	Device Control 3
DC4	Device Control 4
NAK	Negative Acknowledge
SYN	Synchronous Idle
ETB	End of Transmission Block
CAN	Cancel
EM	End of Medium
SUB	Substitute
ESC	Escape
FS	File Separator
GS	Group Separator
RS	Record Separator
US	Unit Separator
DEL	Delete

is called an analog-to-digital (A/D) converter. The output of a computer can be changed from a digital form into an analog form by a device called a digital-to-analog (D/A) converter. The analog output can now be amplified to drive a speaker. The concept of A/D and D/A conversion is built around a principle called Nyquist's Sampling Theorem:

> If an analog signal is uniformly sampled at a rate of least twice its highest frequency content, then the original signal can be reconstructed from the sample.

This principle is illustrated in Fig. 11-41 which shows an analog signal varying with time and the times which the waveform is sampled. Notice that the amplitude does not change very much in the short time.

The process of an A/D conversion is shown in Fig. 11-42. The analog input signal is continuously varying and is sampled at specific times as determined by the sampling rate. The sampled voltage values passes through the A/D and is converted into a unique digital code for each sampled value. The example shown uses an 8-bit code and comes out of the converter in parallel versus serial form. It is this code that is read onto the data bus and into the microprocessor or memory.

There are many different methods of converting an analog signal into a digital representation. A basic binary ramp A/D converter is shown in Fig. 11-43. The analog signal from the sensor is sent to the input of a voltage comparator and is compared with an output reference voltage from the D/A converter. The output of the voltage comparator can have two states—logically 1 or logically 0.

If the analog input voltage is greater than the D/A converter output, then the comparator output will be logically 1. The AND gate will be enabled and pass clock pulses, and the counter will count up. The D/A converted output will rise until it equals the analog voltage from the sensor. At this point, the comparator output switches to a logical 0 and disables the AND gate which inhibits the clock pulses from reaching the counter. The binary representation of the analog signal is taken from the counter. The counter is reset by a computer signal and the conversion process is repeated.

The D/A converter accepts a binary parallel code into a register which affects a number of electronic switches and passes current through a resistive network with an output voltage level that corresponds to the input code. Figure 11-44 shows an 8-bit weighted resistor D/A converter. The binary coded input closes the appropriate solid state switches connected to a constant voltage source which sends current through the corresponding weighted resistor(s) to a summing junction of the amplifier. The resultant signal is amplified by the output amplifier.

During the sample period, the voltage level remains constant and the output has stepped voltage levels. The output voltage is restored very nearly to its original, continuously changing shape by passing the signal through an amplifier and filter as shown in Fig. 11-45.

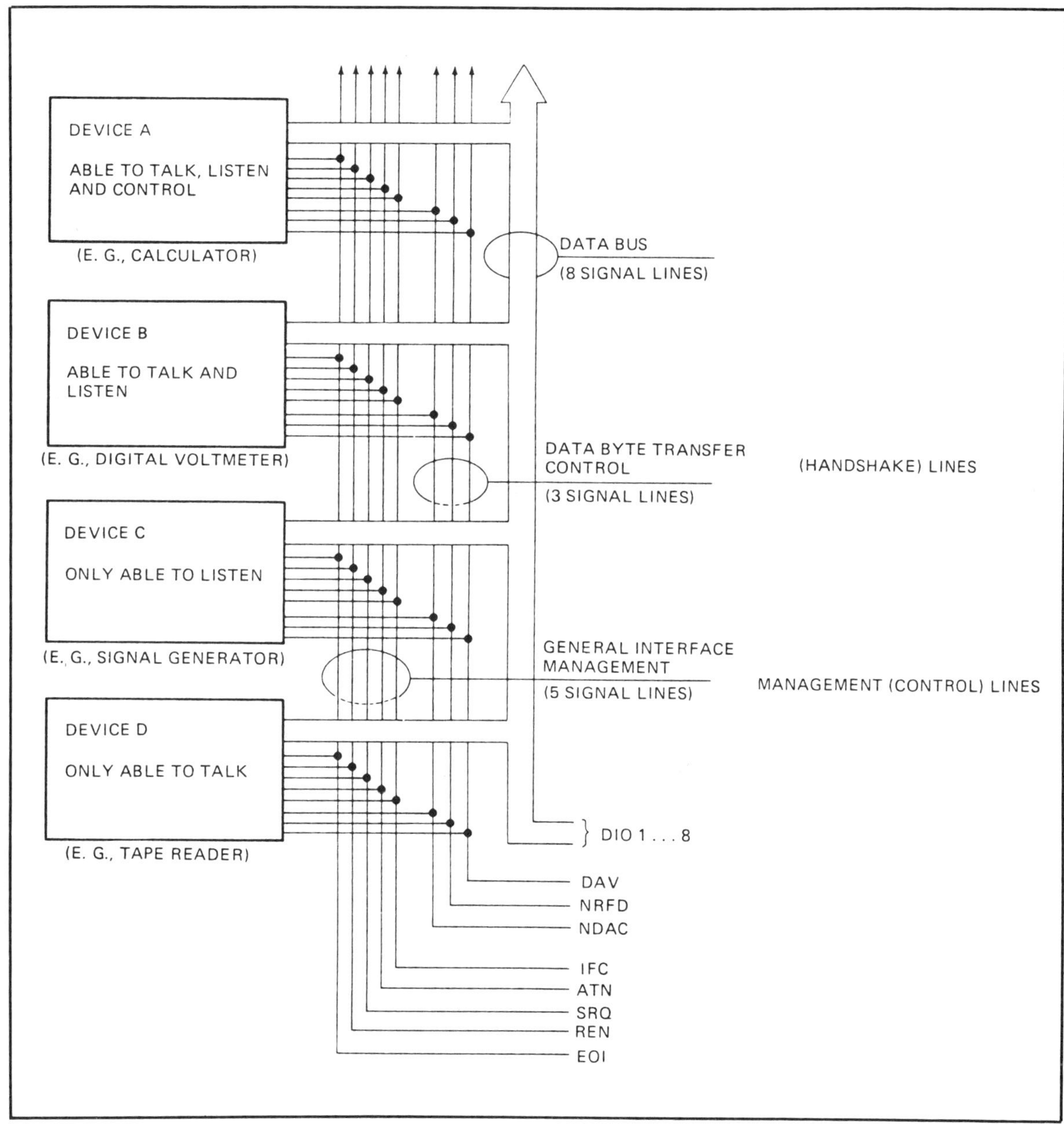

Fig. 11-39. General purpose interface bus (GPIB) structure (courtesy Hewlett-Packard Company).

The more D/A samples taken, the more closely will the output match the original signal. If the original signal was sound and the output was connected properly to a speaker, you would hear the reproduction. An example of a process programmable controller with A/D and D/A converters is shown in Fig. 11-46.

AD/DA Service

To service a D/A converter, perform a master clear and initialize the card to send a signal out. Then, output a data word for the maximum positive voltage. Measure the analog output voltage and, if necessary, adjust a potentiometer for the maximum positive voltage. The

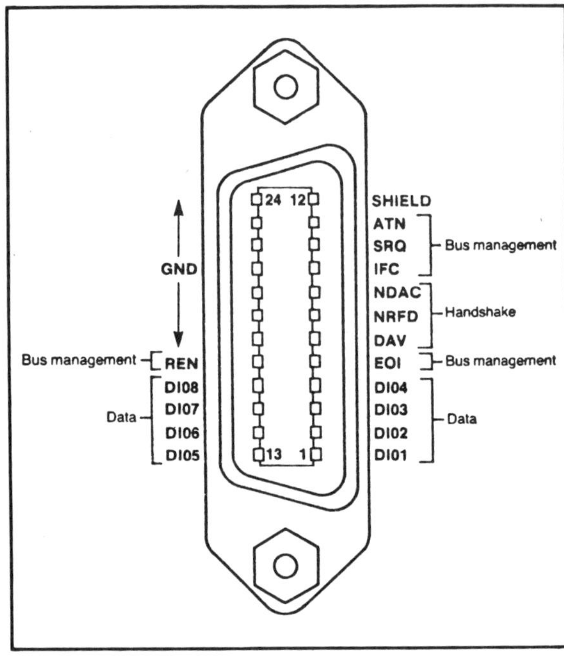

Fig. 11-40. General purpose interface bus (GPIB) connector pin assignment (courtesy Hewlett-Packard Company).

manufacturer's service manual will contain a table with data words and output values; each data word and output value should be verified. If an error occurs, localize the problem using the procedures outlined in Chapter 9.

The A/D converter is serviced or tested by placing the maximum positive voltage on the analog input channel, and having the computer read the data word and display the results. Repeat the process for a range of analog voltages and examine the data word for correct values for each analog voltage.

If you have an A/D and a D/A converter, connect the input and output lines together. Write a program that will output an analog signal on the D/A card and have it read into the computer by the A/D converter. Compare the D/A converter output data word with the A/D converter input data word.

INDUSTRIAL PROGRAMMABLE CONTROLLER

The industrial programmable controller is a microprocessor-based sequential controller. Figure 11-47 illustrates a block diagram of the major components such as the CPU, memory, I/O modules or process cards and programming devices. A total programmable system

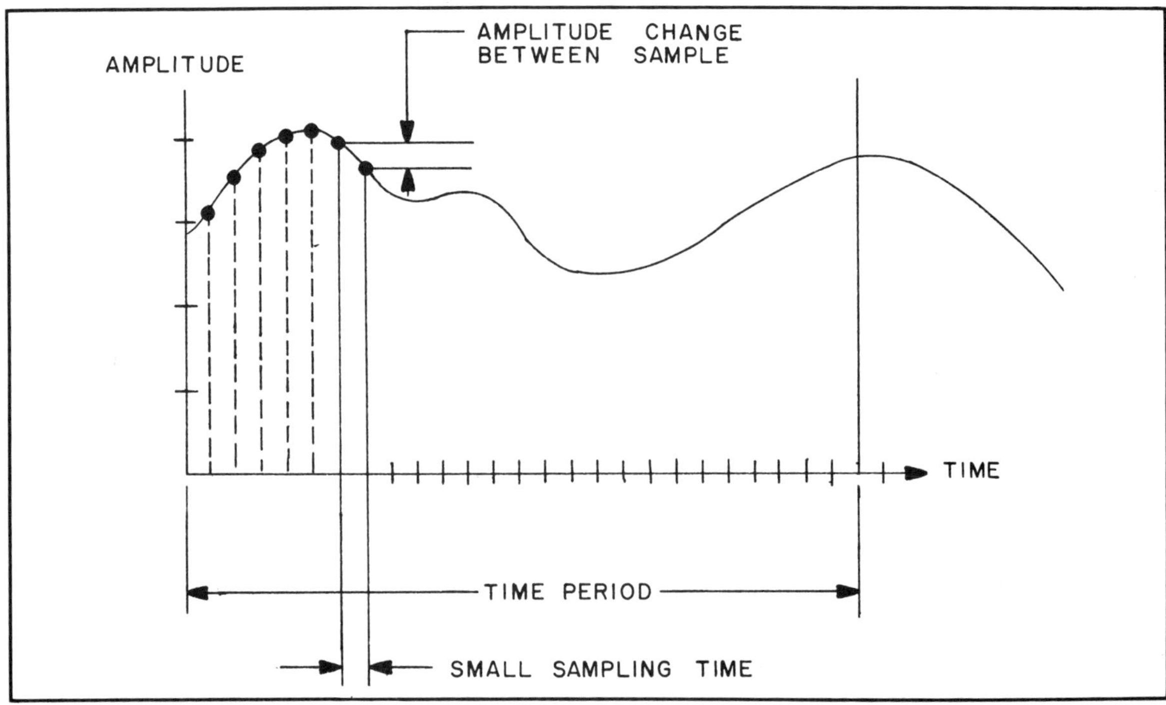

Fig. 11-41. Analog signal sampling.

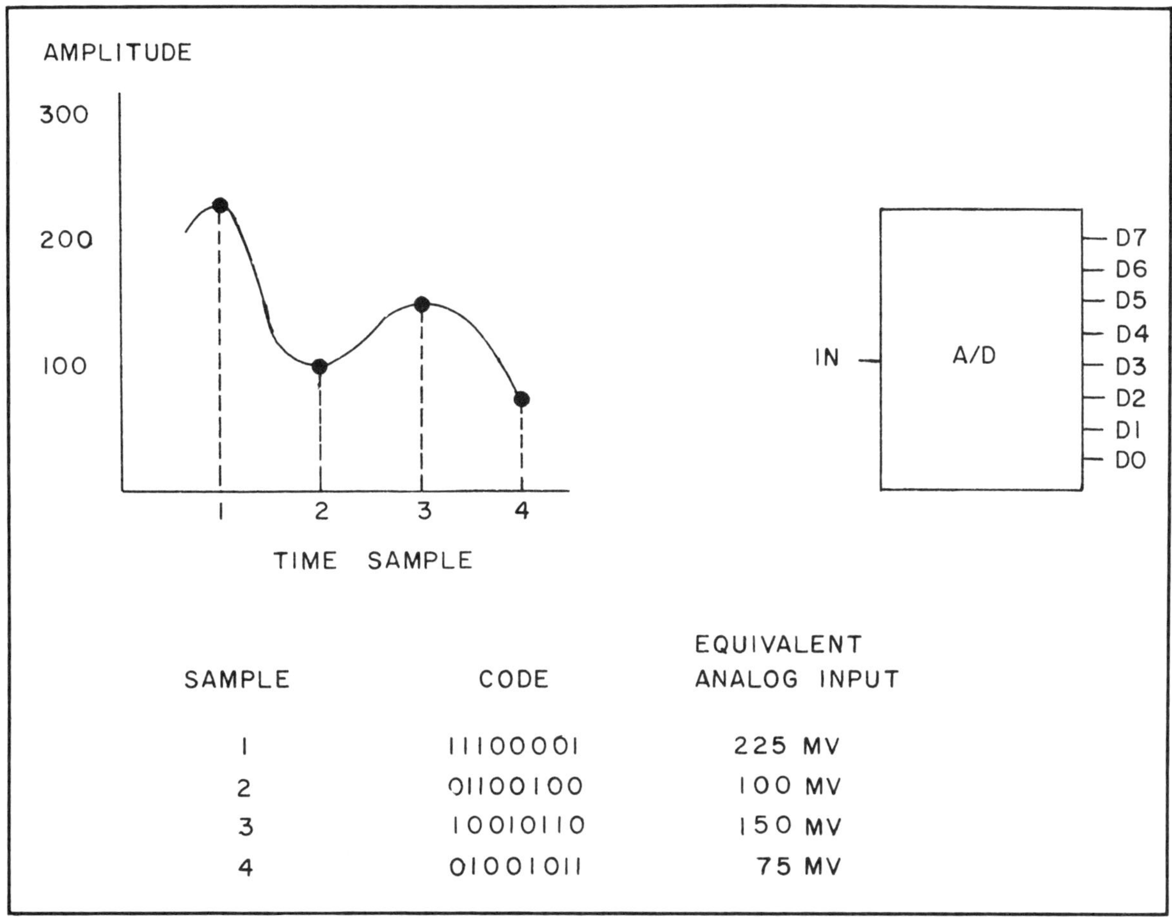

SAMPLE	CODE	EQUIVALENT ANALOG INPUT
1	11100001	225 MV
2	01100100	100 MV
3	10010110	150 MV
4	01001011	75 MV

Fig. 11-42. Analog to digital conversion process.

is shown in Fig. 11-48. Figure 11-49 shows a small programmable system. Notice the different type of program loader or programmer as well as the different style of input/output process cards. The program loader or programmer allows the operator to enter the required operating instructions.

The input process cards are designed to receive a high level ac or dc voltage from switches, contact closures, or photosensors. These process cards can handle single channel or multiple channel inputs. The input process card shown in Fig. 11-50 contains the required circuitry to convert the ac/dc signals of the industrial world into the voltages required by the microprocessor. The output process cards are designed to activate or deactivate a high level ac or dc voltage which controls motors, solenoids, or lamps. The output process card shown in Fig. 11-51 contains the circuitry to convert the microprocessor's signals voltage level to the required value for controlling industrial world devices.

Other process cards commonly used in industrial controllers are: A/D converters, D/A converters, timers, and communications cards for talking between controllers. The memory, CPU, and I/O process cards of a controller mounted on a rack are shown in Fig. 11-52. To allow the microprocessor to identify each rack of input and output cards, a specific reference number is used and set by a group of rack switches. Connection from the O/O process cards to the microprocessor module is made using a 50 conductor flat cable with edge connectors.

Because servicing an industrial controller can consume a great amount of time, manufacturers supply controller diagnostics for locating internal problems within the industrial controller, I/O cards structure, as well as external problems from sensors and activators which feed or are fed by the controller.

A variety of diagnostic techniques are employed by manufacturers. Some use the program loader unit as a diagnostic tool; others require only observing the controller unit; and a few sell a special diagnostic tester. Some controllers have a diagnostic run mode which can be programmed to check a portion of a program or run free after starting. If a problem is encountered, it will stop and the problem location will be indicated on a display. However, the most common method used is LED indicating lights. The LEDs are usually on input and output cards to indicate their status. There are additional LEDs on the I/O cards to indicate that the power is applied to the remote sensor or actuator. LEDs are also on timers to indicate if the timer has timed out.

TROUBLESHOOTING

Troubleshooting a complex digital system such as a microcomputer system requires a fundamental knowledge of microprocesser architecture, memory type and scheme, as well as bus structure and input/output modules. Very often, when armed with a good knowledge of the computer system, a simple description of the problem will quickly suggest a number of probable causes.

Diagnostic Program

Running software diagnostic programs is a way of allowing the intelligent machine to identify its own problems or faults with the system. It is a self-diagnosis. The major requirement in using diagnostic programs is that

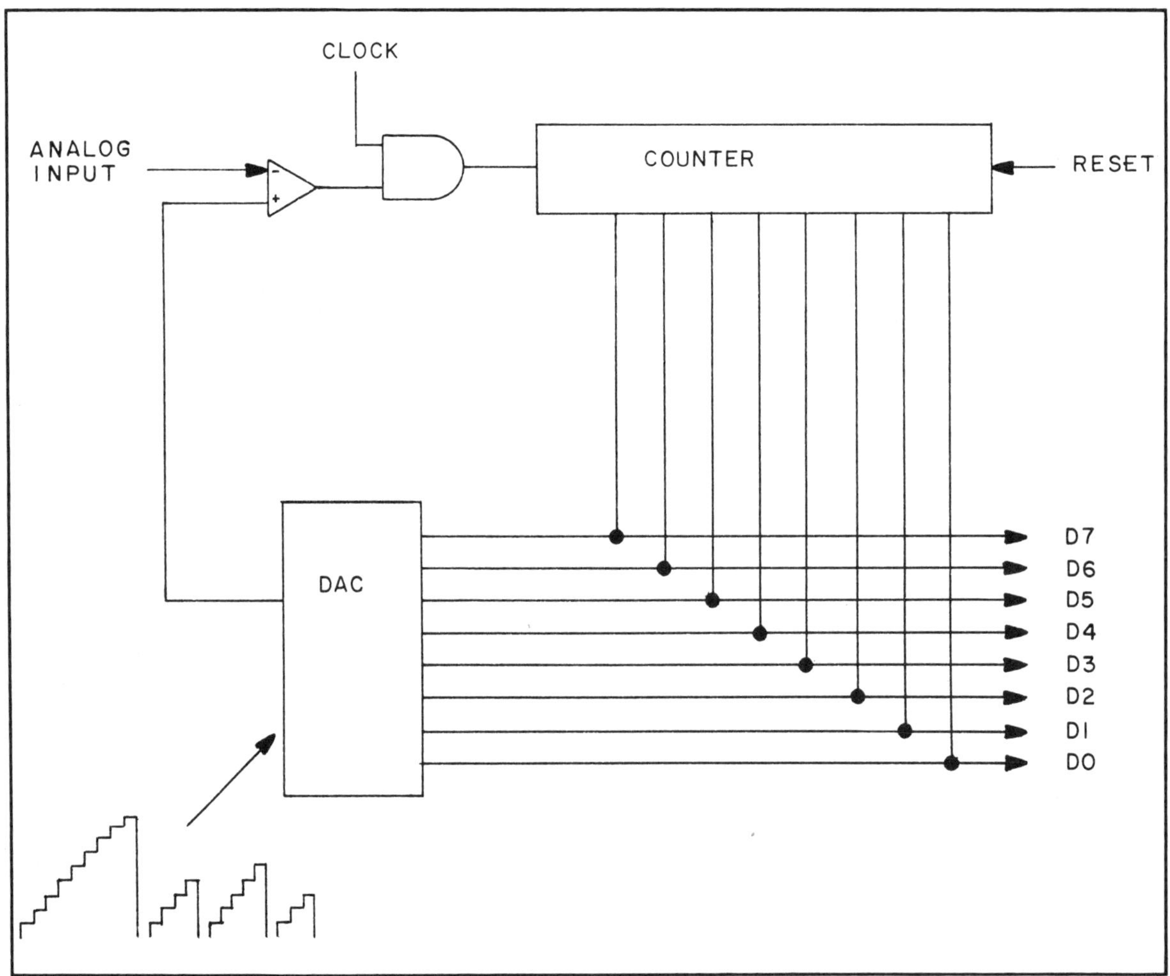

Fig. 11-43. Basic binary ramp A/D converter.

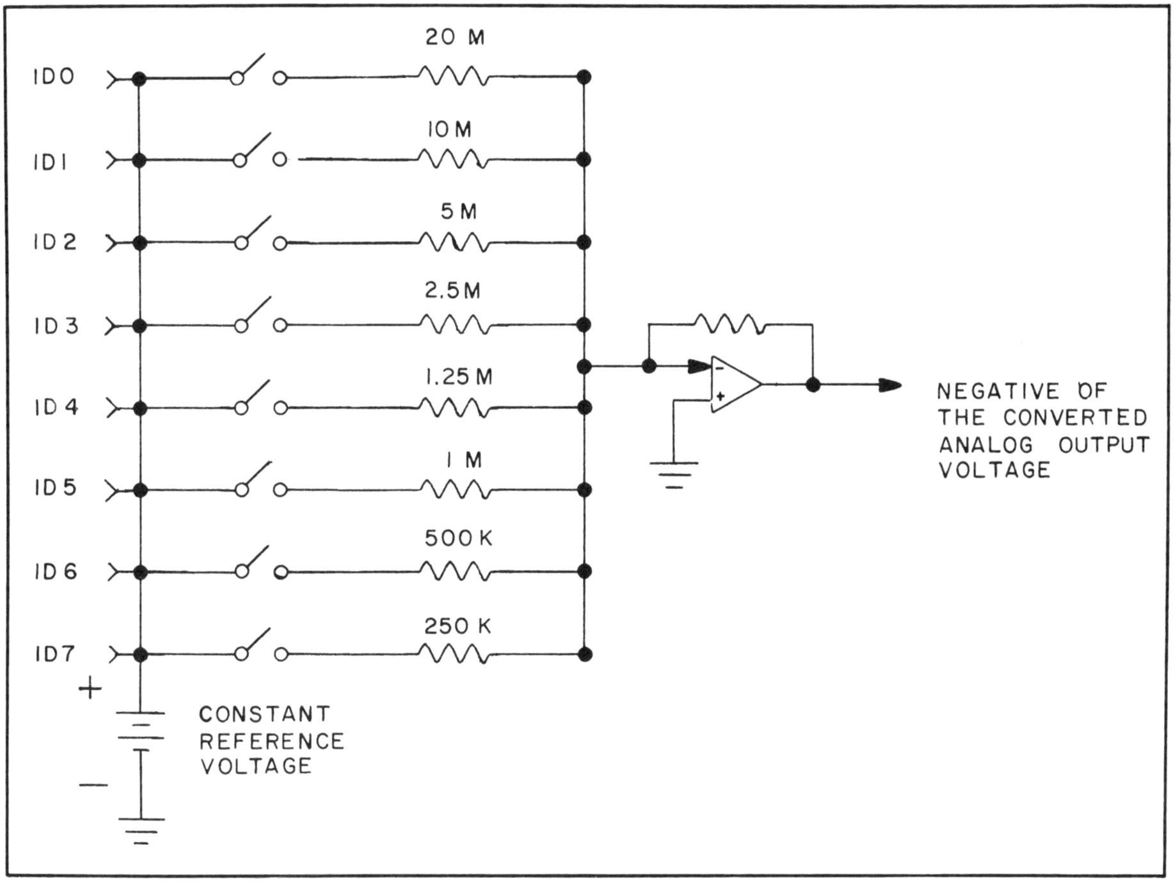

Fig. 11-44. Eight-bit weighted resistor A/D converter.

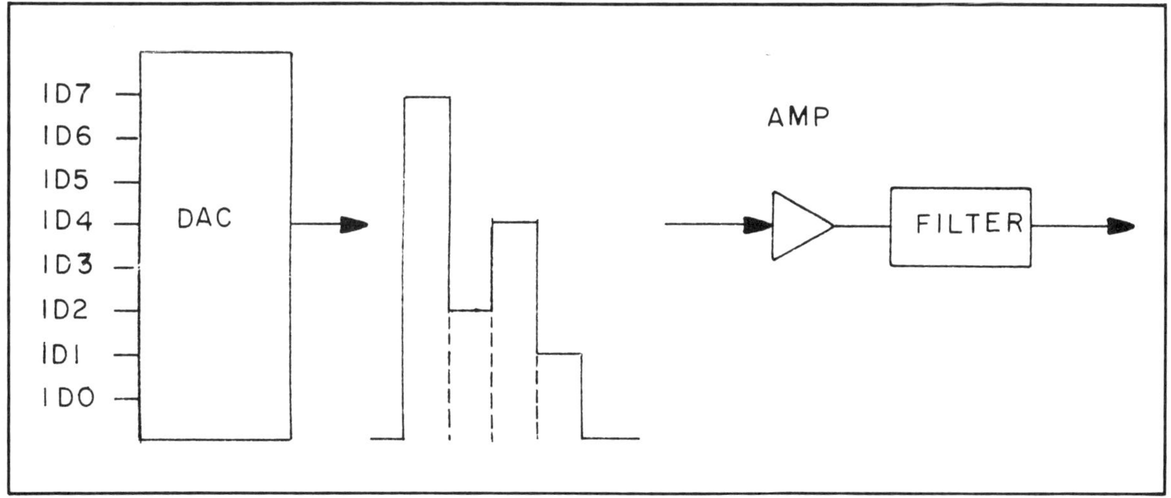

Fig. 11-45. Digital to analog conversion.

Fig. 11-46. Rack mounted process controller with A/D and D/A modules.

those sections of hardware/software systems required by the diagnostic program must be assumed to be operating properly. For example, a diagnostic program which checks for proper operation of the processor instruction set can only validate instructions not used by the diagnostic program. In general, a complete diagnostic program package will include a keyboard check, instruction set check, a RAM or ROM check, a disk speed check, serial and parallel interface check. Often you will need to secure operating manuals from the manufacturers and follow suggested testing of their peripheral card(s).

A popular diagnostic program to test the RAM memory is to have a routine which writes one (1) to each cell under test, reads it back and compares the results. If an error is found, a zero (0) rather than a one is outputted to the monitor or printer. If the RAM passes the ones test, then it is repeated using zeros.

If a fault is found in a memory chip, consult the electrical schematics for the location of the chip on the mother board or extended memory module. Check for appropriate control signals such as chip enable and read/write. Also check for presense of the address sig-

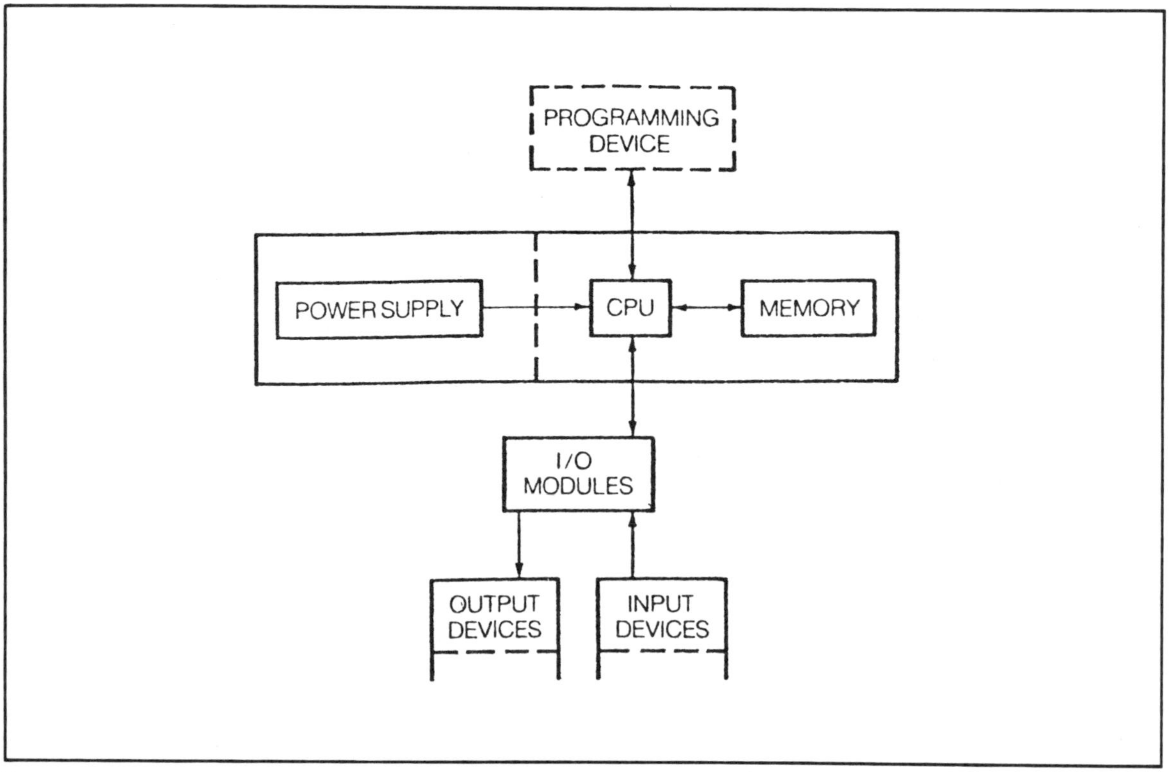

Fig. 11-47 Block diagram of a programmable controller.

Fig. 11-48. Programmable controller system with printer, PC-700, I/O and programmer (courtesy Westinghouse).

Fig. 11-49. A small programmable controller—the EPTAK (R) 200 PC—has been introduced by Eagle Signal Controls, a division of Gulf + Western Manufacturing Company. It is designed to provide digital programmable logic control for small industrial applications currently using hard-wired relay control systems or card logic systems, or to replace other less capable small programmable controllers (courtesy Eagle Signal Control).

Fig. 11-50. Programmable controller input module cutaway (courtesy Allen Bradley).

Fig. 11-51. Programmable controller output module with 16-discrete circuits (courtesy Westinghouse).

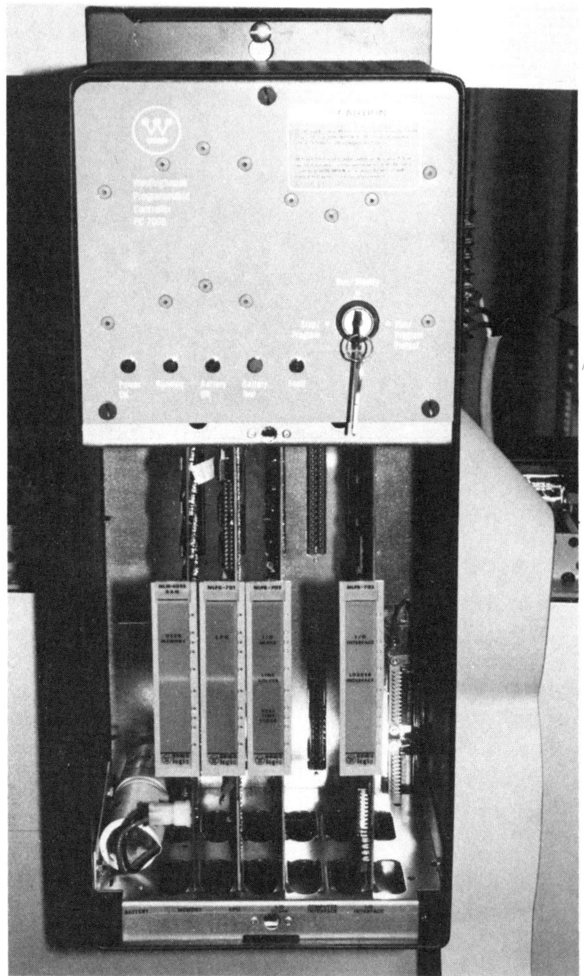

Fig. 11-52. Programmable controller rack mount showing 4 interval printed circuit boards—CPU, memory, I/O boards (courtesy Westinghouse).

nals using an oscilloscope or other device. If the signals are correct, then the memory chip is defective and should be replaced. Figure 11-53 illustrates how a proto clip is connected to an IC and then the oscilloscope probes are connected to it for quick signal tracing.

An extender card can be used when the computer housing does not permit easy access to ICs on a process card or module. Figure 11-54 illustrates the use of an extender card. The extender card plugs into the appropriate slot, and the process card plugs into it. This brings the process card above the surface of the rack so it's easier to monitor the signals.

Before replacing a chip, make certain the power is

Fig. 11-53. Oscilloscope signal tracing using a proto clip.

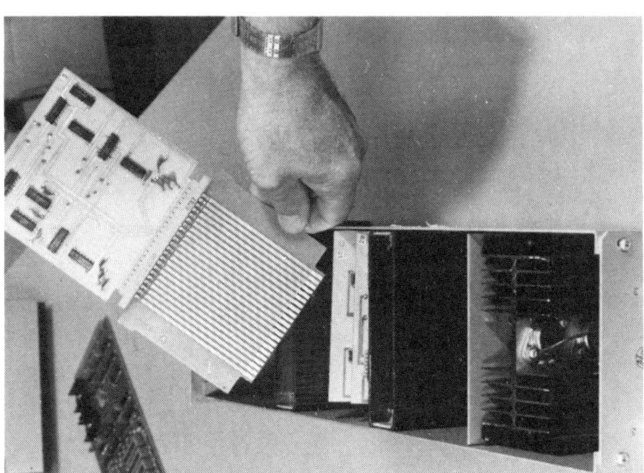

Fig. 11-54. Using an extender to raise the printed circuit above the chassis.

turned off. Discharge any static electricity you may have by touching the power supply chassis. Remove the defective RAM memory chip and insert the new one using methods outlined in Chapter 9.

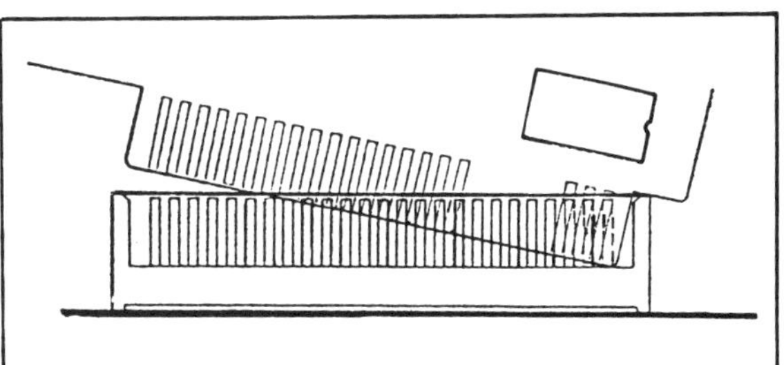

Fig. 11-55. Removing an interface card with power on results in edge connector shorting.

Should the defective RAM chip be located on an extended memory module, make sure the power is turned off before removing the module; otherwise, many additional circuits may be damaged due to shorting. Figure 11-55 illustrates this point. Notice when rocking the module to free it from the edge connector, that the adjacent pins easily cross and short out of the circuits.

Single Stepping

Should the diagnostic program localize the program to a specific module, you may have to analyze the module further to identify which component is defective. Single stepping is a way of slowing down the computer activities for close examination. The single step "freezes" the computer in the wait state of each machine cycle permitting you to directly observe the state of each data transfer operation. Figure 11-56 shows a hardware module for single stepping an Apple computer. The hardware module has LEDs which monitors the address bus and the data bus lines. You can run a diagnostic program or a modified one and single step while comparing the present state with what is expected.

A logic monitor will help you observe the input and outputs of a specific integrated circuit during a single stepping operation. Figure 11-57 shows a 40 pin logic monitor. A voltmeter, an oscilloscope or a logic probe could be used in place of the logic monitor.

Breakpointing

A technique of "freezing" the action of the computer by adding an instruction at predesignated program points is called *breakpoint.* The program runs at normal speed until a breakpoint is reached and allows you to verify the computer operation up to the breakpoint. If the previously run program detects no errors or faults, you may wish to single step further through the program to specifically identify the fault. Typically, a breakpoint is inserted in the program by substituting a break or restart instruction at a point where an error is anticipated to occur. Once again, a logic probe or logic monitor can be used to verify that the appropriate logic state of the ICs has been met. Keep in mind that a suspect hardware fault could really turn out to be an error in software.

Serial Communication Test

A simple way of checking a RS-232 serial communication card is by using a loop test. It consists of writing a

Fig. 11-56. Hardware for single stepping.

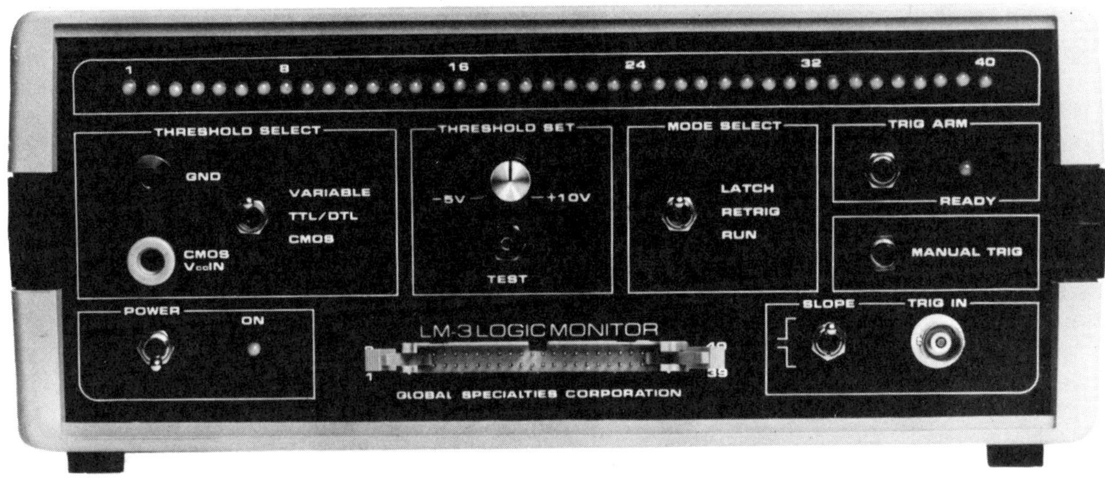

Fig. 11-57. 40-pin logic monitor (courtesy Global Specialties Corporation).

program which transmits a known byte of data, looping the data back to the receiver, reading the data and comparing it to the transmitted data. This test will check out the asynchronous communications chip, clock and line drives. To implement this test, you need a DB-25S female socket or a mating socket. Connect pins 2 and 3 together, which are the data transmit and data receive lines. Connect pins 4 and 5 together, which initiates the request-to-send and clear-to-send lines. Finally, connect pins 8 and 20 together, which initiate receive-line-signal detect and data-terminal-ready lines. If there is no communication, then check the input or output transistor or line drives. Defective line drivers are a common fault with the RS-232 serial communication card. If the test verifies the correct operation of the communication card, then a fault exists on the receiver equipment.

Logic Analyzer

As microcomputer systems become more complex and powerful, the most sophisticated oscilloscope lacks the ability to seek out well-entrenched faults. Consequently, the logic analyzer (LA) is used to test microcomputer system hardware for faults and software debugging. This instrument, shown in Fig. 11-58, is specifically designed to acquire parallel digital words. The logic analyzer operates by performing a data acquisition, and then displays the data acquisition in terms of ones and zeros. The instrument has a threshold adjustment so the data can be acquired from all semiconductor families. The data flows until the RAM memory display is filled. The acquisition data is then compared to a reference memory which contains a good copy the same size as the acquisition data. If they are not equal, the LA highlights the differences, making it easy to identify the error.

The logic analyzer has a trigger/control circuitry which permits the trigger to recognize a qualifier. The simple word recognition feature permits programming the LA to trigger when the microprocessor executes a specific instruction. Then you can observe the display to

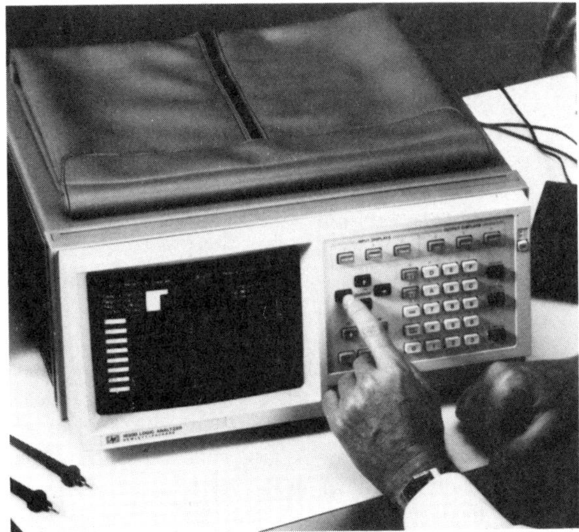

Fig. 11-58. Logic analyzer (courtesy Hewlett-Packard Corporation).

determine if certain sections of the control bus are functioning correctly with respect to timing signals.

The microprocessor base equipment is very complex in terms of the magnitude of the instructional set (machine language), high level language as well as the hardware. Many new digital systems' setup troubles are often blamed on hardware, but the real problem is often software.

PREVENTIVE MAINTENANCE

A systematic preventive maintenance program can be a very useful deterrent to system failures. The program is an aid to both the service person and the user, since detection and prevention of probable failures can reduce down time to a minimum.

As machines are made more intelligent, they employ built-in aids and indicators to help troubleshoot the microprocessor and major units of a computer system as well as I/O modules. Diagnostic programs are used for microcomputer system repair. These programs are generally recommended every 1000 hours for electrical devices and every 500 hours for mechanical devices, or approximately every three months.

Preventive Maintenance Tasks

The following tasks should be performed at least every three months:

1. Clean the exterior and the interior of the equipment cabinet, using a vaccum cleaner and/or a clean cloth.
2. On larger systems clean air filter using a vacuum cleaner to remove accumulated dust or dirt.
3. Visually inspect all wiring and cables for cuts, fraying, deterioration, kinks, strains and mechanical holders. Tape, solder or replace any defective wiring or hooded connectors.
4. Inspect all mechanical security: key switches, indicating lamps, control knobs, fans and data entry keyboards. Tighten or replace as required.
5. Inspect all modules mounted in panels on I/O slots to insure that each module is securely seated in its connector. Remove and clean any module which may have collected excess dust or dirt as shown in Fig. 11-59.
6. Inspect the power supply for proper voltages and check components such as capacitors or resistors for leakage or overheating. Replace any defective components.
7. Check disk drive for proper speed. Adjust speed according to manufacturer's specifications.
8. Clean disk drive heads, magnetic tape heads or optical paper tape reader.
9. Perform all preventive maintenance procedures for each peripheral device included in the system.

Fig. 11-59. Clean edge of modules with a brush.

SELF-EXAMINATION

Select the best answer:

1. A microprocessor is
   A. Another name for a computer.
   B. A CPU integrated circuit.
   C. A name for a calculator chip.
   D. A small scale integrated circuit.
2. Which is NOT one of the five major units of a computer?
   A. Sequencer.
   B. CPU.
   C. Memory.
   D. Control.
3. The system hardward includes
   A. Keyboard.
   B. Memory.
   C. Tape reader.
   D. All of the above.
   E. Only A and B.

4. Which of the following is considered a high level language?
   A. Assembler language.
   B. Machine language.
   C. Pascal.
   D. All of the above.
5. Which of the following is a machine language instruction?
   A. LOAD 5.
   B. IF X THEN Y.
   C. GO TO 280.
   D. 10110110.
   E. All of the above.
6. A program language
   A. Defines the form of instruction.
   B. Is never machine dependent.
   C. Is always machine dependent.
   D. Only A and C.
7. A bus that carries the information signals is
   A. Control.
   B. Address.
   C. Data.
   D. All of the above.
   E. Only A and B.
8. Which of the following steps does NOT occur during a WRITE cycle?
   A. The CPU makes the R/W line a logical 1.
   B. The CPU places a 16-bit address on the address line.
   C. The CPU outputs an 8-bit data onto the data bus.
   D. The selected I/O device or memory cell takes the data from the bus.
9. Which of the following steps does NOT occur during a READ cycle?
   A. The CPU makes the R/W line logical 1.
   B. The CPU places the appropriate address code on the address bus.
   C. The CPU makes the interrupt line a logical 1.
   D. The selected memory location of I/O places a word on the data line.
10. Which memory loses its content when the power is shut off?
    A. ROM.
    B. Fuse link.
    C. EPROM.
    D. RAM.
    E. All of the above.
11. What are two types of memory organization?
    A. Sequential organization.
    B. Bit organization.
    C. Word organization.
    D. Decoder organization.
12. The read/write head of a floppy disk drive is positioned to read a track by
    A. Pulsing a stepping motor.
    B. Activating a dc servo motor.
    C. Activating a power switch.
    D. Controlling the spindle speed.
13. A service procedure to perform on disk drives is
    A. Adjust the stepping motor speed.
    B. Clean the read/write head.
    C. Check the analog card power voltage.
    D. All of the above.
14. If an audio cassette recorder is not reading or writing properly, you should
    A. Clean the read/write head.
    B. Make sure all communication lines are not broken.
    C. Make sure the connecting jacks are firmly seated.
    D. All of the above.
15. Which of the following best describes A/D conversion?
    A. A device to change binary numbers from one form to another.
    B. A device that samples the highest input frequency.
    C. A device that changes an analog signal to a sequence of binary numbers.
    D. Both A and C.
16. If a tape reader performs erratically, you should check
    A. Cables for breaks.
    B. Connectors for proper seating.
    C. Clean the reader head.
    D. All of the above.
17. The audio cassette records data in terms of
    A. Two distinct magnetic patterns.
    B. A 2400 hertz for a logical 1.
    C. A 1200 kilohertz for a logical 0.
    D. A 2400 kilohertz for a logical 0.
18. To test an A/D and D/A converter card, you should
    A. Connect the output of one to the input of another and verify input and output data words.

B. Apply the maximum positive voltage to the analog input of the D/A and compare results with the manufacturer's specifications.
C. Send out a data word to the D/A, measure the analog output, and verify manufacturer's specifications.
D. All of the above.

19. An industrial programmable controller is composed of
A. A CPU.
B. A memory.
C. Input/output devices.
D. All of the above.

20. The serial standard is
A. IEEE 488.
B. GPIB.
C. RS-232C.
D. IEEE 556.

21. Diagnostic tests for a programmable controller consist of
A. Observing LEDs for status and the presence of external power.
B. Using the program loader unit as a diagnostic tool.
C. Using a special diagnostic tester.
D. All of the above.

22. A good preventive maintenance procedure will include
A. Periodic vacuum cleaning of interior and exterior equipment.
B. Scheduled diagnostic testing.
C. Inspection of mechanical connections.
D. Inspection and cleaning of I/O slots.
E. All of the above.

23. An extender card is used to
A. Lengthen the RAM memory of a computer system.
B. Permit easy access to process cards for troubleshooting.
C. Lengthen the ROM memory of a computer system.
D. Only A and C.

24. Before removing any process cards or modules, you should
A. Make sure the cards have warmed up.
B. Make sure to have discharged yourself.
C. Make sure the power is turned off.
D. All of the above.
E. Only B and C.

25. Single stepping
A. Is a way of "freezing" the computer so that you directly observe the state of each data transfer.
B. Uses a hardware module for controlling the machine cycle of the computer.
C. Is an instruction inserted in the program at predesignated points to interrupt the computer organization.
D. All of the above.
E. Only A and B.

Appendix A
Motor Troubleshooting Guide

Probable Causes of Motor Troubles

	Motor Type					
	AC Single Phase				AC Polyphase (2 or 3 Phase)	Brush Type (Universal, Series, Shunt or Compound)
Trouble	Split Phase	Capacitor Start	Permanent Split Capacitor	Shaded Pole		
Will not start.	1, 2, 3, 5	1, 2, 3, 4, 5	1, 2, 4, 7, 17	1, 2, 7, 16, 17	1, 2, 9	1, 2, 12, 13
Will not always start, even with no load, but will run in either direction when started manually.	3, 5	3, 4, 5	4, 9		9	
Starts, but heats rapidly.	6, 8	6, 8	4,8	8	8	8
Starts, but runs too hot.	8	8	4,8	8	8	8
Will not start, but will run in either direction when started manually—overheats.	3, 5, 8	3, 4, 5, 8	4, 8, 9		8, 9	
Sluggish—sparks severely at brushes.						10, 11, 12, 13, 14
Abnormally high speed—sparks severely at the brushes.						15
Reduction in power—motor gets too hot.	8, 16, 17	8, 16, 17	8, 16, 17	8, 16, 17	8, 16, 17	13, 16, 17
Motor blows fuse, or will not stop when switch is turned to off position.	8, 18	8, 18	8, 18	8, 18	8, 18	18, 19
Jerky operation—severe vibration.						10, 11, 12, 13, 19

* Probable causes correspond to the numbers in the table.

1. Open in connection to line.
2. Open circuit in motor winding.
3. Contacts of centrifugal switch not closed.
4. Defective capacitor.
5. Starting winding open.
6. Centrifugal starting switch not opening.
7. Motor overloaded.
8. Winding short circuited or grounded.
9. One or more windings open.
10. High mica between commutator bars.
11. Dirty commutator or commutator is out of round.
12. Worn brushes and/or annealed brush springs.
13. Open circuit or short circuit in the armature winding.
14. Oil-soaked brushes.
15. Open circuit in the shunt winding.
16. Sticky or tight bearings.
17. Interference between stationary and rotating members.
18. Grounded near switch end of winding.
19. Shorted or grounded armature winding.

courtesy Bodine Electric Co.

Appendix B

Motor Control Troubleshooting Guide

Motor Control Troubles

Trouble	Cause	Remedy
OVERLOAD RELAYS Tripping	1. Sustained overload.	1. Check for grounds, shorts, or excessive motor currents and correct cause.
	2. Loose connection on load wires.	2. Clean and tighten.
	3. Incorrect heater.	3. Heater should be replaced with correct size.
MAGNETIC AND MECHANICAL PARTS Noisy Magnet	1. Broken shading coil.	1. Replace magnet and armature.
	2. Magnet faces not mating.	2. Replace magnet and armature.
	3. Dirt or rust on magnet faces.	3. Clean.
	4. Low voltage	4. Check system voltage and voltage dips during starting.
Failure to Pick-up and Seal	1. Low voltage.	1. Check system voltage and voltage dips during starting.
	2. Coil open or shorted.	2. Replace
	3. Wrong coil.	3. Replace.
	4. Mechanical obstruction.	4. With power off check for free movement of contact and armature assembly.
Failure to Drop-Out	1. Gummy substance on pole faces.	1. Clean pole faces.
	2. Voltage not removed.	2. Check coil circuit.
	3. Worn or rusted parts causing binding.	3. Replace parts.
	4. Residual magnetism due to lack of air gap in magnet path.	4. Replace magnet and armature.
PNEUMATIC TIMERS Erratic Timng	1. Foreign matter in valve.	1. Replace timing head complete or return timer to factory for repair and adjustment.
Contacts Do Not Operate	1. Maladjustment of actuating screw.	1. Adjust as per instruction in service bulletin.
	2. Worn or broken parts in snap switch.	2. Replace snap switch.
LIMIT SWITCHES Broken Parts	1. Overtravel of actuator.	1. Use resilient actuator or operate within tolerances of the device.
MANUAL STARTERS Failure to Reset	1. Latching mechanism worn or broken.	1. Replace starter.
COMPENSATORS (MANUAL) Welding of Contacts on Starting Side	1. Inching, jogging and operating handle slowly.	1. Excessive inching and jogging not recommended (caution operator) move handle swiftly and surely to start position.
Welding of Contacts Running Side	1. Moving handle slowly to run position.	1. Move handle swiftly and surely to run position as motor approaches full speed.
	2. Lack of sufficient spring pressure.	2. Replace contacts and contact springs.
Damaged or Burned Transformer	1. Repeating inching and jogging.	1. Excessive inching and jogging not recommended (caution operator).

	2. Holding handle in start position for long periods.	2. Hold handle in start position only until motor approaches full speed.
CONTACTS Contact Chatter	1. Broken shading coil. 2. Poor contact in control circuit. 3. Low voltage.	1. Replace magnet and armature. 2. Replace the contact device or use holding circuit interlock (3 wire control). 3. Correct voltage condition. Check momentary voltage dip during starting.
Welding or Freezing	1. Abnormal inrush of current. 2. Rapid jogging. 3. Insufficient tip pressure. 4. Low voltage preventing magnet from sealing. 5. Foreign matter preventing contacts from closing. 6. Short circuit.	1. Check for grounds, shorts or excessive motor load current or use larger contactor. 2. Install larger device rated for jogging service. 3. Replace contacts and springs, check contact carrier for deformation or damage. 4. Correct voltage condition. Check momentary voltage dip during starting. 5. Clean contacts with Freon, contactors, starters, and control accessories used with very small current or low voltage, should be cleaned with Freon. 6. Remove short or fault and check to be sure fuse or breaker size is correct.
Short Tip Life or Overheating of Tips	1. Filing or dressing. 2. Interrupting excessively high currents. 3. Excessive jogging. 4. Weak tip pressure. 5. Dirt or foreign matter on contact surface. 6. Short circuits. 7. Loose connection. 8. Sustained overload.	1. Do not file silver tips. Rough spots or discoloration will not harm tips or impair their efficiency. 2. Install larger device or check for grounds, shorts or excessive motor currents. 3. Install larger device rated for jogging. 4. Replace contacts and springs, check contact carrier for deformation or damage. 5. Clean contacts with Freon. 6. Remove short or fault and check to be sure fuse or breaker size is correct. 7. Clean and tighten. 8. Check for excessive motor load current or install larger device.
COILS Open Circuit	1. Mechanical damage.	1. Handle and store coils carefully.
Roasted Coil	1. Over voltage or high ambient temperature. 2. Incorrect coil. 3. Shorted turns caused by mechanical damage or corrosion. 4. Under voltage, failure of magnet to seal in. 5. Dirt or rust on pole faces increasing air gap.	1. Check application, circuit, and correct. 2. Install correct coil. 3. Replace coil. 4. Correct system voltage. 5. Clean pole faces.

courtesy Square D. Co.

Appendix C
Radio/Stereo Troubleshooting Guide

Radio & Stereo Troubles

Symptom	Section	Probable Cause
Needle skids on record.	Tone arm/support.	Adjust needle pressure.
Tone arm incorrect placement.	Tone arm/support.	Adjust tone arm height and set down point.
Turntable too slow.	Idler wheel, motor, drive surfaces.	Clean idler wheel, turntable rims, lubricate bearings.
Incorrect or inoperative dial indicator.	Dial chord.	Replace/restring dial cord.
Weak sound, very distorted output.	Output amplifiers/drivers.	Replace output, check circuit.
No output or extremely low output.	Cartridge/driver circuits, outputs.	Check needle, cartridge, drivers, outputs.
Hum, low volume.	Power supply.	Filter.
Poor selectivity/sensitivity.	Tuner/alignment.	Check converter, align receiver.
Noisy, rattling vibration.	Speakers.	Check connections, speaker, etc.
Motorboating, squealing.	Tuner/power supply.	Filter, tuner, check connections.
Tubes fail to light.	All circuits/power supply.	Check tube, heaters, fuse, switch, rectifier, filter, etc.
Intermittent operation.	All circuits.	Check connections, amps, etc.
Station drift.	Local oscillator.	Check oscillator/leaky capacitors.
Volume varies.	AGC.	Check AGC circuit.
One channel dead and/or highly distorted with low volume.	Outputs/driver circuits, etc.	Check outputs, coupling, bias, circuits, check AM/FM switch.
Good fm, but low volume, distorted AM.	AM circuits.	Outputs, etc.
Poor treble/bass fidelity.	Preamp crossover system.	Check circuits.
Noisy volume or balance control.	Control potentiometer.	Clean or replace control.
Faulty balance control.	Balance control circuit.	Check circuit.

Appendix D
Tape Player Troubleshooting Guide

Tape Player Troubles

Trouble	Probable Cause
Speed is sluggish or slow, capstan does not rotate.	Defective motor, drive belt, idler, flywheel, capstan, or mechanical binding.
High distortion, poor treble, low output.	Dirty head (clean with isopropyl alcohol), defective or misadjusted head, defective amplifier circuit (damagnetize head).
Bad erase.	Defective or dirty erase head (clean or replace head).
Rewind or fast forward inoperative.	Defective belt, idler motor, or housing.
Bad tape drive, flutter.	Dirty head and drive mechanisms; flywheel, idler, motor, bearing (may need to lubricate with light machine oil).
Take-up reel drags.	Dirty or defective belts, pulleys, motor mechanical binding (may need cleaning and lubricating).
Ejection button does not work properly.	Check spring, mechanical wear, or misalignment.

Appendix E
Appliance Troubleshooting Guide

Appliance Troubles

Symptom/Problem	Probable Cause
Fails to heat.	Fuse, switch, heating element, rectifier, overload, cord, connections.
Fails to operate.	Fuse, switch, rectifier, overload, cord, connections, plug.
Shocks obtained.	Check for grounds, loose or pinched wires.
Erratic, intermittent operation.	Check connections, contacts, etc., clean appliance.
Any type of electric motor problem.	See Appendix C.
Leaky water, air, oil, etc.	Replace gasket, tighten loose bolts, connections, etc.
Pump does not spray.	Clean/check plunger, valve, spring, jets, nozzles, etc.
Overload keeps tripping.	Check for shorts/replace overload.
Noisy, rattling, vibration.	Check for loose connections.
Appliance will not shut off.	Check switch, thermostat.
Cord overheats.	Check for plug corrosion, check for shorts, replace cord.
Toaster will not pop bread up.	Clean/replace thermostat, check/adjust projecting lever.
Toaster does not lower bread.	Check for binding, trip lever, thermostat blade.
Toaster does not heat uniformly.	Check/replace heating elements.
Pilot lamp does not light.	Check for power, replace bulb.
Timing cycle does not operate.	Replace timer, thermostat.
Insufficient heat.	Check voltage, connections, contacts, temperature, setting switch.

Appendix F

TV Troubleshooting Guide

Television Troubles

Symptom	Section	Probable Cause
Dead set.	Power supply.	Line cord, fuse, circuit breaker, rectifier, filter, switch, tube, filaments.
Set fails to shut off.	Power switch.	Check diode/switch.
Low B+ voltage.	Power supply or short in another stage.	Rectifier, filter.
Horizontal black and white hum bars or bend.	Power supply.	Filter.
No sound, no picture, tubes lit.	Power supply.	Tube, fuse, filter, switch, circuit breaker, rectifier.
Tuner jitters the picture.	Tuner.	Clean, lubricate tuner, contacts.
Snowy picture.	Tuner.	Check rf amp/converter, circuit.
Weak picture, weak sound or no picture, no sound.	Tuner, i-f, video detector, AGC.	Check amps, circuit.
Smear, poor detail, grainy picture.	I-f.	Check amps, circuit.
Wormy picture, herringbone weave.	Video detector, video amp.	Check sound trap, circuit.
Sound level varies.	Audio i-f detector, output amp.	Check volume control, amps, etc.
Buzz, distorted sound.	Audio i-f, detector, output amp.	Adjust sound, i-f can, circuit.
Low sound.	Audio i-f, detector, output amp.	Check amps, circuit.
No sound.	Audio i-f, detector, output amp.	Amps, circuit.
Intermittent sound.	Audio i-f, detector, output amp.	Check connections, amps, circuit.
No contrast, weak picture.	Video detector, amp.	Check amp, circuit.
Overloaded picture.	AGC	Adjust control, circuit.
Wavy, weaving picture with distorted sound.	AGC	Adjust control, circuit.
Low brightness.	Horizontal, CRT	Check hv; check CRT.
Silvery picture, blooming.	CRT.	Check CRT.
Dark, burnt spots on CRT.	CRT.	Screen burn, replace crt.
Lost one color in picture.	CRT, color gun, demodulator.	Check gun, CRT.
Vertifical jitter, horizontal line only, shrunken picture top to bottom.	Vertical oscillator/amp.	Adjust linear, height, hold controls, check amps, circuit.

Symptom	Section	Probable Cause
Vertical roll.	Vertical/sync.	Check oscillator, amps, circuits.
No raster, sound normal.	Horizontal circuits, hv rectifier, flyback, CRT.	Check hv, CRT, horizontal circuits.
Picture pulled in side to side.	Horizontal.	Horizontal oscillator, damper, output.
High pitched ringing.	Horizontal.	Check output.
White vertical bars, fluttering picture, brightness weak.	Horizontal.	Check damper, etc.
Triangular shape, shrunken, unnatural shaped picture.	Deflection yoke, pincushion correction.	Replace yoke.
Tilted picture, picture tearing horizontally.	Deflection yoke, horizontal sync.	Adjust yoke, check horizontal oscillator, sync,
Confetti.	Color killer, color detector.	Adjust color killer, check circuits.
Weak color.	Chroma amp.	Check amp.
Washed out, white, loss of color.	Demodulator.	Check demodulators, amps.
Drifting color bars.	Reactance, afpc.	Check circuits.
One color too strong.	Color amps.	Adjust color drives/screens.
Incorrect facial tones.	Color oscillator.	Adjust tint control.
Color fades.	Color circuits.	Connections, demodulators, amps, etc.
Blooming, poor focus.	TV regulator, focus rectifier.	Check connections, adjust TV, check circuits.
All colors too strong.	Chroma amp.	Adjust color control, check chroma amp.
Hv arcs, snapping, etc.	Focus/hv rectifier, regulator.	Check hv lead dress connections, circuits.
Unclear color picture, fringing, color positions do not hold.	Convergence.	Perform convergence setup, check convergence rectifier.

Appendix G

General Troubleshooting Guide

Suspected Problem	Suggested Techniques*
Fuse	1,2,13
Filter (capacitor)	2,4,11,16
Transistor	2,4,6,7,10
Diode	2,3,6,10
Circuit board	1,2,5,6,7,8,12
IC	1,2,3,6
Tube	1,2,3,5,15
Coil	2,3,13,16
Transformer	1,2,3
Resistor	1,2
Switch	1,2
Lamp	2,3
Wire/cable	1,2,17
Motor winding	13,14,17
Armature	14
Electronic circuit	1,2,5,6,8,9,10,12,16

***Options** (correspond to numbers on table)

1. Voltage measurement.
2. Resistance measurement
3. Substitution.
4. Bridging.
5. Tapping.
6. Heat/freeze.
7. Transistor cut off.
8. Signal tracing.
9. Oscilloscope.
10. Diode/transistor tester.
11. Spark test.
12. Resoldering.
13. Test lamp.
14. Growler.
15. Tube tester.
16. Other component testers.
17. Megohmmeter.

Appendix H

TV/FM Reception Troubleshooting Guide

Ghosts, Extra Images, and Smear

A. Possible Multi-Path Reception

- ☐ Reposition antenna by turning it to the left or right or up and down.
- ☐ Increase spacing between your TV antenna and any other antennas on the mast.
- ☐ Replace antenna with a highly directional one.
- ☐ Be sure lead-in wire runs in as direct a line as possible.
- ☐ Twist 300-ohm twinlead one turn per foot coming down from antenna.
- ☐ Switch 300-ohm twinlead to coaxial cable.

B. Imperfect Impedance Match

- ☐ Check if TV is at fault. Substitute another TV.
- ☐ In strong signal areas, introduce an attenuator pad between downlead and the TV (e.g., 12 db pad).
- ☐ Check lead-in connections.

Snow or Grainy Picture

A. TV Set at Fault

- ☐ Check setting of fine tuning controls.
- ☐ Check setting of AGC control.
- ☐ Check tuner and i-f circuit.
- ☐ Clean tuner (tap on channel selector; if picture flickers, tuner needs cleaning/servicing).

B. Bad Lead-In

- ☐ Check lead-in connections.
- ☐ Replace lead-in.

C. Antenna/Pre-Amplifier at Fault

- ☐ Change direction of antenna.
- ☐ Increase height of antenna.
- ☐ Replace antenna.
- ☐ Replace pre-amplifier.

Noise and Interference

A. Automobile Ignition, Motors, Generators, Etc.

- ☐ Replace 300-ohm lead-in with coaxial cable.
- ☐ Increase height of antenna.

B. FM Radio, Citizens Band, AM Radio and Ham Radio, Airplane Flutter

- ☐ Add high pass filter ahead of TV set.
- ☐ Add another high pass filter by antenna.
- ☐ Eliminate long horizontal runs of lead-in wire.
- ☐ Install better quality coaxial cable.
- ☐ Consider stacking your antennas in a vertical array (airplane flutter).

C. High-Power Line, X-Ray, Diathermy, etc.

- ☐ Replace 300-ohm lead-in with 75-ohm coaxial cable.
- ☐ Relocate antenna.

D. Flashing in Picture

- ☐ Check for poor or broken lead-in connections.
- ☐ Make sure antenna elements are tight and installed properly
- ☐ Check TV set (e.g., tuner).

E. Poor Reception in Rainy or Snowy Weather

- ☐ Replace lead-in wire.
- ☐ Change 300-ohm lead-in to 75-ohm coaxial cable.
- ☐ Reposition elements so that they do not touch each other.
- ☐ Check/replace amplifiers, couplers, etc.

*Adapted: courtesy Winegard Company.

Appendix I
Microcomputer Troubleshooting Guide

Section	Symptom	Probable Cause/ Remedy
Data transfer between disk drive	Unable to transfer data accurately between disk drive	1. Clean read/write head 2. Check drive speed
Data transfer between computer and disk drive	Unable to store or retrieve data from a disk drive	1. Check cable connection 2. Check for proper insertion of diskette 3. Programming error 4. Check tristate chip on drive analog card
Keyboard to display	Inappropriate character displayed on screen	1. Check encoder chip for proper seating 2. Check encoder chip
Keyboard to processor	Error between the keyboard and processor	1. Check band rate 2. Check processoı 3. Check keyboard contacts

Section	Symptom	Probable Cause/ Remedy
Magnetic tape communication	Unable to establish or maintain communication with tape recorder	1. Check cable connection 2. Check tape read switch 3. Check for properly seated tape 4. Check write protect tab 5. Clean head 6. Check clock frequency on interface card
Paper tape communication	Unable to establish or maintain communication with tape reader	1. Check cable connection 2. Check tape read switch 3. Insure tape is properly seated 4. Check reader head for cleanliness
Input Process Card	Single input point inoperative	1. Check for proper seating of card in socket 2. Check for appropriate input voltage 3. Input card bad 4. Mounting base bad 5. Programming error
Output Process Card	Single output point inoperatve	1. Check for proper seating of card in socket 2. Logic interface module bad on programmable controller 3. Mounting base bad 4. Sequencer bad on programmable controller 5. Programming error
Analog-to-digital card	No input value change	1. Check for appropriate input voltage 2. Check for end of conversion signal, if absent replace A/D chip

Section	Symptom	Probable Cause/ Remedy
		3. Check for proper seating of card in its mount 4. Check for appropriate addressing signal
Digital-to analog card	No output voltage change	1. Check for proper output connection 2. Check for proper seating of card in its mount 3. Check for appropriate address signals
Display	No display	1. Check brightness control and adjust as required 2. Power down, then power up again

Answers to Self-Examinations

Chapter 1
1. E
2. D
3. A
4. C
5. B
6. D
7. B
8. C
9. A
10. D
11. B
12. A
13. C
14. A
15. E
16. B
17. B
18. B
19. B
20. D
21. A
22. B
23. C
24. C
25. C
26. A
27. C
28. C
29. B
30. A

Chapter 2
1. C
2. D
3. E
4. B
5. E
6. D
7. D
8. E
9. E
10. D

Chapter 3
1. D
2. D
3. C
4. B
5. A
6. D
7. E
8. A
9. D
10. A
11. C
12. A
13. A
14. C
15. D

Chapter 4
1. C
2. A
3. B
4. A
5. C
6. A
7. C
8. C
9. B
10. B
11. A
12. C
13. D
14. B
15. B
16. A
17. B
18. C
19. B
20. C

Chapter 5
1. C
2. D
3. D
4. C
5. C
6. C
7. A
8. D
9. D
10. E
11. A
12. C
13. D
14. C
15. C
16. D
17. D
18. B
19. B
20. C

Chapter 6
1. C
2. D
3. C
4. D
5. A
6. B
7. A
8. C
9. A
10. C
11. A
12. C
13. B
14. C
15. D

Chapter 7
1. B
2. D
3. D
4. B
5. A
6. B
7. E
8. C
9. C
10. A
11. E
12. C
13. C
14. B
15. B
16. C
17. D
18. B
19. C
20. A
21. D
22. B
23. D

Chapter 8
1. A
2. B
3. B
4. C
5. B
6. B
7. C
8. B
9. D
10. B
11. D
12. B
13. C
14. C
15. B
16. D
17. C
18. A
19. B
20. C

Chapter 9
1. E
2. B
3. B
4. A
5. C
6. B
7. C
8. C
9. D
10. D
11. B
12. D
13. D
14. E
15. C
16. D
17. E
18. B
19. A
20. D
21. E

Chapter 10
1. E
2. C
3. C
4. B
5. D
6. C
7. B
8. B
9. C
10. D
11. D
12. D
13. B
14. E
15. C
16. D
17. B
18. E
19. B
20. E
21. D
22. C
23. A
24. E
25. B

Chapter 11
1. B
2. A
3. D
4. C
5. D
6. A
7. C
8. A
9. C
10. D
11. E
12. A
13. B
14. D
15. C
16. D
17. B
18. A
19. D
20. C
21. D
22. E
23. B
24. E
25. B

Index

OTHER POPULAR TAB BOOKS OF INTEREST

Transducer Fundamentals, with Projects (No. 1693—$14.50 paper; $19.95 hard)
Second Book of Easy-to-Build Electronic Projects (No. 1679—$13.50 paper; $17.95 hard)
Practical Microwave Oven Repair (No. 1667—$13.50 paper; $19.95 hard)
CMOS/TTL—A User's Guide with Projects (No. 1650—$13.50 paper; $19.95 hard)
Satellite Communications (No. 1632—$11.50 paper; $16.95 hard)
Build Your Own Laser, Phaser, Ion Ray Gun and Other Working Space-Age Projects (No. 1604—$15.50 paper; $24.95 hard)
Principles and Practice of Digital ICs and LEDs (No. 1577—$13.50 paper; $19.95 hard)
Understanding Electronics—2nd Edition (No. 1553—$9.95 paper; $15.95 hard)
Electronic Databook—3rd Edition (No. 1538—$17.50 paper; $24.95 hard)
Beginner's Guide to Reading Schematics (No. 1536—$9.25 paper; $14.95 hard)
Concepts of Digital Electronics (No. 1531—$11.50 paper; $17.95 hard)
Beginner's Guide to Electricity and Electrical Phenomena (No. 1507—$10.25 paper; $15.95 hard)
750 Practical Electronic Circuits (No. 1499—$14.95 paper; $21.95 hard)
Exploring Electricity and Electronics with Projects (No. 1497—$9.95 paper; $15.95 hard)
Video Electronics Technology (No. 1474—$11.50 paper; $16.95 hard)
Towers' International Transistor Selector—3rd Edition (No. 1416—$19.95 vinyl)
The Illustrated Dictionary of Electronics—2nd Edition (No. 1366—$18.95 paper; $26.95 hard)
49 Easy-To-Build Electronic Projects (No. 1337—$6.25 paper; $10.95 hard)
The Master Handbook of Telephones (No. 1316—$12.50 paper; $16.95 hard)
Giant Handbook of 222 Weekend Electronics Projects (No. 1265—$14.95 paper)
Introducing Cellular Communications: The New Mobile Telephone System (No. 1682—$9.95 paper; $14.95 hard)
The Fiberoptics and Laser Handbook (No. 1671—$15.50 paper; $21.95 hard)
Power Supplies, Switching Regulators, Inverters and Converters (No. 1665—$15.50 paper; $21.95 hard)
Using Integrated Circuit Logic Devices (No. 1645—$15.50 paper; $21.95 hard)
Basic Transistor Course—2nd Edition (No. 1605—$13.50 paper; $19.95 hard)
The GIANT Book of Easy-to-Build Electronic Projects (No. 1599—$13.95 paper; $21.95 hard)
Music Synthesizers: A Manual of Design and Construction (No. 1565—$12.50 paper; $16.95 hard)
How to Design Circuits Using Semiconductors (No. 1543—$11.50 paper; $17.95 hard)
All About Telephones—2nd Edition (No. 1537—$11.50 paper; $16.95 hard)
The Complete Book of Oscilloscopes (No. 1532—$11.50 paper; $17.95 hard)
All About Home Satellite Television (No. 1519—$13.50 paper; $19.95 hard)
Maintaining and Repairing Videocassette Recorders (No. 1503—$15.50 paper; $21.95 hard)
The Build-It Book of Electronic Projects (No. 1498—$10.25 paper; $18.95 hard)
Video Cassette Recorders: Buying, Using and Maintaining (No. 1490—$8.25 paper; $14.95 hard)
The Beginner's Book of Electronic Music (No. 1438—$12.95 paper; $18.95 hard)
Build a Personal Earth Station for Worldwide Satellite TV Reception (No. 1409—$10.25 paper; $15.95 hard)
Basic Electronics Theory—with projects and experiments (No. 1338—$15.50 paper; $19.95 hard)
Electric Motor Test & Repair—3rd Edition (No. 1321 $7.25 paper; $13.95 hard)
The GIANT Handbook of Electronic Circuits (No. 1300—$19.95 paper)
Digital Electronics Troubleshooting (No. 1250—$12.50 paper)